MEP 016B 3KW Generator
Operators Manual
and Organizational Maintenance
TM 5-6115-615-12

Conversion Instructions
for
Repower from MEP 016B to MEP 016E

Diesel Generator Set
Skid Mounted, Tactical Quiet

also applies to
MEP 021B
MEP 026B

The black marks on the page edges
are to assist in locating the manuals
in this book. They are in vertical order of appearance

in the book with the first manual at the top.

edited by
Brian Greul

The MEP series of Military Generators are rugged, durable and incorporate proven diesel engine technology. This book is the operators manual and incorporates organizational maintenance instructions. A second publication is included at the end that addresses the re-powering of the diesel engine. This conversion results in the unit becoming a MEP 016EIt is being republished to assist enthusiasts, restorers, and aftermarket owners who use or wish to use these generators outside of military use.

An 8.5x11 3 hole punched loose leaf copy may be purchased for your 3 ring binder. Email books@ocotillopress.com for current information.

Should you have suggestions or feedback on ways to improve this book please send email to Books@OcotilloPress.com

Edited 2021 Ocotillo Press
ISBN 978-1-954285-23-1

Printed in the United States of America

Ocotillo Press
Houston, TX 77017
Books@OcotilloPress.com

Disclaimer: The user of this book is responsible for following safe and lawful practices at all times. The publisher assumes no responsibility for the use of the content of this book. The publisher has made an effort to ensure that the text is complete and properly typeset, however omissions, errors, and other issues may exist that the publisher is unaware of.

MARINE CORPS TECHNICAL MANUAL   TM 05926B/06509B-12/1 ARMY TECHNICAL MANUAL   TM 5-6115-615- 12 NAVY PUBLICATION   NAVFAC P-8- 646-12 AIR FORCE TECHNICAL ORDER   TO 35C2-3-386- 31

TECHNICAL MANUAL

OPERATOR AND ORGANIZATIONAL
MAINTENANCE MANUAL

GENERATOR SET, DIESEL ENGINE DRIVEN, TACTICAL, SKID MOUNT- ED, 3 KW, 3 PHASE 120/208 AND SINGLE PHASE 120/240 VOLTS AC AND 28 VOLTS DC

DOD MODEL CLASS M O D E NSN

MEP-016B	UTILITY 60 HZ	6115-01-150-4140
MEP-021B UTILITY	400 HZ	6115-01-151-8126
MEP-026B	UTILITY 28 VDC	6115-01-150-0367

PUBLISHED UNDER THE AUTHORITY OF HEADQUARTERS U.S. MARINE CORPS,

JULY 1987

PCN 184 059267 00

MARINE CORPS TM 05926B/06509B-12/I
ARMY TM 5-6115-615-12
NAVY NAVFAC P-8-646-12 AIR
FORCE TO 35C2-3-386-31

C1

CHANGE⎫
       ⎬
No. 1 ⎭

Headquarters, U.S. Marine Corps
Department of the Army, Navy & Air Force
WASHINGTON, D.C.,   31 May 1989

Operator and Organizational Maintenance Manual GENERATOR SET, DIESEL ENGINE DRIVEN, TACTICAL, SKID MOUNTED 3KW, 3 PHASE 120/208 and SINGLE PHASE 120/240 VOLTS AC AND 28 VOLTS DC

DOD MODEL	CLASS	MODE	NSN
MEP-016B	UTILITY	60 HZ	6115-01-150-4140
MEP-021B	UTILITY	400 HZ	6115-01-151-8126
			6115-01-150-0367

Approved for public release; distribution is unlimited.

1. Change 1 to this joint technical manual TM 05926B/06509B-12/1/TM 5-6115-615-12/NAVFAC P-8-646-12/TO 35C2-3-386-31 is for Army use only. It is effective upon receipt and provides operator and organizational maintenance instructions, and a Maintenance Allocation Chart (MAC) for Generator Set, Engine Driven, Tactical, Skid-Mounted, 3 KW DOD Model MEP 701A, 60 Hz, NSN 6115-01-234-5966. Model 701A is a modified model MEP 016B with the addition of the Acoustic Suppression Kit (ASK). The ASK is intended to suppress the high noise level inherent in a diesel-driven generator.

2. Reporting of discrepancies or suggested change should be submitted on a DA Form 2028 directly to:

Commander
U.S. Army Troop Support Command
ATTN: AMSTR-MCTG
4300 Goodfellow Boulevard St.
Louis, MO 63120-1798

3.

Army Users: Remove and insert the following pages as indicated below. Remove Pages

Insert Pages

v through xiv	v thru xv/ xvi
--	5-1 thru 5-45/ 5-46 C-7 thru
C-9/ C-10	C-7 thru C-10
Index 1 thru Index 5/ Index 6	Index 1 thru Index 6

4.

MARINE CORPS TM 05926B/06509B-12
ARMY   TM 5-6115-615-12
NAVY   NAVFAC P-8-646-12
AIR FORCE   TO 35C2-3-386-31

By Order of the Secretaries of the Army, the Navy,and the Air Force (Including the Marine Corps)

CARL E.VUONO
General, United States Army
Chief of Staff

Official:

WILLIAM J. MEEHAN, II
BrigadierGeneral, United States Army
The Adjutant Genenal

B. F. MONTOYA
Rear Admiral CEC, US Navy
Commander
Naval Facilities Engineering Command

LARRY D. WELSH, General *USAF*
Chief of Staff

Official:

ALFRED G. HANSEN
General, USAF, Commander, Air Force
Logistics Command

C. N. PASTINO
Colonel, U.S. Marine Corps
Deputy Commander

DISTRIBUTION:

To be distributed in accordance with DA Form 12-25A, , Operator Maintenance requirements for Generator Set, Gas Driven, 120/208V, 28V DC, 3 KW, 60/400 HZ, 3PH (MEP-016A, MEP-021A, MEP 026A) and Unit 120/240V,

MARINE CORPS TECHNICAL MANUAL

DEPARTMENT OF THE ARMY TECHNICAL MANUAL TM 5-6115-615-12

DEPARTMENT OF THE NAVY PUBLICATION NAVFAC P-8-646-12

DEPARTMENT OF THE AIR FORCE TECHNICAL ORDER

TM 05926B/06509B-12/1

TO 35C2-3-386-31

HEADQUARTERS U.S. MARINE CORPS, DEPARTMENTS
OF THE ARMY, NAVY, AND AIR FORCE
WASHINGTON, D.C. (July 1987)

OPERATOR AND ORGANIZATIONAL
MAINTENANCE MANUAL

GENERATOR SET, DIESEL ENGINE DRIVEN, TACTICAL, SKID MOUNTED, 3 KW, 3 PHASE
120/208 AND SINGLE PHASE 120/240 VOLTS AC AND 28 VOLTS DC

DOD MODEL	CLASS	MODE	NSN
MEP-016B UTILITY		60 HZ	6115-01-150-4140
MEP-021B	UTILITY	400 HZ	6115-01-151-8126
MEP-026B UTILITY		28 VDC	6115-01-150-0367

Marine Corps TM 05926B/06509B-12/1
Army TM 5-6115-615-12
Navy NAVFAC P-8-646-12 Air
Force TO 35C2-3-386-31

DEPARTMENT OF THE NAVY Headquarters, U.S. Marine Corps
Washington, D.C. 20380-0001

31 July 1987

1. **This Manual is effective upon receipt and contains operator and organizational maintenance instructions, and a Maintenance Allocation Chart (MAC) for Generator Set, Engine Driven, Tactical, Skid Mounted, 3 KW, DOD Models MEP 016B, 60 HZ, NSN 6115-01-150-4140, MEP 021B, 400 HZ, NSN 6115-01-151-8126, and MEP 026B, 28 VDC, NSN 6115-01-150-0367.**

2. Notice of discrepancies or suggested changes:refer to paragraph 1-4 titled, Reporting of Errors, of this Manual for applicable Servicesform number and forwarding address.

BY DIRECTION OF THE COMMANDANT OF THE MARINE CORPS

OFFICIAL:
J. G. O'NEILL
Head, Materiel Acquisition Support Branch
Materiel Division
Installations and Logistics Department

CARL E. VUONO
General, United

States Army
Chief of Staff

R. L. DILWORTH
Brigadier General, United States Army
The Adjutant General

B. F. MONTOYA, RADM, CEC USN

LARRY D. WELSH, General
USAF, Chief of Staff

ALFRED G. HANSEN, General, USAF
Commander, Air Force Logistics Command

DISTRIBUTION :AGB/L77/L82

Copy to:7000161 (2)

LIST OF EFFECTIVE PAGES

INSERT LATEST CHANGED PAGES. DESTROY SUPERSEDED PAGES.

NOTE: The portion of the text affected by the changes is indicated by a vertical line in the outer margins of the page. Changes to Illustrations are indicated by miniature pointing hands. Changes to wiring diagrams are indicated by shaded areas.

Date of issue for original and changed pages are:
Original . . . 0 . . . July 1987

TOTAL NUMBER OF PAGES IN THIS PUBLICATION IS 234 CONSISTING OF THE FOLLOWING:

Page No.	* Change No.
Cover	0
Blank	0
Title Block	0
Blank	0
A	0
Blank	0
i thru xiv	0
1-1 thru 1-13	0
Blank	0
2-1 thru 2-19	0
Blank	0
3-1 thru 3-43	0
Blank	0
4-1 thru 4-113	0
Blank	0
A-1	0
Blank	0
B-1	0
Blank	0
C-1 thru C-9	0
Blank	0
Index-1 thru Index-5	0
Blank	0
FO-1	0
Blank	0
FO-3	0
Blank	0

* Zero in this column indicates an original page.

MARINE CORPS TM 05926B/06509B-12
ARMY   TM 5-6115-615-12
NAVY NAVFAC P-8-646-12
AIR FORCE   TO 35 C2-3-386-31

TABLE OF CONTENTS

TABLE OF CONTENTS - Continued

TABLE OF CONTENTS - Continued

MARINE
CORPS
ARMY

TM 05926B/06509B-l2
TM 5-6115-615-12
NAVFAC P-8-646-12
TO 35C2-3-386-31

TABLE OF CONTENTS - Continued

TABLE OF CONTENTS - Continued

TABLE OF CONTENTS - Continued

LIST OF ILLUSTRATIONS

MARINE CORPS TM 05926B/06509B-12
ARMY   TM 5-6115-615-12
NAVY   NAVFAC P-8-646-12

LIST OF ILLUSTRATIONS - Continued

Figure	Title	Page

LIST OF ILLUSTRATIONS - Continued

LIST OF TABLES

SAFETY SUMMARY

The following are general safety precautions that are not related to any specific procedures and therefore do not appear elsewhere in this publication. These are recommended precautions that personnel must understand and apply during many phases of operation and maintenance.

KEEP AWAY FROM LIVE   CIRCUITS
Operating personnel safety   must at all times observe all
regulations. adjustments to this Do not replace components or make equipment
                                 with the high voltage

supply turned on.Under certain conditions, dangerous potentials may exist when the power control is in the off position, due to changes retained by capacitors. To avoid casualties, always remove power and discharge and ground a circuit before touching it.   Always remove
rings, watches and other jewelry before servicing this equipment.

DO NOT SERVICE OR ADJUST ALONE
Under no circumstances should any person reach into this equipment for the purpose of servicing or adjusting this equipment except in the presence of someone who is capable of rendering aid.

RESUSCITATION
Personnel working with or near high voltages should be familiar with modern methods of resuscitation.   Such
information may be obtained from the Bureau of Medicine and Surgery.

SECURE LOOSE CLOTHING
Personnel working on this equipment should secure all loose fitting clothing to prevent clothing from catching in moving parts.

KEEP COMPRESSED AIR AWAY FROM SKIN
Personnel using compressed air should not exceed 15 psi nozzle pressure when drying parts, and should not direct compressed air toward skin.Personal injury could result.

OPERATE EQUIPMENT IN A WELL VENTILATED AREA
Exhaust discharge contains noxious and deadly fumes.
Do not operate this equipment in an enclosed area unless exhaust discharge is properly vented to the outside.When using cleaning solvents, clean parts in a well ventilated area and avoid inhalation of solvent
fumes .

WEAR EAR PROTECTION
The noise level of this generator damage.   To   set can cause hearing always wear
avoid hearing damage, protectors, as recommended   ear medical or safety equipment.
by the officer, when operating near this

USE CAUTION WHEN WORKING AROUND FLAMMABLES
Do not smoke, use open flame or use excessive heat in the vicinity of the generator
set when refueling, working around the cleaning solvents. which could result
                              battery, or working with flammable Doing so could
                              cause an explosion in severe personal injury or death.

The following em-   warnings appear in   the text in this volume and are repeated here for

WARNING

All personnel who operate or maintain the generator sets should become thoroughly
familiar with the safety precautions prior to performing operation or maintenance
procedures. Paragraph 2-1.

WARNING

Personnel should not attempt any of the following operating procedures without first
grounding the unit as outlined in paragraph 2-5.1.Failure to do so could result in serious electrical shock or death from electrocution.   Paragraph 2-1.
Operation of this equipment presents a noise hazard to personnel in the area. The noise
level exceeds the allowable limits for unprotected personnel. Wear ear

WARNING

muffs or ear plugs. Paragraph 2-1.
Battery electrolyte is an acid solution that gives off flammable fumes.Do not smoke
or use open flame when working around battery.Doing so can cause an explosion that

WARNING

could result in serious personal injury. If skin is exposed to electrolyte, flush exposed
area with water immediately.   If eyes are exposed to electrolyte, flush them
with water and seek immediate medical attention.Paragraphs 2-10, 3-15, 3-16, 4-19,
4-22, Tables 3-2, 3-3.

WARNING

Prior to connecting load cables, be certain that all switches and circuit breakers
are in the OFF or OPEN position, and that the generator set is not running. Failure
to do so can result in death from electrocution. Paragraphs 2-2, 4-2.
not hold main circuit breaker in ON position.   Shock death from

WARNING

Do   electrocution can result. Table 3-3.
or

WARNING

Disconnect load before con-   switching load terminal death from electro-
nections.   Shock result.   or
Table 3-3.

WARNING

Shut down generator connec-   cution can
tions.   Shock result.
Table 3-3.

WARNING

Unless manual directs otherwise, do not attempt any of the following maintenance
procedures when generator set is operating.0o not touch exposed electrical
connection when a source of power such as utility power or another generator set is
connected to the load terminals.   Severe electrical shock or death by electrocution can result. Section V, Section VIII, Section IX, Section X, Section XI, Section
XII, Section
XIII, Section XIV.
Do not use a lifting device with a capacity of less than 600 lbs (272 kg).Do not allow
the crated generator to swing while it is suspended.   Failure to observe

WARNING

this warning can result in serious injury or death to personnel.   Paragraph 4-1.

MARINE CORPS TM 05926B/06509B-12
ARMY   TM 5-6115-515-12
NAVY   NAVFAC P-8-646-12
AIR FORCE   TO 35C2-3-386-31

WARNING

Do not operate the generator set in an enclosed area unless the exhaust gases are piped to the outside. Inhalation of exhaust fumes will result in serious illness or death.Paragraph 4-2.
Do not remove oil filler cap when engine is running.

WARNING

Hot oil can splash up and cause burns.Paragraph 4-7.
Keep feet clear when tilting and working around tilted generator set.Failure to do

WARNING

so can result in personal injury.   Paragraph 4-7.
Avoid prolonged contact and inhalation of fumes of dry cleaning solvent. Use dry

WARNING

cleaning solvent only in a well ventilated area.Table 4-1, Paragraph 4-52.

WARNING

Do not   direct pressurized air toward skin.Personal could result.Paragraph
injury

WARNING

4-52.
Disconnect battery cables before servicing generator components. The high current output of the DC electrical system can cause arcing and/or burns if a short

WARNING

circuit occurs.Paragraph 4-22.
Make sure muffler is completely cooled before

WARNING

When changing the position of the output reconnection switch, make sure that the
grounding jumper is connected to the correct output terminal as indicated in paragraph
2-5.l.b.   Serious damage may result if the ground jumper
is incorrectly connected. Paragraph 2-5.1.
In the 240 volt single phase mode, the output
receptacle"NEUTRAL" lead is NOT GROUNDED. When L-2 is grounded in the

---
WARNING
---

240 volt mode, both leads to the receptacle will measure a potential with respect to
ground. Serious damage may result if the output receptacle is grounded.Paragraph
2-5.1.
The output receptacle is connected to a floating ground system.   The set MUST be
grounded as specified in
paragraph 2-5.1 to effectively ground the receptacle. Failure to do so may cause

---
WARNING
---

severe injury or death.
Paragraph 3-35.

CHAPTER 1

INTRODUC-

Section I. GENERAL.

1-1.   SCOPE .   This manual is for your use in operating and maintaining the 3KW DED Generator Set, Type I (Tactical), Class 2 (utility) skid mounted Models MEP-016B, MEP-021B, and MEP-026B.   It contains information on operation, lubrication, preventive maintenance checks and services, troubleshooting, operator/crew maintenance and organizational maintenance. Thoroughly familiarize yourself with the unit before operating or servicing.

1-1.1.   <u>Appendices.</u>   Appendix A contains a list of reference publications applicable to this manual.Appendix B contains a Components List.Appendix C contains the Maintenance Allocation Chart (MAC) which determines the level of maintenance responsibility for Army users.

1-2. LIMITED APPLICABILITY. Some portions of this publication are not applicable to all services.These portions are prefixed to indicate the services to which they pertain: (A) for Army, (AF) for Air Force, (N) for Navy, and (MC) for Marine Corps. Portions not prefixed are applicable to all services.

1-3.   MAINTENANCE FORMS AND RECORDS.

1-3.1.   (MC) Maintenance forms and records used by Marine Corps personnel are prescribed by TM 4700-15/1.

1-3.2. <u>(A)</u> Maintenance forms and records used by Army personnel are prescribed by DA PAM 738-750.

1-3.3.   (AF) Maintenance forms and records used by Air Force personnel are prescribed in AFM 66-1 and the applicable 00-20 Series Technical Orders.

1-3.4. <u>(N)</u> Navy users should refer to their service peculiar directives to determine applicable maintenance forms and records to be used.

1-4. REPORTING OF ERRORS. Reporting of errors, omissions, and recommendations for improvement of this publication by the individual user is encouraged.   Reports should be submitted as follows:

1-4.1. (MC) By NAVMC form 10772 directly to:Commanding General, Marine Corps Logistics Base (Code 850), Albany, Georgia 31704-5000.
1-4.2. <u>(A)</u> DA Form 2028 directly to: Commander, US Army Troop Support Command, ATTN: AMSTR-MCTS, 4300 Goodfellow Boulevard, St. Louis, MO 63120-1798.

1-4.3. <u>(AF)</u> AT<u>TN:</u>   AFTO Form 22 directly to: Commander, Sacramento Air Logistics Center, McClellan Air Force Base, CA. 95652 in accordance with TO-00-5-1.

1-4.4. (N) By Control   letter directly to: Commanding Officer, U.S. Navy, Ships Parts ATTN :   Code 783, Mechan-
Center,   icsburg, PA 17055.

1-5.1. (MC) Marine Corps users shall refer to the Repair Parts List.

1-5.2. (A) Army users shall refer to the Maintenance Allocation Chart (MAC) tasks and levels of maintenance to be   for
performed.

1-5.3. (AF) Air Force users shall accomplish maintenance at the user level consistent their capability in accordance
with AFM 66-1.

1-5.4. (N) Navy users shall determine their maintenance levels in accordance with their service directives.

1-6. (MC, A) DESTRUCTION OF ARMY MATERIEL TO PREVENT ENEMY USE. Demolition of
materiel to prevent enemy use shall be in accordance with the requirement of TM-750-244-3 (Procedures for Destruction of
Equipment to Prevent Enemy Use for U.S. Army).

1-7.   ADMINISTRATIVE STORAGE.

1-7.1.   (MC, N) Refer to individual service directives for requirements.

1-7.2. (A) Refer to TM 740-90-1 (Administrative Storage).

1-7.3. (AF) Refer to TO 35-1-4 (Processing and Inspection of Aerospace Ground Equipment

1-8. PREPARATION FOR SHIPMENT AND STORAGE.

1-8.1.   (MC) Refer to MCO P4450.7.

1-8.2. (A) Refer to TB 740-97-2 and TM   item generator sets and TO 38-1-5 for

1-8.3. (AF) Refer to TO 35-1-4 for end installed engine.
                                                 directives for requirements.
1-8.4. (N) Refer to individual service
740-90-1.

1-8.5. (A) REPORTING EQUIPMENT IMPROVEMENTRECOMMENDATIONS (EIR). EIR will be
prepared using DA form 2407, Maintenance Request.Instructions for preparing EIR's are provided in DA PAM 738-750, The
Army Maintenance Management System.EIR' S should be mailed directly to: U.S. Army Troop Support Command, ATTN: AM-
STR-QX, 4300 Goodfellow Boulevard, St. Louis, MO 63120-1798.A reply will be furnished directly to you.

1-8.6. (MC) REPORTING EQUIPMENT IMPROVEMENTRECOMMENDATIONS (EIR). Submit Quality Assurance
Report on standard form 368 in accordance with MCO 4855.10.

Section II.   DESCRIPTION AND DATA.

1-9.   DESCRIPTION. The generator sets are self contained, frame mounted portable units.   They are powered by a single cylinder diesel engine that is directly coupled to the generator.   See Figures 1-1 and 1-2 for locations of major compo-

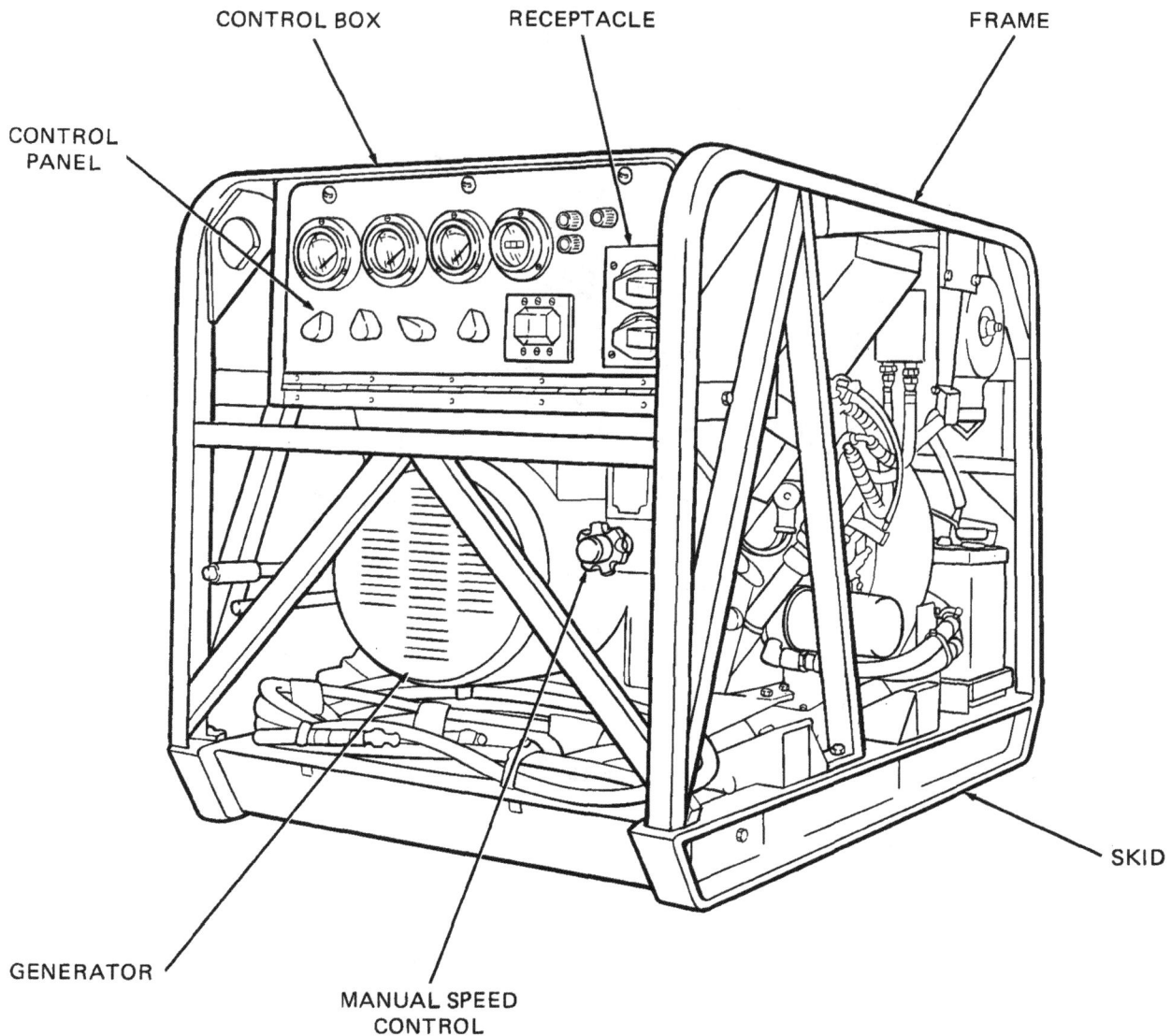

Figure 1-1. Generator Set, Right Front, Three Quarter View.

CONTROL BOX

FUEL TANK

AIR CLEANER

SLAVE RECEPTACLE

BATTERY

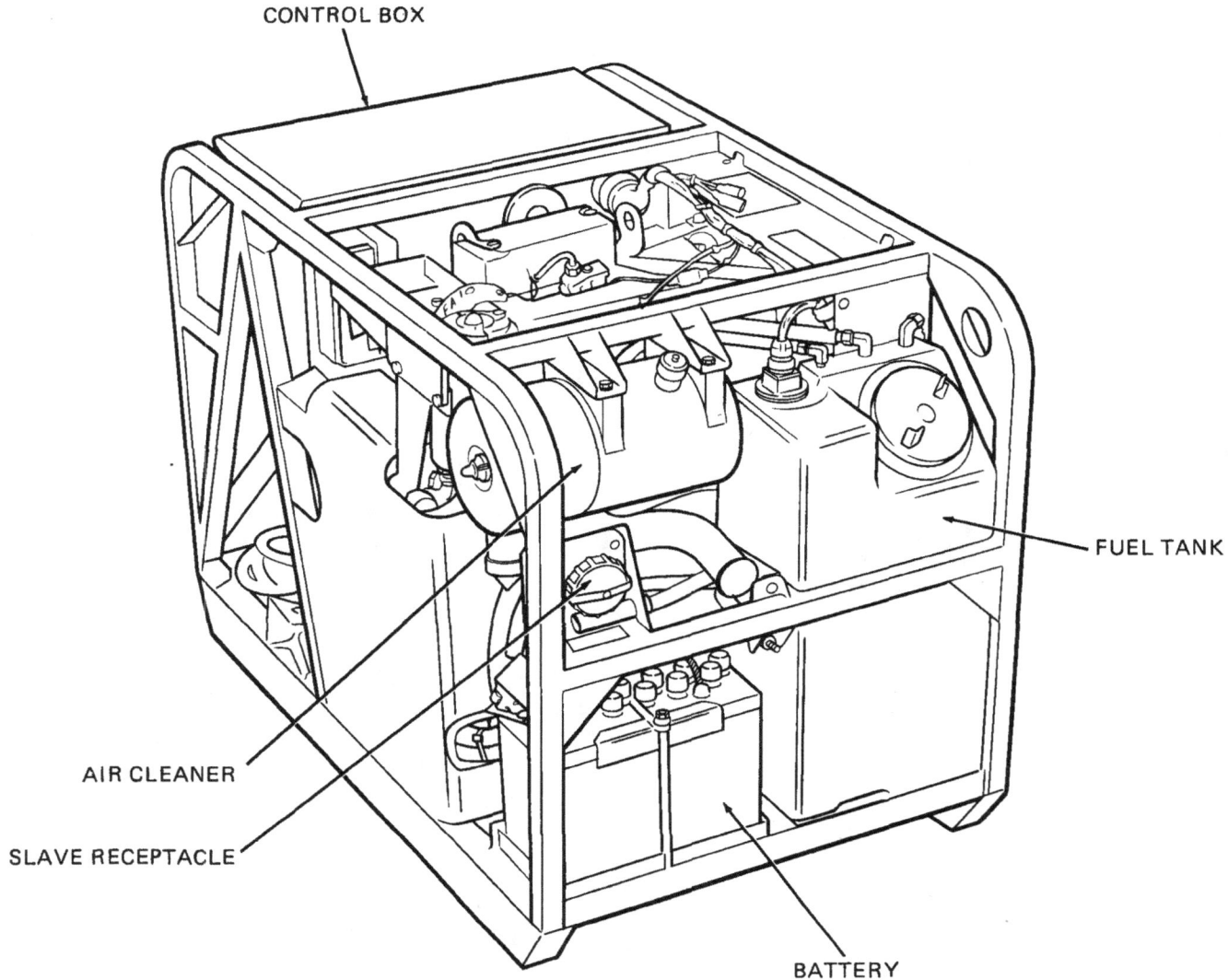

Figure 1-2. Generator Set, Left Rear, Three Quarter View.

1-9.1._____Engine. The engine is a single cylinder, air cooled, direct injection
diesel with a displacement of 36.80 cubic inches (603cc).A mechanical governor is used to maintain engine speed under rated load
conditions.

1-9.1.1. Fuel System. Fuel is supplied from the unit's self-contained fuel tank. The fuel is filtered by an integral strainer/filter and
water separator. A mechanical fuel transfer pump delivers fuel to the fuel injection pump which pumps fuel at high pressure to
the fuel injection nozzle in the cylinder head. The fuel transfer pump has a manual priming lever that is used to prime the fuel
system when necessary.   The unit can also be run from an auxiliary fuel source. An adapter connects the auxiliary fuel source
to the electric auxiliary fuel pump mounted to the skid base.The auxiliary fuel pump transfers the fuel into the unit's main fuel tank
allowing fuel to be delivered to the engine in the normal manner.

1-9.1.2.   Electrical. electric   One 24-volt "wet cell" battery supplies power for the 24 volt glow plug located in the cylinder
starter, the auxiliary fuel pump.   head, and for the electric An alternator automatically recharges the battery when the
engine is running.

1-9.2.   Generators.

      MEP-061-B.
a self-excited, 60 hertz alternating current generator. The generator output is 120/240 volts, single phase; 120 volt, 3 phase; or
120/208 volt, 3 phase, 4 wire. It is rated at 3 kilowatts at 8000 feet altitude when operating on diesel fuel.
When operating with JP4 fuel, the altitude operation is limited by temperature (see Figure 1-3).

1-9.2.2.   MEP-021B.   The generator provided with the Model MEP-021B generator set is a self-excited, 400 hertz
alternating current generator.   The generator output is 120/240 volt, single phase; 120 volt, 3 phase; or 120/208 volt, 3 phase,
4 wire. It is rated at 3 kilowatts at 8000 feet altitude when operating on diesel fuel. When operating with JP4 fuel, the altitude
operation is limited by temperature (see Figure 1-3).

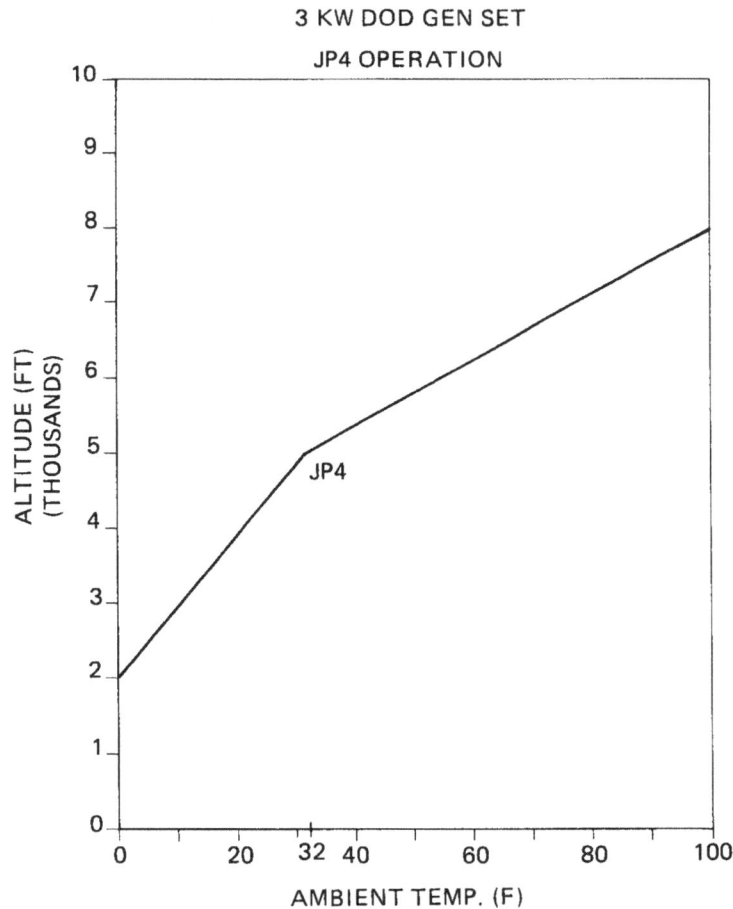

Figure 1-3. JP4 Operation.

1-9.2.3. MEP-026B. The generator provided with the Model MEP-026B generator set is a self-excited, 28 volt direct current generator. It is rated at 3 kilowatts at 8000 feet altitude when operating on diesel fuel. When operating with JP4 fuel, the altitude operation is limited by temperature (see Figure 1-3).

1-9.3. Controls. (See Figures 1-2, 1-4 and 1-5). All of the controls necessary for operation of the generator set are located on the main control panel, with the exception of the output selection switch and the engine speed control. The output selection switch is located inside the control box (easily accessible by opening the front panel of the control box).   The engine manual speed control is located directly below the control box.

1-10. TABULATED DATA.
1-10.1. Identification and Instruction Plates. All identification and instruction plates are explained in Table 1-1.
1-10.2. Tabulated Data. The tabulated data for the generator sets are contained in

Table 1-2.

1-10.3. Torque Data. Torque data applicable to operator/crew and organizational maintenance is found in Table 1-3.

1-10.4. Installation Plans and Schematics.   Installation plans and schematics are shown in Figures 1-4,1-5, 1-6, FO-1 and FO-2.

Table 1-1. Identification and Instruction Plates.

Location	Type	Description
Inside of hinged control box cover (60,400 Hz)	Schematic	Electrical schematic
Top of control box (28VDC)	Instruction	Electrical schematic/operating instructions
Top of control box (60,400 Hz)	Identification	Operating instructions
Back of control box	Identification/ caution	Load terminals
Load terminal cover	Identification	
		Identifies load terminals and has voltage caution
Front of control box (28VDC) Schematic/ instruction		Circuit breaker

MARINE CORPS TM 05926B/06509B-12
ARMY   TM 5-6115-615-12
NAVY   NAVFAC P-8-646-12

Table 1-1. Identification and Instruction Plates, Continued.

Location	Type	Description
Muffler heat shield plate	Caution	Instructions for disconnecting and connecting the battery
Muffler heat shield plate	Diagram	U.S. Department of Defense Data Plate Contains model, serial number and rating information for the set
	Identification	Hearing protection required
Muffler heat shield plate	Instruction	Block diagram of the fuel system
Muffler heat shield plate	Identification -slave receptacle	Identifies lifting and tie down points
Muffler heat shield plate	Identification -speed adjustment control	Instructions for connecting load cables
Engine guard panel (60,400 Hz)	Identification -auxiliary fuel connection	Identifies the slave receptacle
On frame	Identification -ground	Identification and instructions for the engine manual speed control
Bracket under control box	Data plate	Identifies the auxiliary fuel connection point
on skid base	Data plate	Identifies the ground stud
On skid base	Warning	Manufacturer's identification
Engine cooling air outlet		Contains the manufacturer's identification and rating
Top of generator housing		Danger- rotating fan can cause severe injury
Both sides of generator housing Instruction		
Identification		

Table 1-2. Tabulated Data.

Generator Set Manufacturer. . . . . .ONAN Corporation, Minneapolis, MN.

MODEL	HERTZ/MODE	CLASS
MEP-016B	60 Hz	UTILITY
MEP-021B	400 Hz	UTILITY
MEP-026B	28 VDC	UTILITY

Operating Temperature Range
without external heat
MEP-016B. . . . .   -25°F to +125°F (-32°C to +52°C)
MEP-021B. . . . .   -25°F to +125°F (-32°C to +52°C)
MEP-026B. . . . .   -25°F to +125°F (-32°C to +52°C)

Voltage Output
MEP-016B (60Hz). . . . .   120/240 VAC, single phase
                                                            120 VAC, 3 phase
                                                            120/208 VAC, 3 phase, 4 wire

MEP-021B (400Hz). . . . .   120/240 VAC, single phase
                                                            120 VAC, 3 phase
                                                            120/208 VAC, 3 phase, 4 wire

MEP-026B 28VDC

Power Factor
MEP-016B. 0.8
MEP-021B. 0.8
MEP-026B. . . . .   N/A

Capacities
Fuel System . . . . 4.5 gal. (17.4 Liters)
Lubricating Oil. . . . .   3.0 qts. (2.8 Liters)

Dimensions and Weights
Overall Length. . . . .   39" (99 cm)
Overall Width. . . . .   23.75" (60.3 cm)
Overall Height. . . . .   28" (71.1 cm)
Net Weight Empty. . . . .   440 lbs. (199 kg)
Net Weight Filled.   498 lbs. (226.3 kg)
Shipping Weight.   500 Ms. (227.6 kg
Cubage. . . . . .19.5 ft³(552.3 dm³)

Table 1-2. Tabulated Data, Continued.

Engine

Manufacturer . . . . .Onan Corporation, Minneapolis, MN

Model Q106D

Type. . . . .   Diesel, direct injection,
                                                                              4 stroke/cycle

Number of Cylinders. 1

Displacement . . . . .36.80 Cubic inches (603 cc)

Horsepower . . . . .8 BHP at 3600 rpm (as installed)

Rotation . . . . .ccw (viewed from generator end)

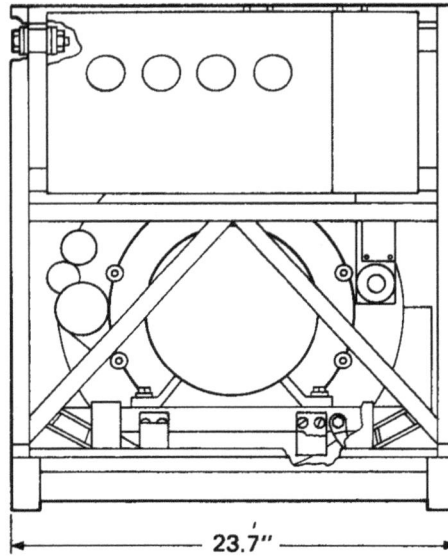

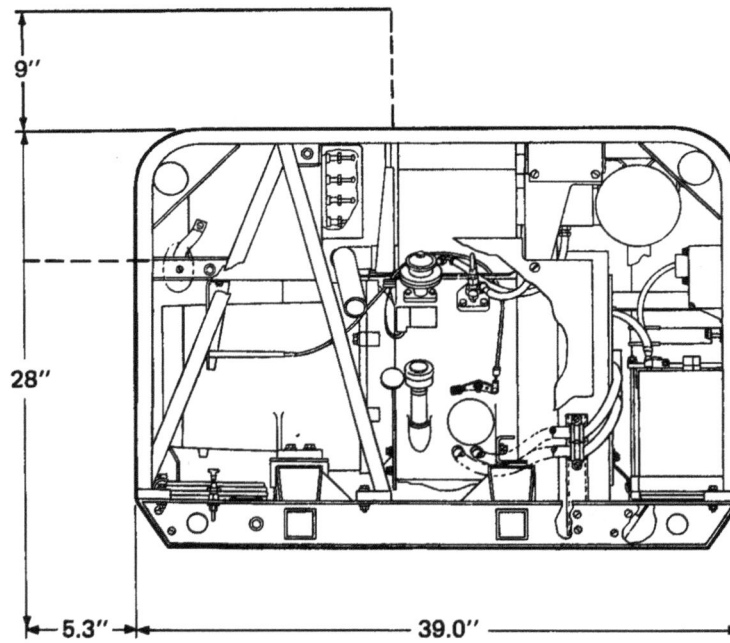

Figure 1-4.Installation Plans.

LIFTING AND TIE DOWN DIAGRAM

CG 19   11.85 CG

LIFTING
3400 LB CAP

TRANSPORT
IN ARROW
DIRECTION

ENGINE
END

CG

13.25

TOWING
2125 LB CAP

(4) TIE DOWN
HOLES
1275 LB CAP

(4) .560 MTG HOLES

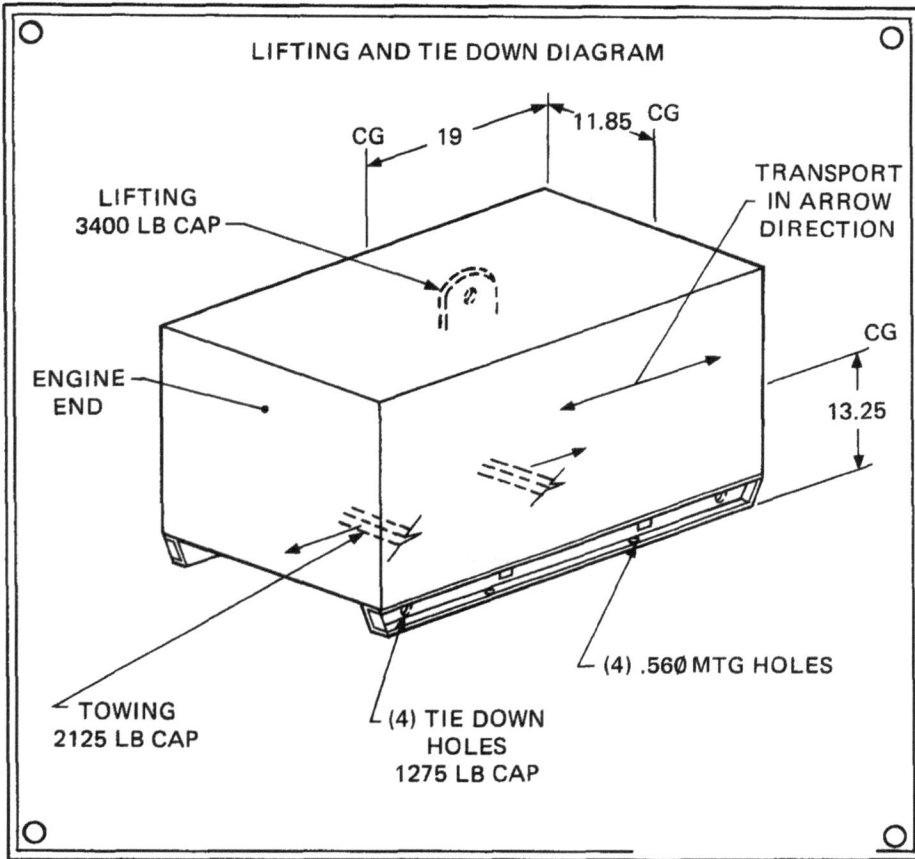

Figure 1-5. Tie Down, Lifting and Towing provisions.

FUEL SYSTEM DIAGRAM

PART OF ENGINE

COMBINATION PRIMARY
AND SECONDARY FILTER
AND WATER SEPARATOR

DRAIN

FUEL INJ PUMP

TO INJECTORS

RETURN

MECHANICAL FUEL PUMP

RETURN FROM PUMP

AUX FUEL PUMP

AUX FUEL SUPPLY

FUEL LEVEL SWITCH

FILLER ASSY

FUEL TANK ASSY

DRAIN

- VENT ON UNDERSIDE OF FILLER CAP MUST
  BE IN OPEN POSITION WHEN SET IS RUNNING
- FOR OPERATION FROM SET "MASTER
  SWITCH" MUST BE IN "RUN" POSITION
- FOR OPERATION FROM EXTERNAL FUEL
  SUPPLY "MASTER SWITCH" MUST BE IN
  "RUN AUX FUEL" POSITION
- VENT ON UNDERSIDE OF FILLER CAP MUST
  BE IN OPEN POSITION WHEN SET IS RUNNING

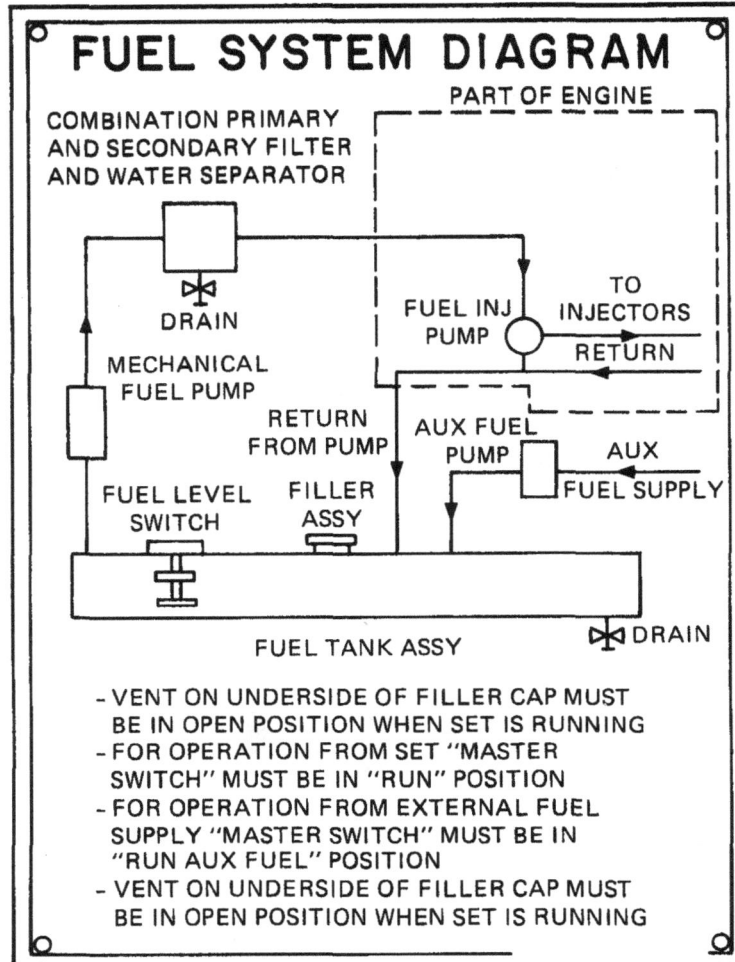

Figure 1-6. Fuel System Schematic.

1-11. DIFFERENCES BETWEEN MODELS. This manual covers DOD Models MEP-016B, MBP-021B and MEP-026B. The main differences between these models is the frequency, output voltage, and current that each delivers. The output characteristics are listed in Table 1-3. The other noticeable differences are in the control panels for each of the sets. These differences are thoroughly discussed in Chapter 2 of this manual.

Table 1-3. Output Characteristics.

Model MEP-016B	Model MEP-021B	Model MEP-126B
120 or 240 VAC single phase	120 or 240 VAC single phase 60 Hz	28 VDC 400 Hz
120 VAC three phase 400 Hz	120 VAC three phase 60 Hz	
120/208 VAC three phase, 4 wire Hz	120/208 VAC three phase, 4 wire 60 Hz 400	

CHAPTER 2

OPERATING INSTRUC-

Section I. OPERATING PROCEDURES.

2-1. GENERAL. This chapter contains instructions for starting, operating, and stopping the generator sets.   Location and functions of all controls and indicators are provided, as well as safety precautions for operating and maintaining the generator sets.

WARNING

All personnel who operate or maintain the generator sets should become thoroughly familiar with the safety precautions prior to performing operation or maintenance procedures.

Personnel should not attempt any of the following operating procedures without first grounding the unit as outlined in paragraph 2-5.1. Failure to do so could result in serious electrical shock or death from electrocution.

Operation of this equipment presents a noise hazard to personnel in the area. The noise level exceeds the allowable limits for unprotected personnel. Wear ear muffs or ear plugs.

WARNING

Prior to connecting load cables, be certain all the OFF or switches and circuit breakers are in position, and that the generator set OPEN is not running. Failure to do so can result in death from electrocution.

Models MEP-016B and MEP-021B can be switched to provide voltage outputs of either 120/240 volts, single phase; 120 volts, 3 phase or 120/208 volts, 3 phase, 4 wire. The output selection switch is located inside the control box (see Figures 2-1 and 2-2) and is accessible by turning the three "quarter turn" screws on the front of the control panel. It is important that the load is connected before starting the unit, and that the load connections and the setting of the output selection switch match the load requirements.

2-3.EQUIPMENT RECONNECTION. Before operating the unit, be voltage and frequency rating of the load matches that of the output voltage of the generator set must be changed to match rotate the output selection switch (see Figures 2-1 and 2-2) setting before starting the generator set. certain that the generator set.If the a particular load, to the desired output

2-4. CONTROLS AND INSTRUMENTS.

2-4.1. Model MEP-016B.   Controls and instruments for Model MEP-016B are illustrated in Figure 2-1 and described in Table 2-1.

2-4.2. Model MEP-021B.   Controls and instruments for Model MBP-021B are illustrated in Figure 2-2 and described in Table 2-2.

2-4.3. Model MEP-026B.   Controls and instruments for Model MEP-026B are illustrated in Figure 2-3 and described in Table 2-3.

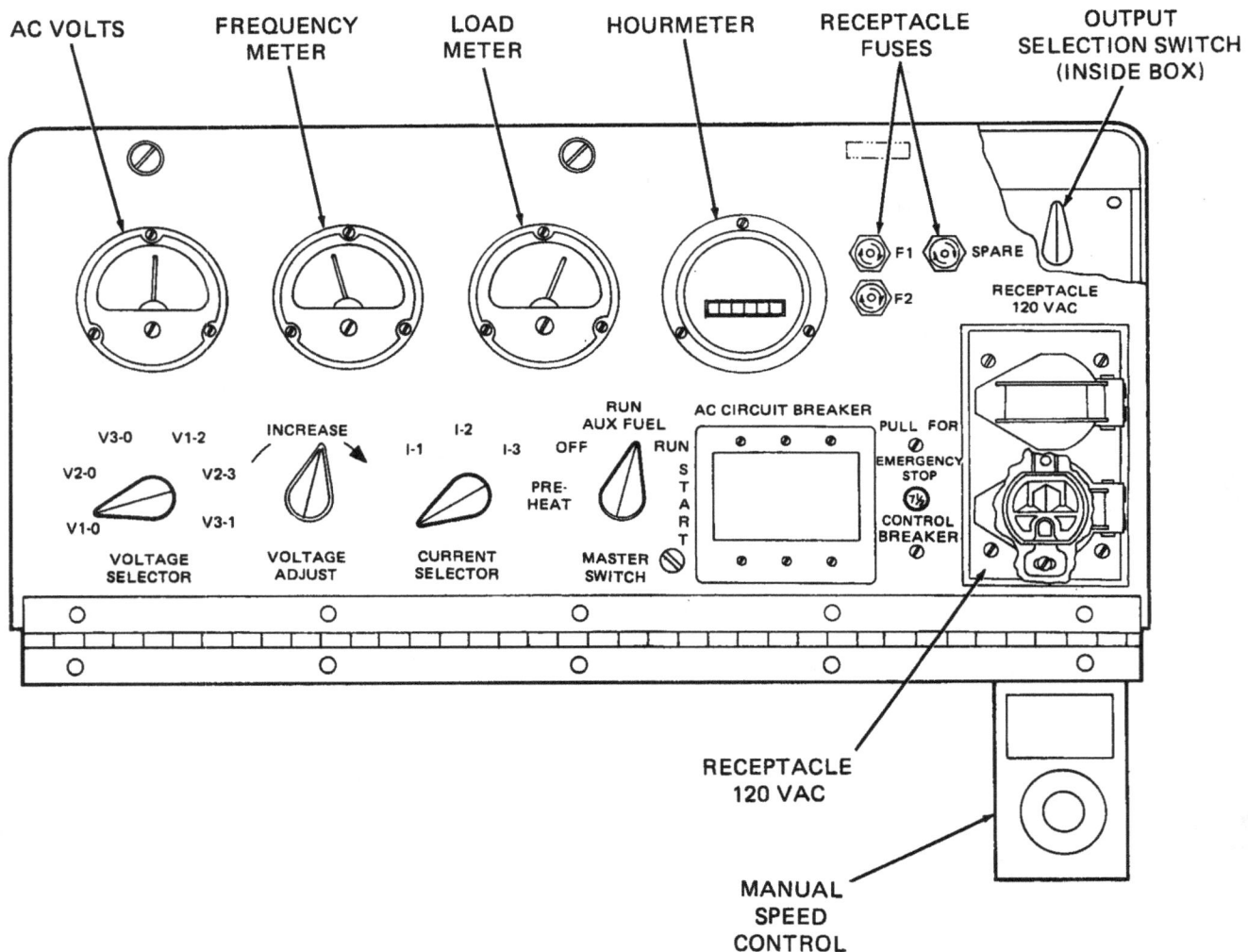

Figure 2-1. Controls and Instruments, Model MEP-061B (60 Hz).

AC Volts	50 to 250 volt scale in 10 volt increments. Red marks at 120, 208 and 240 volts .	Indicates output frequency.Normal reading is 60 Hz (red line). Indicates output current as percentage of load.
Frequency Meter	Range 55 to 65 Hz. in 0.5 Hz scale divisions. Red mark at 60 Hz.	Load is not to exceed 100%.
Load Meter	0 to 125 percent range. Red band from 100 to 125 percent.	Records total engine operating time in hours.
Hourmeter	0 to 9999.9 hours	

Indicates output voltage.Normal reading is 120, 208 or 240 volts.

Voltage Selector	Rotary switch	When this switch corresponds to the load terminal connections, the voltmeter will indicate output voltage for the selected load.
Voltage Adjust	Rheostat	Adjusts generator set output voltage. Clockwise to increase; counterclockwise to decrease.
Current Selector Rotary Switch		When this switch corresponds to the load, the load meter will indicate output current for the selected load terminal connections.
Master Switch	Rotary Switch	Controls the Preheat, Start, and Stop functions of the engine. Also used for switching between main and auxiliary fuel sources.
Main Circuit Breaker		Used to connect the load and protect the generator against
DC Control	7.5 amp breaker	Protects DC circuitry in case of short.Also used for emergency stopping (pull out to stop engine).

Control	Description	Function
Receptacle, 120 Provides power for lighting, tools or	Standard 120 VAC	Convenience outlet (120 volts AC, 60 Hz). VAC receptacle appliances.
Fuse, F1 12 amp fuse	Protects convenience outlet against short circuits.	
Fuse, F2	12 amp fuse	Protects convenience outlet against short circuits.
Fuse, Spare	12 amp fuse	Spare fuse for replacement of F1 or F2.
Output Selection Rotary switch Switch located inside	or 240 volts single phase; 120 volts, 3 control box.	Selects output voltage from generator: 120 phase; or 120/208 volts, 3 phase, 4 wire.
Manual Speed Control		Controls engine speed.

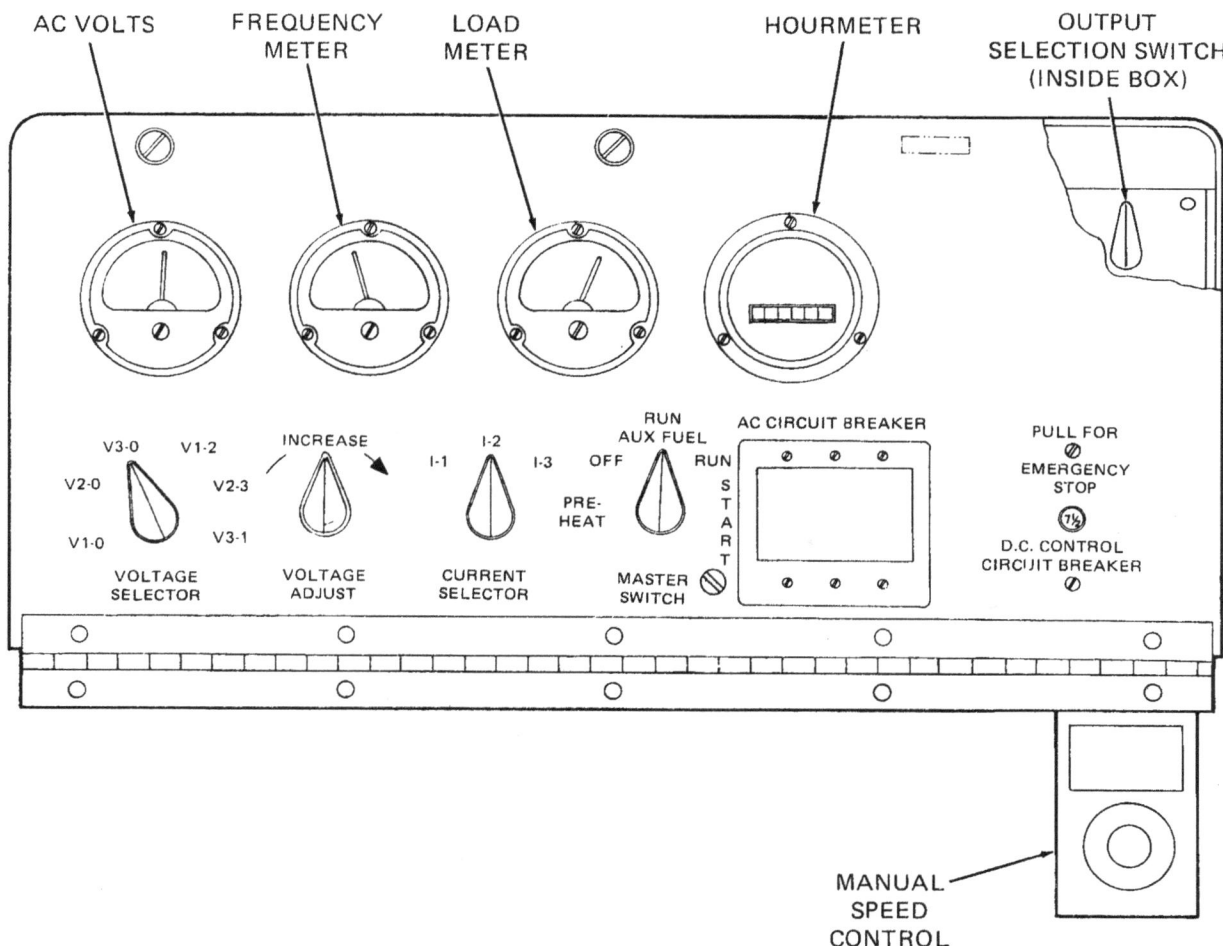

Figure 2-2. Controls and Instruments, Model MEP-021B (400 Hz).

Table 2-2. Controls and Instruments, Model MEP-021B (400 Hz).

Control	Description	Function
AC Volts	50 to 250 volt scale in 10 volt increments. Red marks at 120, 208 and 240 volts .	Indicates output voltage. Normal reading is 120, 208 or 240 volts.
Frequency Meter	Range 380 to 420 Hz.in 0.5 Hz scale divisions. Red mark at 400 Hz.	Indicates output frequency. Normal reading is 400 Hz (red line).

MARINE CORPS TM 05926B/06509B-12
ARMY   TM 5-6115-615-12
NAVY   NAVFAC P-8-646-12
AIR FORCE   TO 35C2-3-386-31

Control	Description	Function
Load Meter	0to 125 percent range. Red band from 100 to 125 percent.	Indicates output current as percentage of load. Load is not to exceed 100%.
Hourmeter	0to 9999.9 hours	Records total engine operating time in hours.
Voltage Selector	Rotary switch Rheostat	When this switch corresponds to the load terminal connections, the voltmeter will indicate output voltage for the selected load terminal.
Voltage Adjust	Rotary switch	Adjusts generator set output. Clockwise to increase, counterclockwise to decrease.
Current Selector	Rotary switch 7.5 amp breaker	When this switch corresponds the load meter to the load, output will indicate for the selected load. current
Master Switch		Controls the Preheat, Start, functions of the engine. Also used for switching between main and auxiliary fuel sources.
Main Circuit Breaker		Used to connect the load and to protect the generator against shorts in the load.
DC Control Circuit Breaker	Rotary switch located inside control box.	Protects DC circuitry in case of short. Also used for emergency stopping (pull out to stop engine).
Output Selection Switch		Selects output voltage from generator: 120 or 240 volts, single phase; 120 volts, 3 phase; or 120 or 208 volts, 3 phase, 4 wire.
Manual Speed Control		Controls engine speed.

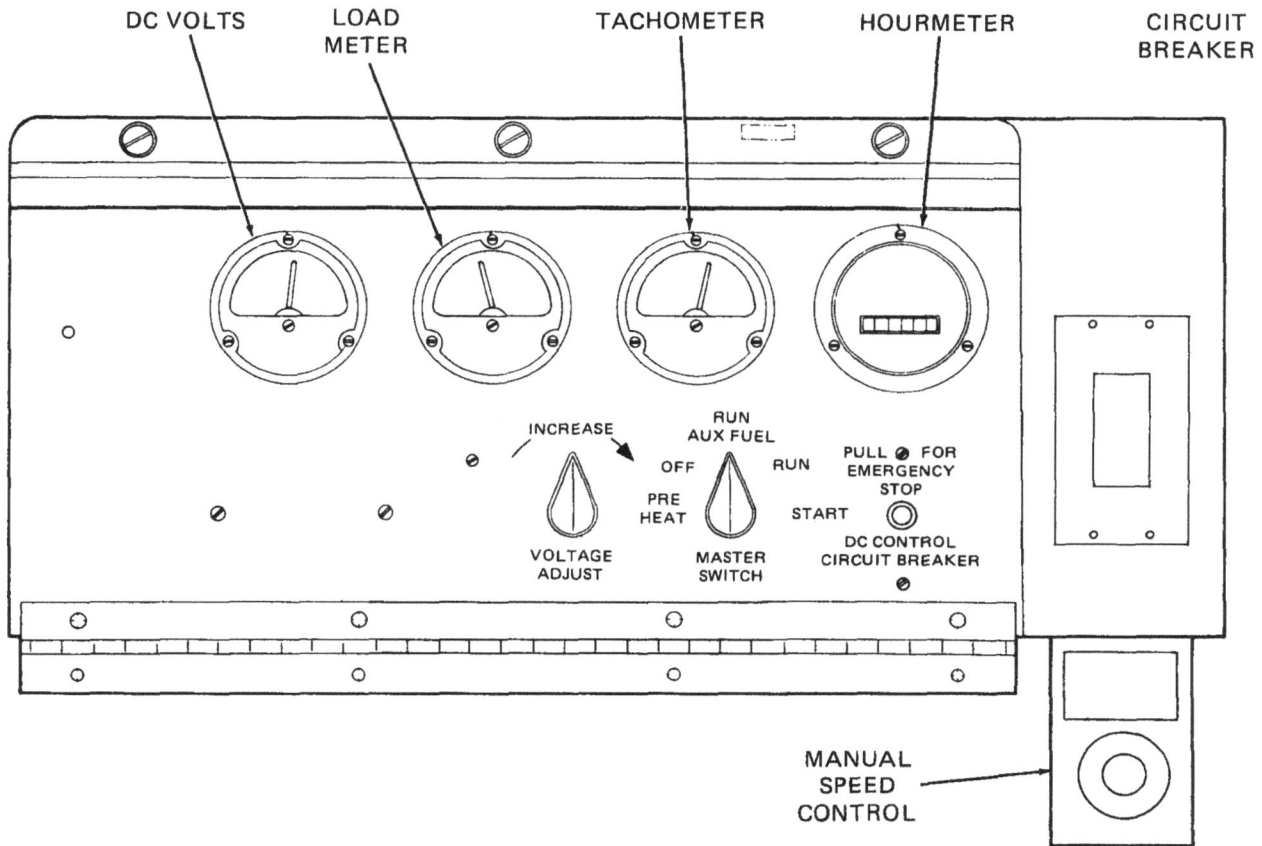

Figure 2-3. Controls and Instruments, Model MEP-026B (28 VDC).

Table 2-3.Controls and Instruments, Model MEP-02613 (28 VDC).

Control	Description	Function
DC Volts	0 to 50 volt range in 1 volt increments. Red mark at 28 volts .	Indicates output voltage.Normal reading indicates 28 volts DC.
Load Meter	0 to 125 percent range. Red band from 100 to 125 percent.	Indicates output current as percentage of load. Load is not to exceed 100 percent.
Tachometer	Range from 3300 to 3900 rpm in increments of 20 rpm. Red mark at 3600 rpm.	Indicates engine speed in revolutions per minute (RPM). Do not exceed 3600 RPM. Normal reading is 3600 RPM.
Hourmeter	0 to 9999.9 hours	Records total engine operating time in hours.

Control	Description	Function
Voltage Adjust	Rheostat	Adjusts generator set output voltage. Clockwise to increase; counterclockwise to decrease.
Master Switch	Rotary switch	Controls the Preheat, Start, Run and Stop functions of the engine. Also used for switching between main and auxiliary fuel sources.
DC Control Circuit Breaker	7.5 amp breaker	Protects DC circuitry in case of short. Also used for emergency stopping (pull out to stop engine).
Main Circuit Breaker		Used to connect the load and to protect the generator against a short in the load.
Manual Speed Control		Controls engine speed.

2-5. OPERATING PROCEDURES. The instructions in this section are provided for the information and guidance of the personnel responsible for the operation of the generator sets. The operator must know how to perform every operation of which the generator sets are capable.   This section gives instructions on starting and stopping the generator sets and regulating them to perform the specific tasks for which the equipment is designed.

2-5.1. Grounding. The generator set must be properly grounded before operation.

a. System Description. The AC alternator stator consists of three individual windings which are brought out of the generator housing by six wires, T-1 and T-4, T-2 and T-5, T-3 and T-6, which are connected to the output reconnect switch. By positioning the output switch to the desired position, the alternator output leads, T-1 through T-6 are connected to give the desired output voltage configuration. As wired, the generator output is always floating (not connected to ground); safety requirements dictate that the system must be grounded. This is accomplished by placing a #6 American Wire Gauge (AWG) stranded conductor between the proper output terminal (paragraph 2-5.l.b) and the generator set ground terminal located on the generator frame. To complete the grounding requirement, the ground terminal must also be connected to a proper grounding electrode system (ground rod, ground plate, etc.).

b. Set Grounding Instructions.   When grounding the 60 Hz and 400 Hz sets, grounding connections MUST be made to both the output terminal and frame terminal (see Figure 2-4.1). When grounding the 28 VDC set, only the frame (ground terminal stud) need be grounded (see Figure 2-4.2).

MARINE CORPS TM 05926B/06509B-12
ARMY   TM 5-6115-615-12
NAVY   NAVFAC P-8-646-12

(1)   For 120 volt single phase,120 volt three phase, or 240 volt single phase modes, connect a #6 AWG stranded conductor between the output terminal stud L-2 and frame (ground terminal stud) (see Figures
2-4.1 and 2-4.3).

(2)   For a 120/208 volt three phase mode, connect a #6 between output termi-   AWG stranded conductor terminal nal stud LO and frame (ground Figures 2-4.1 and 2-4.3).

WARNING

When changing the position of the output reconnection switch, make sure that the grounding jumper is connected to the correct output terminal as indicated in paragraph 2-5.1.b.Serious damage may result if the ground jumper is incorrectly connected.

For ALL modes, ground the frame (ground terminal stud) to an appropriate ground electrode system as described in paragraph 2-5.l.d.

(3)   c. Output Receptacle Grounding.   The leads to the output receptacle are connected to alternator leads T-1 and T-4. The receptacle is floating (not grounded).When a grounding jumper is installed between the specified output terminal (paragraphs 2-5.1.b (1) and 2-5.1.b (2)), the receptacle will automatically be grounded with proper polarity EXCEPT when the output switch is in the 240 volt single phase mode.In the 240 volt single phase mode, the output receptacle is always floating (not grounded).

In the 240 volt single phase mode, the output receptacle "NEUTRAL" lead is NOT GROUNDED.When L-2 is grounded in the 240 volt mode, both leads to the receptacle will measure a potential with respect to ground . Serious damage may result if the output receptacle is grounded.

WARNING
---

d.   Grounding Electrode (Rod) System.

(1) The generator set must be grounded before operation. The ground can be, in order of preference: (1) a underground metallic water piping; (2) a driven metal rod; (3) a buried metal plate (see Figure 2-4.2).

NOTE

A ground rod is furnished with the unit.

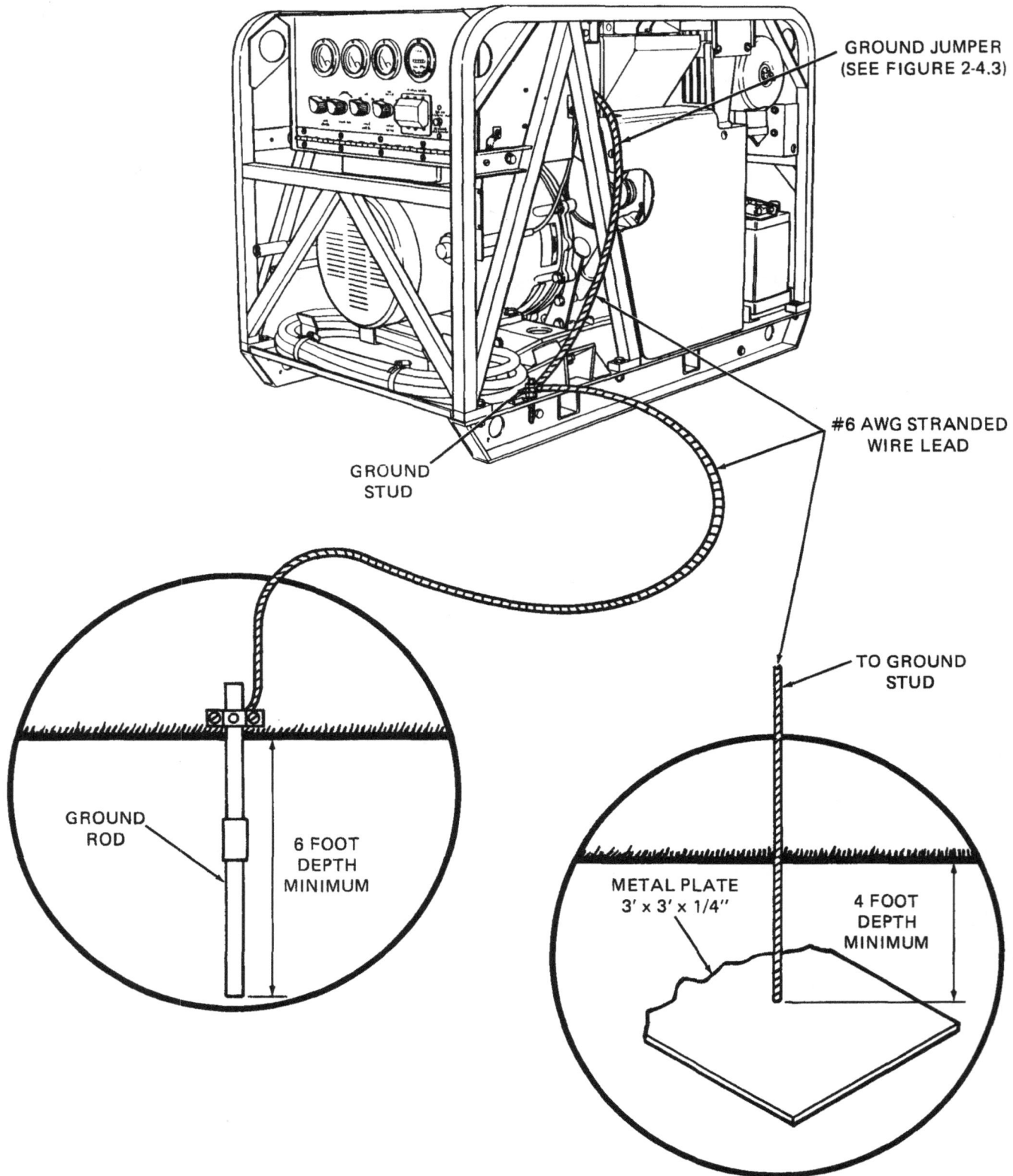

Figure 2-4.1. Grounding the GeneratorSet (60 Hz and 400 Hz).

GROUND
STUD

#6 AWG STRANDED
WIRE LEAD

TO GROUND STUD

GROUND
STUD

MIN
1/4" THICK

4 FT DEPTH
MINIMUM

METAL PLATE
9 SQ FT MIN

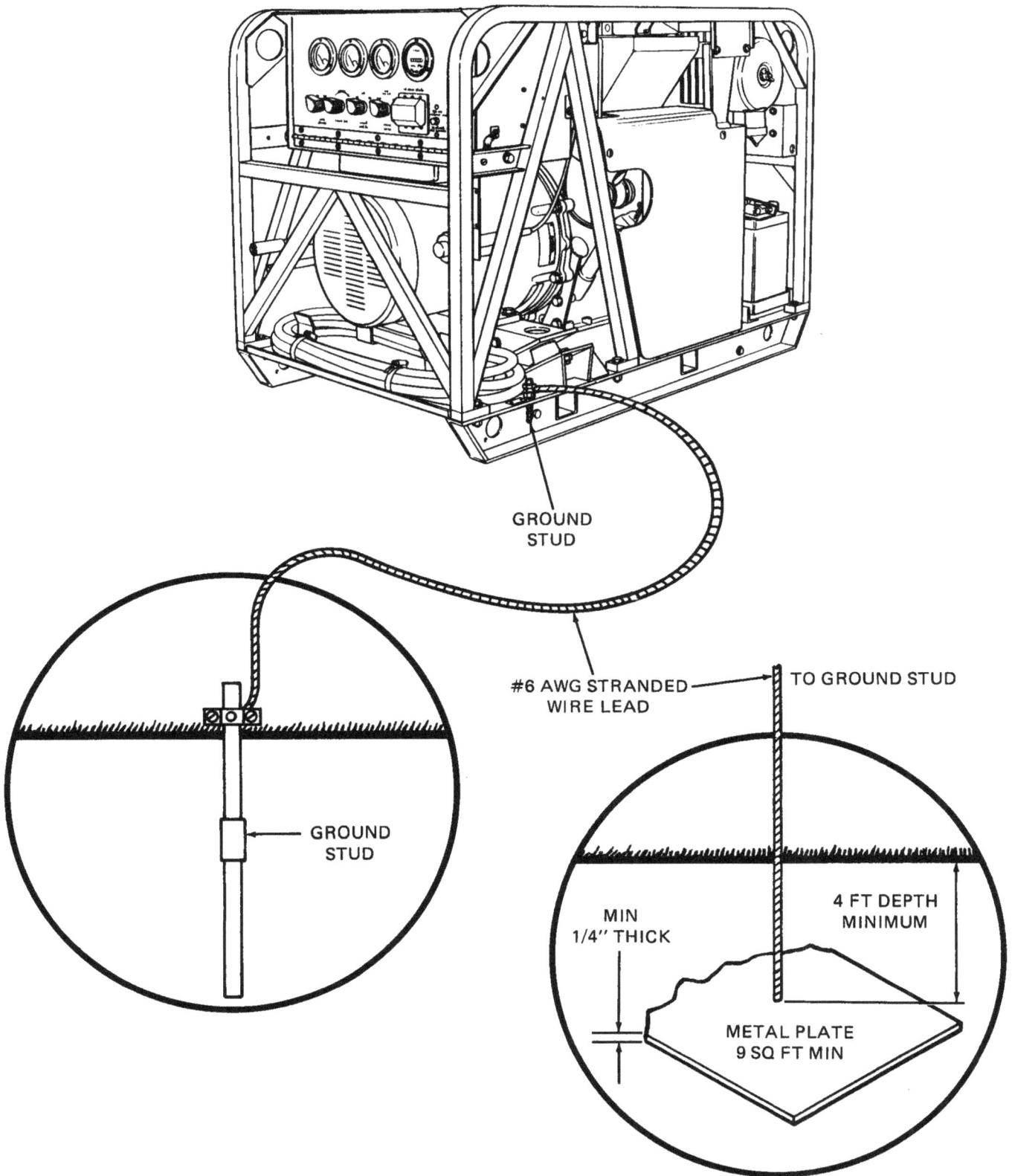

Figure 2-4.2. Grounding the Generator Set (28 VDC).

(2) If the effectively grounded portion of the buried metallic water pipe is less than 10 feet (3.0 meters) because of insulated sections or joints, this preferred grounding method must be supplemented by an additional driven metal ground rod or a buried metal plate.A driven metal ground rod must have a minimum diameter of 5/8 inch (16 mm) if solid, or 3/4 inch (20 mm) if pipe. The rod must be driven to a minimum depth of 6 feet (2 meter).   A buried metal plate must have a minimum area of nine square feet
(1 square meter), a minimum thickness of 1/4 inch (7 mm), and be buried to a minimum depth of four feet (1.5 meters).   The ground lead must be at least a #6 AWG or
thicker stranded copper wire.

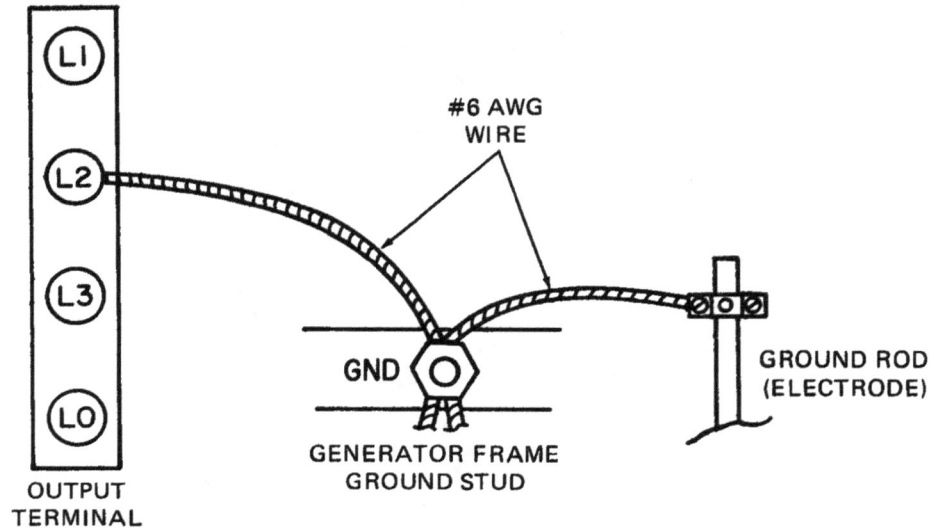

GROUNDING CONNECTION FOR 120V 1 PHASE,
120V 3 PHASE AND 240V 1 PHASE

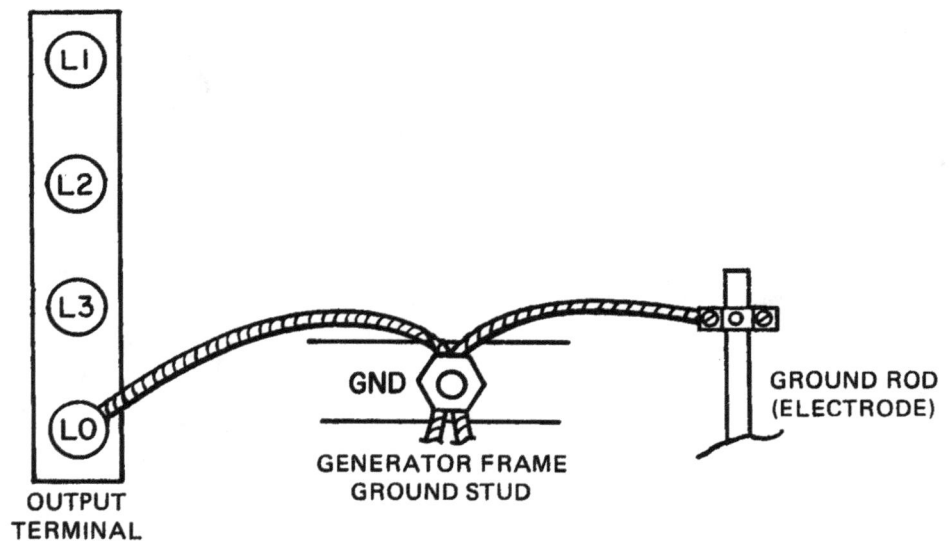

GROUNDING CONNECTION FOR 120/208V 3 PHASE

Figure 2-4.3.Grounding Connections (60 Hz and 400 Hz).

MARINE CORPS TM 05926B/06509B-12
ARMY   TM 5-6115-615-12
NAVY NAVFAC P-8-646-12
AIR FORCE   TO 35C2-3-386-31

2-6.   STARTING.

2-6.1. Preparation For Starting.

a.   Be sure set is adequately grounded (refer to paragraph 2-5.1).

b. Perform the daily preventive maintenance services (see Table 3-2).

c. Be certain load is connected, and that setting of output selection switch corresponds to load requirements.

d.   Switch Main circuit breaker to the OFF position.

e.   If fuel filter has been removed for any reason, or if bleed screw on top of filter was opened to drain the filter, be sure bleed screw is tight and use manual priming lever on fuel transfer pump to prime fuel system (see Figure 2-5).

CAUTION

Do not attempt to prime fuel system fuel filter assembly by adding fuel to permanently. Doing so can fuel injection pump.

NOTE

Manual priming lever must be depressed several times before fuel system is primed. Priming is complete when fuel drains back into fuel tank through return line (observable by removing fuel filler cap).

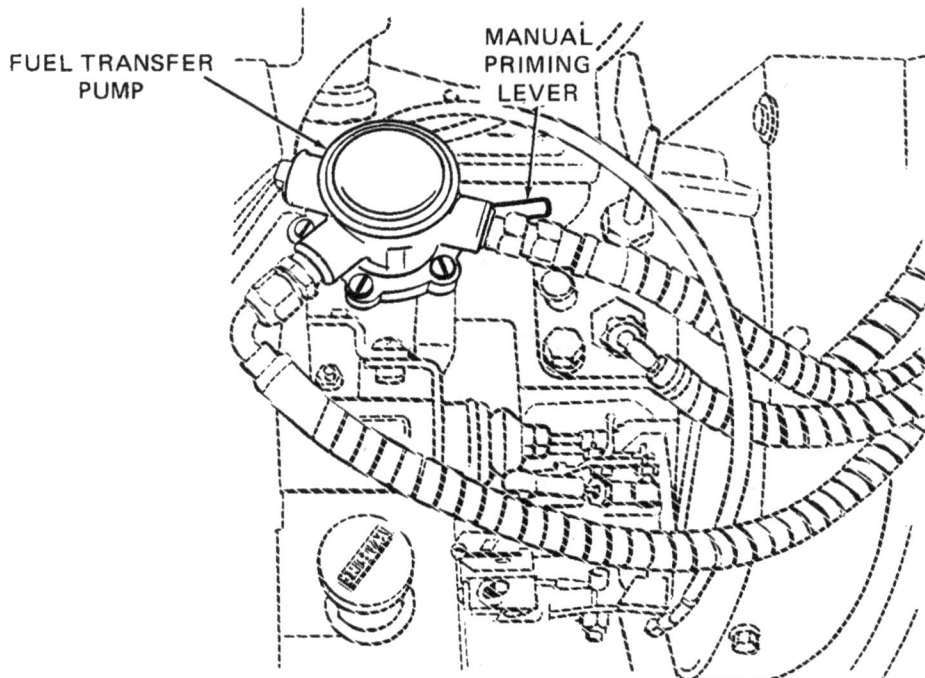

Figure 2-5. Manual Priming Lever and Fuel Transfer Pump Connection.

g. If an auxiliary fuel source is to be used, connect fuel line to auxiliary fuel pump (see Figure 2-6).

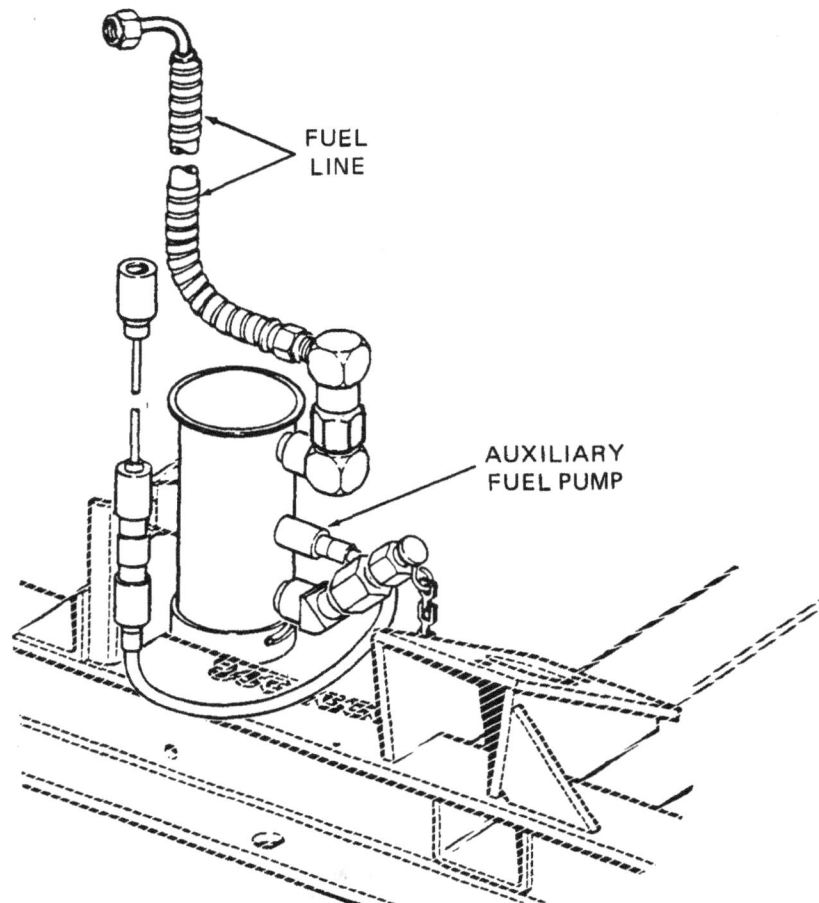

Figure 2-6. Auziliary Fuel Pump.

h. Vent on underside of fuel filler cap must be in the open position whenever the set is running (see Figure 2-7).

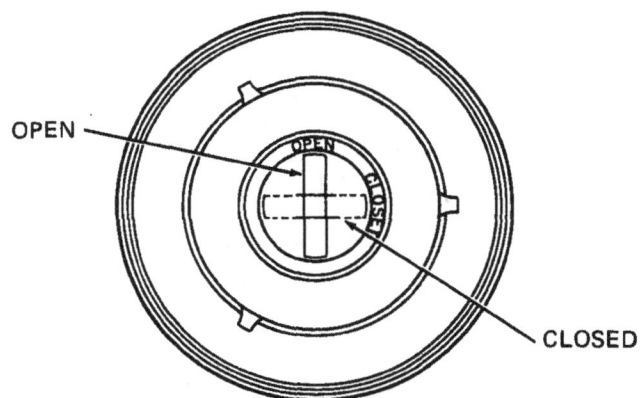

Figure 2-7. Fuel Filler Cap.

2-6.2.   Starting and Operation (all models).

a.   Make sure that the DC control circuit breaker is in the ON position.

b.   Rotate master switch to PREHEAT position and hold for 30 seconds.

c.   Roate master switch to START position and hold until voltmeter reads normal operation. Release master switch. It will automatically go to the RUN position.

NOTE

If engine does not start within 30 seconds, repeat start sequence above.   If engine will not start after a total of five attempts, turn master switch to OFF position and allow the starter to cool for 60 seconds before attempting to start engine again.   Repeat steps (b) and (c).   If engine fails to start, refer to troubleshooting (Table 3-3).

If running from an auxiliary fuel source,rotate the master switch to the RUN AUX FUEL position.

d.   Models MEP-016B and MEP-021B.Rotate the current selector to the desired position corresponding to the load requirements.Place the AC circuit breaker in the ON position.Use the engine manual speed control (located directly below the control panel) to adjust frequency.Turn the knob clockwise to increase, counterclockwise to decrease.

e.   Model MEP-026B.Use the engine manual speed control (located directly below the control panel) to adjust the engine speed to 3600 rpm.Turn the knob clockwise to increase engine speed, counterclockwise to decrease.

Models MEP-016B and MEP-021B.Place the AC circuit breaker in the ON position.

Models MEP-026B.Place the DC control circuit breaker in the ON position.

Rotate the voltage selector to the desired position corresponding to the load requirements.   Use the voltage adjust knob to obtain the desired output voltage: 120, 208 or 240 for Models MEP-016B and MEP-021B or 28 volts DC for Model MEP-026B.

f.

g.

h.   After unit has reached normal output current   operating temperature recheck output voltage, and engine RPM,and and frequency,

2-7.   **STOPPING THE UNIT.**

a.   Models MEP-016B and MEP-021B.   Switch   the main (AC) circuit breaker to the

Model MEP-026B. Switch the main (DC)   circuit breaker to the OFF position.

b. Rotate voltage adjust knob counterclockwise until it stops.

c. Rotate engine manual speed control knob counterclockwise to the in position
    to bring engine to idle.

d. Rotate the master switch to the OFF position.

e. After operation of the generator set, perform the "after operation" inspection and service procedures as out-
    lined in Table 3-2, Operator Preventive Maintenance Checks and Services.

2-8.   **EMERGENCY STOPPING.**   To stop the generator set in an emergency, pull out the
                                   (see Figures 2-1, 2-2 and 2-3).

MARINE CORPS TM 05926B/06509B-12
ARMY   TM5-6115-615-12
NAVY   NAVFAC P-8-646-12
AIR FORCE   TO 35C2-3-386-31

Section II. OPERATION UNDER UNUSUAL CONDITIONS.

2-9.   OPERATION IN BXTREME COLD (BELOW -25 degrees Fahrenheit, -32   degrees Celsius).

do the following:

a.   Use OEA grade of lubricating oil in the engine crankcase for arctic conditions (see Table 3-1 and Table 3-1, item 7). Change oil only when the engine is warm.

b.   Use arctic grade diesel fuel (see Table 3-1).

c.   Keep batteries in a well charged condition (between 1.280 and 1.300 on hydrometer reading).   If possible, remove battery from unit and store in a heated area when not in use.

d.   Keep fuel tank as full as possible to prevent accumulation of moisture or condensation.

e.   Remove any ice or snow which may have accumulated on the engine, generator, or wiring.

2-10.   OPERATION IN BXTREME HEAT.

To ensure satisfactory operation under conditions of extreme heat, do the following:

a.   Be sure that nothing obstructs air flow to or from the unit.

b.   Keep cooling fins clean.

WARNING

Battery electrolyte is an acid solution that gives off flammable fumes.Do not smoke or use open flame when working around battery.   Doing so can cause an explosion that could result in serious personal injury. If skin is exposed to electrolyte, flush exposed area with water immediately.   If eyes are exposed to electrolyte, flush them with water and seek immediate medical attention.

c.   Inspect the battery electrolyte level daily.Add distilled water if necessary to keep the electrolyte level over the plates.

d.   Keep the generator free of dirt and grime.Be sure ventilating screens are free of obstructions.

e.   Inspect load meter frequently to make sure that generator is not overloaded.

2-11.   **OPERATION IN DUSTY OR SANDY AREAS.**

To ensure that the generator set will operate satisfactorily in dusty or sandy areas, do the following:

a.   Shield generator from dust.Take advantage of natural barriers that offer protection from wind and dust.

b.   Clean the generator set as required.

c.   Service air cleaner as required.Check air restriction indicator daily. If red signal is visible , service the air cleaner.Be certain that all air cleaner and intake manifold connections do not leak.   Be certain oil filler cap fits tightly.

Change crankcase oil and oil filter every 60 hours or as needed.Clean area around oil filler cap and filter connection before inspecting or servicing engine.

d.   Store oil and fuel in dust free containers. Be certain that foreign matter does not enter fuel tank when refueling.

e.

f.   Make sure   that the generator set ground connections are free of dust and that connections are tight before
      sand, and   starting unit.

2-12.

To ensure that the geneator set will operate satisfactorily the following: in wet or humid areas,
do

a.   Keep the unit covered when not operating.Remove the periods.   cover during dry

b.   Keep the main and auxiliary fuel tanks as full as possible to protect against moisture and condensation accumu-
      lation.

c.   Keep electrical components and wiring clean and dry.Humid conditions can cause corrosion and deterioration of electri-
      cal components.

2-13.   **OPERATION IN SALT WATER AREAS.**

a.   Wipe the generator set with a clean cloth dampened with clean, fresh water at frequent intervals. Use care not to con-
      taminate the fuel supply or damage the electrical system with water.

b.   Use care to prevent salt water from entering the engine when adding or changing oil.

c.   Paint all exposed nonpolished surfaces.   Coat exposed parts of polished steel or other ferrous
      metal with standard issue rustproofing material if available, or cover exposed parts with a light coating of grease (Table
      3-1, no. 17).

2-14. OPERATION AT HIGH ALTITUDE.

a. Be sure that oper-
ation.

b. the air flow is not obstructed to and from the generator during
Keep the cooling fins clean.

c. Keep generator free of dirt and grime.　　　　　　Be sure ventilating screens are clean and free of obstructions.

d. The generator sets are rated at 3 kilowatts up to 8,000 feet altitude. To calculate specific generator set output above
8,000 feet, use the following formula:

$$\frac{7\% \times (\text{actual altitude} - 8000) \times 3kw}{1000} = kw \text{ derating}$$

Example:　　　　　1,000　　　　$= 1.05$ kw derating

3kw- 1.05 kw =1.95 kw derated power at 13,000 feet.

CHAPTER 3

OPERATOR/CREW MAINTENANCE INSTRUC-

Section I. CONSUMABLE OPERATING AND MAINTENANCE   SUPPLIES.

3-1. CONSUMABLE SUPPLIES. Table 3-1 contains all consumable   supplies and the
quantities necessary for operating and maintaining this generator set.

Table 3-1. Consumable Operating and Maintenance Supplies.

Item tion	component Number	National Stock Description	Required	Required For Initial Operation	Qty 8 Hours No. Operation	Qty	Applica-
1	Tank, fuel	9130-00-256-8613		JP-4, MIL-J-5624 Bulk *	1.0 Gal.	4.8 Gal.	
				FUEL OIL, DIESEL as follows:			
2		9140-00-286-5294		Regular Grade, DF2	1.0 Gal.		4.0 Gal.
3		9140-00-286-5286		Winter Grade, DF1	1.0 Gal.	4.0 Gal.	
4		9140-00-286-5283		Arctic Grade, DFA	1.0 Gal.	4.0 Gal.	

˙For Emergency use only

	Crankcase		OIL, LUBRICATING # five gallon can as follows:				
5		9150-00-188-9858	Grade OE/HDO 30	3.0 Qt.			
6		9150-00-186-6668	Grade OE/HDO 10	3.0 Qt.			
7		9150-00-402-2372	Grade OEA/APG-PD-l	3.0 Qt.			

Refer to Figure 3-1.

Table 3-1. Consumable Operating and Maintenance Supplies, Continued.

Item	Component	National Stock	Required	Required For Initial	Qty 8 Hours	Qty No.	Applica-
8	Battery	6810-00-249-9354		Electrolyte	0.7 Gal.		
9	Misc.	6850-00-264-9037		Dry cleaning solvent, P-D-680	As req'd		
10	Misc.	8030-01-025-1692		Sealing compound, MIL-S-46163-A Type II, Grade N, Removable	As req'd		
11	Misc.	8030-00-148-9833		Sealing compound, MIL-S-46163-A Type III, Grade R Removable	As req'd		
12	Misc.	8030-00-148-9833		Sealing compound, MIL-S-46163-A Type I, Grade K	As req'd		
13	Misc.	8030-00-133-3164		Sealing compound, MIL-S-22473-E Grade HVV	As req'd		
14	Misc.	8030-00-148-9833		Sealing compound, MIL-R-40082 Type I	As req'd		
15	Misc.	8030-00-058-5398		Loctite Superflex Ultra Blue Silicone sealant	As req'd		
16	Misc.	8030-00-059-2761		Anti-seize lubricant MIL-A-907	As req'd		
				Grease			
17	Mics.	xxxx-xx-xxx-xxxx		Plastic tie wraps	As req'd		
18	Misc.	5975-00-074-2072			As req'd		

Section II. LUBRICATION INSTRUCTIONS.

3-2. GENERAL. The generator utilizes sealed bearings; lubrication of the generator is not required.For general lubrication information on the engine; Army, Marine Corps and Navy users should refer to LI 05926B/06509B-12/5 reproduced in this section. Air Force users will use the lubrication section of applicable T.O. series workcards.

3-3. LUBRICATION ORDER.

Refer to Figure 3-1 for a reproduction of the Lubrication Order.

LUBRICATION INSTRUCTION

MARINE CORPS LI 05926B/06509B-l2/5
ARMY LO 5-6115-615-12
NAVY P-8-646-LO

GENERATOR SET, DIESEL ENGINE DRIVEN,
TACTICAL SKID MOUNTED, 3 KW

DOD MODEL   CLASS   MODE   NSN
MEP-016B UTILITY 50/60 HZ 6115-01-150-4140
MEP-021B UTILITY 400 HZ 6115-01-151-8126
MEP-026B UTILITY 28 VDC 6115-01-150-0367

Intervals (on condition or hard time) and Clean areas to be lubricated. Clean parts **the related man-hour times are based on** with dry cleaning solvent (SD), type II or normal operation. The man-hour time speci-   equivalent. Dry before lubricating. Dotted fied is the time you need to do all the   arrow points indicate lubrication on both services prescribed for a particular interval.   sides of the equipment.

On-condition (OC) oil sample intervals Level of maintenance. The lowest level of shall be applied unless changed by the maintenance authorized to lubricate a Army Oil Analysis Program (AOAP) labora-   point is indicated by one of the following tory, Change the hard time interval if your   symbols as appropriate: Operator/Crew lubricants are contaminated or if you are   (C); and Organizational Maintenance(O). operating the equipment under adverse operating conditions, including longer-than-usual operating hours. The hard time inter-val may be extended during periods of low activity. If extended, adequate preservation precautions must be taken. Hard time inter-vals will be applied in the event AOAP laboratory support is not available.

TOTAL MAN-HOURS		TOTAL MAN-HOURS	
INTERVAL HOURS	MAN-	INTERVAL HOURS	MAN-
D		125	0.5
0.2			
Q			
0.1			

Figure 3-1. Lubrication Instruction/Lubrication Order (Sheet 1 of 3).

LI 05926B/06059B-12/5

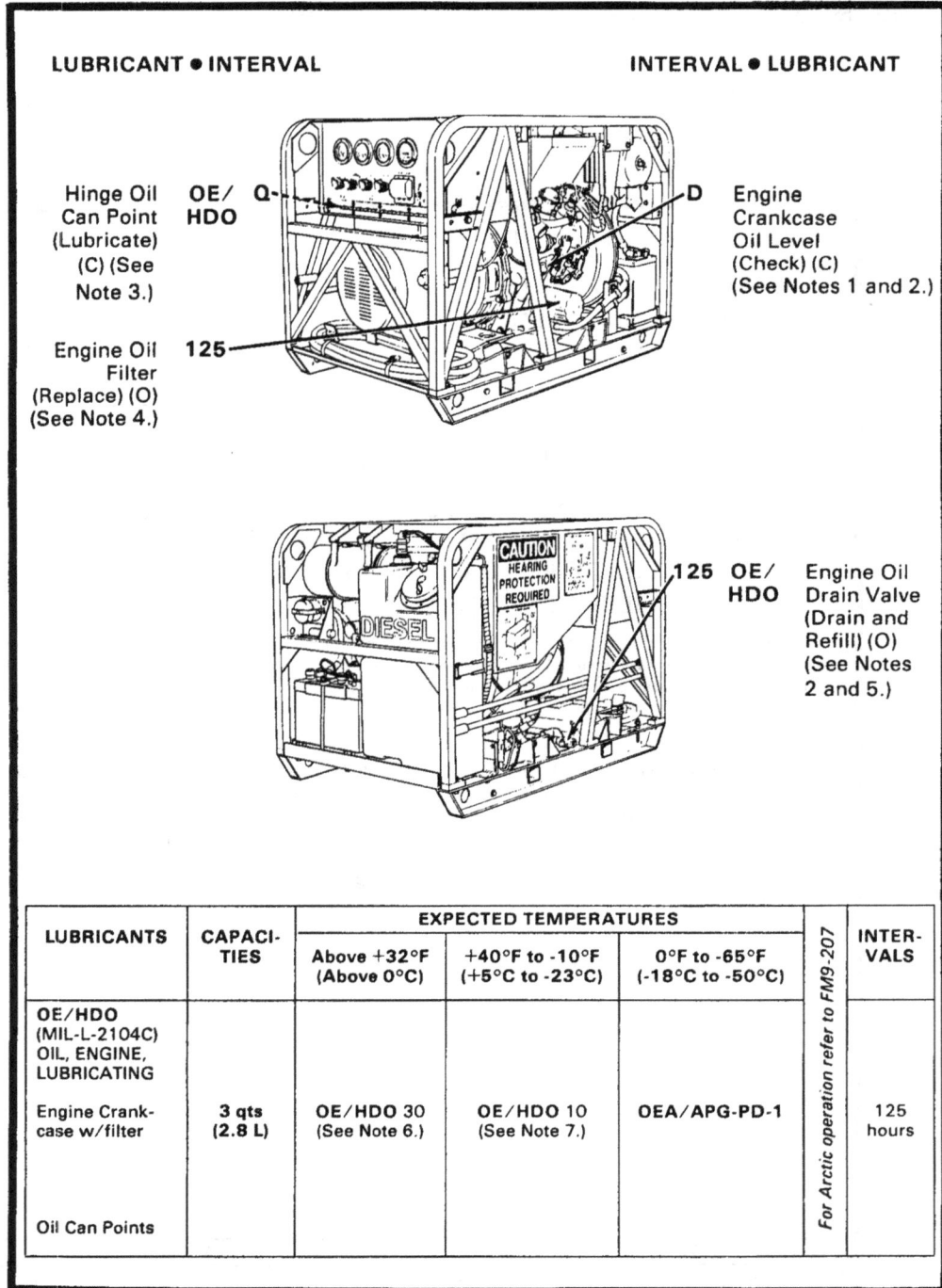

| LUBRICANT ● INTERVAL | | INTERVAL ● LUBRICANT |

Hinge Oil Can Point (Lubricate) (C) (See Note 3.) — OE/HDO — Q

Engine Oil Filter (Replace) (O) (See Note 4.) — 125

D — Engine Crankcase Oil Level (Check) (C) (See Notes 1 and 2.)

125 OE/HDO — Engine Oil Drain Valve (Drain and Refill) (O) (See Notes 2 and 5.)

LUBRICANTS	CAPACITIES	EXPECTED TEMPERATURES			For Arctic operation refer to FM9-207	INTERVALS
		Above +32°F (Above 0°C)	+40°F to -10°F (+5°C to -23°C)	0°F to -65°F (-18°C to -50°C)		
OE/HDO (MIL-L-2104C) OIL, ENGINE, LUBRICATING						
Engine Crankcase w/filter	3 qts (2.8 L)	OE/HDO 30 (See Note 6.)	OE/HDO 10 (See Note 7.)	OEA/APG-PD-1		125 hours
Oil Can Points						

CARD 2 of 3

Figure 3-1. Lubrication Instruction/Lubrication Order (Sheet 2 of 3).

LI 05926B/06509B-12/5

NOTES:

WARNING

Do not remove oil filler cap when engine is running. Hot oil can splash

1. ENGINE CRANKCASE OIL LEVEL. Start engine and allow to run until normal operat-achieved (about 5 min-utes). Check for leaks. Stop the unit and allow to sit for one minute. Remove the access panel on the left side of the unit. Remove filler cover/dipstick and wipe with

a clean, dry cloth. Insert dipstick and im-mediately remove. Oil level should be be-oil as needed and recheck oil level.

2. ENGINE CRANKCASE OIL. If a sample of engine crankcase oil is to be sent to an AOAP laboratory for analysis, refer to TB43-0210 for complete sampling require-ments.

3. OIL CAN POINTS. Clean and lightly coat hinge, fasteners, control cable and all ex-posed adjusting threads. DO NOT lubricate governor linkages.

4. ENGINE OIL FILTER. Drain engine oil (Note 5). Remove filter by turning counter-clockwise and clean base with a clean, dry cloth. Apply a thick film of clean engine oil to new filter gasket. Install new filter until **gasket contacts base, then turn an ad-ditional 3/4 turn clock-wise.** Add engine oil and check level (Note 1).

5. ENGINE OIL DRAIN VALVE. Start engine and allow to run until normal operating tempera-ture is achieved (about 5 minutes).

Stop the unit and open drain valve. Allow all oil to drain into suitable container. Change oil filter (Note 4). Refill crankcase and check oil level (Note 1). **up and cause burns.**

6. Grade 15W/40 (OE/HDO 15/40) may be used when expected temperatures are above +5°F (-15°C). ing temperature is

7. If OEA lubricant is required to meet the low "expected temperature" range, OEA

lubricant is to be used in place of OE/HDO-

10 lubricant for all expected temperature ranges where OE/HDO-10 is specified in the key. tween marks on dipstick. Add or drain

Copy of this Lubrication Order will remain with the equipment at all times; instructions contained herein are mandatory.

CARD 3 OF 3

Figure 3-1. Lubrication Instruction/Lubrication Order (Sheet 3 of 3).

Section III. PREVENTIVE MAINTENANCE CHECKS AND SERVICES (PMCS).

3-4. GENERAL. To ensure that the generator set is ready for operation at all times, it must be inspected systematically so that defects may be discovered and corrected before they result in serious damage or failure. The necessary preventive maintenance checks and services to be performed by the operator personnel are listed and described in paragraph 3-8.

3-5. CORRECTING AND REPORTING DEFICIENCIES. Defects discovered during operation will be noted for future correction.   Stop operation immediately if a deficiency is noted which could damage the equipment or present a safety hazard. All deficiencies will
Marine Corps users should refer to current issue of TM 4700-15/1.
       be recorded together with the corrective actions taken on the applicable form.

(MC)   Army users should refer to current issue of DA PAM 738-750.

(A)   Air Force users should refer to current issue of AFM 66-1 and the applicable 00-20 Series Technical Orders.

(AF)   Navy users should refer to their service peculiar directives to determine applicable maintenance forms and records to
       be used.
              DETERMINING PMCS INTERVAIS. Certain Operator PMCS on this unit should be performed on a "per

(N)   hours of operation" basis. The hourmeter on the control panel should be used to determine the generator set operating
       time.

3-6.

3-7. PMCS FOR UNITS IN CONTINUOUS OPERATION. For PMCS performed on an operating time basis, the PMCS should be performed as close as possible to the time intervals indicated. For units in continuous operation, perform PMCS before starting operation if continuous operation will extend beyond the service interval indicated.

3-8. OPERATOR PREVENTIVE MAINTENANCE CHECKS AND SERVICES. Table 3-2 contains a tabulated listing of PMCS to be performed by the Operator personnel. The item numbers are listed consecutively and indicate the sequence of minimum requirements.

Table 3-2.Operator Preventive Maintenance Checks and Services.
Interval B - Before Operation A - After Operation

Total M/H 1.3

Operator Daily	D-During Operation Daily- 8 Hours Items To Be Inspected	Equipment Is Not Ready Or Available

GENER-

1 9 15	Make a visual inspection of the entire generator set for cleanliness, any obvious deficiencies such as loose or missing hardware, and for any bent, cracked or broken parts. Inspect all wires and terminals for damage and loose connections.	Damaged components, loose or missing hardware, or leaking fluids are found.
2 10 16	Check fuel supply.Make sure that quantity of fuel in tank or in auxiliary fuel supply is enough for operation.	
3 11	Open drain valves on fuel tank and filter. Remove any water or sediment. Make sure vent on underside of fuel filler cap is open whenever set is running.	Fuel tank is empty. Water is present in fuel, or filler cap vent is closed.

WARNING

Battery electrolyte is an acid solution that gives off flammable fumes.   Do not smoke or use open flame when working around battery.   Doing so can cause an explosion that could result in serious personal injury. If skin is exposed to electrolyte, flush exposed area with water immediately.   If eyes are exposed to electrolyte flush them with water and seek immediate medical attention.

| 4 | Inspect battery electrolyte level. | Electrolyte level is below plates. |

ENGINE

| 5 | 17 | Inspect oil in crankcase   oil as necessary.
for proper level. Add | Oil level in |

Table 3-2.Operator Preventive Maintenance Checks and Services, Continued.
Interval B - Before Operation A - After Operation   **Total M/B 1.3**

Operator Daily BDA	Items To Be Inspected Inspection Procedures	Equipment Is Not Ready If:
6	Inspect air filter element.Clean if necessary. Refer to paragraph 3-29.	Air filter is damaged or clogged.
7 12 18	Inspect air flow indicator. If indicator shows red, clean or replace air cleaner element. indicator.Clean dust valve.Refer to paragraph 3-29.	Air flow indicator shows red. Reset
13	Inspect for unusual noises or operation, too Any of the listed much vibration, lack of power, excessive condi-tions exist. smoke or engine failing to respond to the controls. Refer to troubleshooting, Table 3-3.	Shut down genera-tor set under such conditions.
14	Inspect controls and instruments for proper Controls and/or operation. and instruments do not 2-3 for descriptions and ranges. Replace aged components. Refer to paragraph 3-35.	Refer to Tables 2-1, 2-2, operate properly. any da

GENERATOR

| 8 | Inspect the ground connections to make certain all are clean and tight. connections are loose | Ground stud t |

Section IV. TROUBLESHOOTING.

3-9. GENERAL. This section contains troubleshooting information for locating and correcting operating troubles which may develop in the generator set. Each
malfunction for an individual component, unit or system is followed by a list of tests or inspections which will help you determine the probable causes and corrective actions to take. You should perform the tests/inspections and corrective actions in the order listed.

3-10. MALFUNCTIONS NOT CORRECTBD BY THE USE OF THE TROUBLESHOOTING TABLE. This
manual cannot list all malfunctions that may occur, nor all tests or inspections and corrective actions.   If a malfunction is not listed, or cannot be corrected by listed corrective actions, notify your supervisor.

NOTE

Before you use the troubleshooting table, be sure you have performed all the applicable Operator Preventive Maintenance Checks and Services (refer to Table 3-2).

Table 3-3. Troubleshooting.

Malfunction
   Test Or Inspection
      Corrective Action

---

1. ENGINE FAILS TO CRANK WHEN MASTER SWITCH IS HELD IN THE START POSITION.

   Step 1.   Inspect to see that the DC control circuit breaker on the control
         panel is depressed.

            Depress button to set DC control circuit breaker. If breaker will not stay depressed, notify higher echelon of maintenance.

   Step 2.   Inspect for empty fuel tank.

            If fuel tank is empty or low, refill tank or switch to auxiliary fuel source.

MARINE CORPS TM 05926B/06509B-12
ARMY   TM 5-6115-615-12
NAVY   NAVFAC P-8-646-12

Table 3-3. Troubleshooting, Continued.

Malfunction
> **Test or Inspection**
>> **Corrective Action**

1. ENGINE FAILS TO CRANK WHEN MASTER SWITCH IS HELD IN THE START POSITION (CONT ' D).

    Step 3. Inspect for loose, corroded *or* broken battery cables or starter ground cable.

        If loose, corroded or broken battery cables and/or starter ground cable are found, notify higher echelon of maintenance.

WARNING

Battery electrolyte is a acid solution that gives off flammable fumes.   Do not smoke or use open flame when working around battery.Doing so can cause an explosion that could result in serious personal injury. If skin is exposed to electrolyte, flush exposed area with water.   If eyes are them with water and seek

        exposed to electrolyte, flush immediate medi-

    Step 4.   Inspect to see that   electrolyte (liquid) level in each battery cell the plates. is above the top of

        If electrolyte level is below top of plates, add distilled water.

2. ENGINE CRANKS NORMALLY BUT FAILS TO START.

    Step 1.   Inspect for empty fuel tank.

        If fuel tank is empty or auxiliary fuel

MARINE CORPS TM 05926B/06509B-12
ARMY   TM 5-6115-615-12
NAVY   NAVFAC P-8-646-12

Table 3-3. Troubleshooting, Continued.

Malfunction
    Test Or Inspection
        Corrective Action

2. ENGINGE CRANKS NORMALLY BUT FAILS TO START (CON'T).

    Step 2.   Inspect for sediment or water in fuel filter.

        Open drain on bottom of fuel filter and water.   If   drain sediment and   and
        necessary, drain fuel system clean fuel supply.

    Step 3.   Inspect for loose fuel fittings or bad fuel lines.

        If fuel lines are loose, cracked or show signs of leaking, notify higher echelon of maintenance.

    Step 4.   Inspect governor linkage for obstructions or binding (see Figure 3-2).

        Remove any obstructions or notify higher echelon of maintenance.

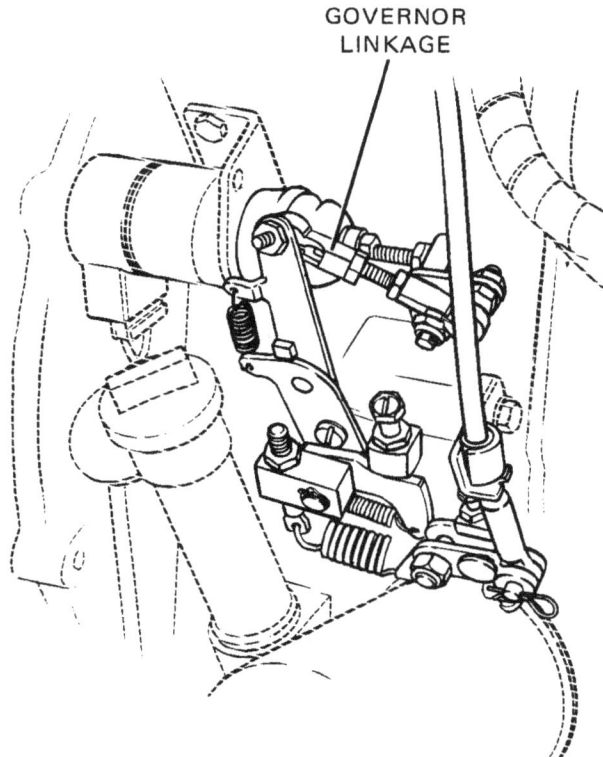

Figure 3-2 Governor Linkage.

MARINE CORPS TM 05926B/06509B-12
ARMY   TM 5-6115-615-12
NAVY   NAVFAC P-8-646-12

Table 3-3. Troubleshooting, Continued.

Malfunction

    Test Or Inspection

        Corrective Action

      Step 5.   Inspect Stop/Run solenoid to be sure it is fully engaged.
                  Inspect for loose or broken wiring or connections.

                    If Stop/Run solenoid is not fully engaged, or wiring is damaged, notify higher echelon of maintenance.

      Step 6.   Inspect to see if glow plug lead connection is clean and tight.

                    Clean and tighten connection as needed, or notify higher echelon of maintenance to repair or replace wiring.

3. ENGINE STARTS BUT DOES NOT RUN SMOOTHLY (MISFIRES, KNOCKS OR MAKES UNUSUAL NOISES).

      Step 1.   Refer   to Steps 2 and 3 under ENGINE CRANKS NORMALLY BUT FAILS TO above.

                  Perform corrective action as necessary.

      Step 2.   Inspect exhaust muffler assembly for obstructions.

                  Remove obstructions if possible or notify higher echelon of maintenance.

4. ENGINE STARTS AND RUNS NORMALLY BUT SUDDENLY STOPS.

      Step 1.   Inspect for empty fuel tank.

                    If fuel tank is empty, refill tank or switch to auxiliary fuel source.

              Inspect to see that vent on underside of fuel filler cap is open.
      Step 2.

                    Clear vent hole.

              EMITS BLACK SMOKE IN EXHAUST.
5. ENGINE RUNS BUT
              Inspect for restricted air intake. Red signal on air flow indicator should not be visible under
      Step 1.   normal conditions.

                    Remove any restrictions from intake port. Inspect air cleaner element and notify higher echelon of maintenance replacement is necessary.   Reset air flow indicator by pushing reset button.   if

Table 3-3. Troubleshooting, Continued.

Malfunction
    Test Or Inspection
        Corrective Action

Step 2.   Inspect load on generator by checking percent rated current meter on control panel.

If meter indicates more immediate than 100 percent load, notify supervisor.

Step 1.   Inspect for oil leaks, especially at front and rear oil seals, at oil pan gasket and at dipstick cap.

If oil is leaking, notify higher echelon of maintenance.

Step 2.   Inspect for white smoke or oil coming from exhaust pipe.

If exhaust pipe emits white smoke or oil, notify higher echelon of maintenance.

7. GENERATOR SUPPLIES NO VOLTAGE TO LOAD.

Step 1.   Inspect to be sure that the panel is in the main circuit breaker on the control MEP-
(Models MEP-016B and

WARNING

Do not hold main circuit breaker in ON position.   Shock
or

CAUTION

Do not hold AC circuit breaker in ON position.   Damage
to equipment can result.

Place main circuit breaker in the ON position.   If breaker will
not stay in the ON position, notify higher echelon of maintenance.

Table 3-3. Troubleshooting, Continued.

Malfunction

   Test Or Inspection

        Corrective Action

WARNING

Disconnect load before connec-   result.
tions.   Shock or

Step 2.   Inspect load terminal board.

             Make certain that load leads are attached to the correct load terminals. Make certain
             that connections are clean and tight.

8. GENERATOR SUPPLIES IMPROPER (UNDER OR OVER) VOLTAGE/FREQEUNCY TO LOAD.

        Step 1.   Inspect voltage selector switch located inside control box.
                     (Models MEP-016B and MEP-021B only).

                     Make certain switch is in the proper position.

WARNING

Shut down generator connec-   set before switching load terminal or death from
tions.   Shock

        Step 2.   Inspect for restricted air intake. Red signal on air flow
                     indicator should not be visible under normal conditions.

                     Remove any restrictions from intake port. cleaner element and   Check air
                     notify higher echelon replacement is necessary.Reset air flow   of maintenance indica- if
                     pushing reset button.   tor by

        Step 3.   Inspect load terminal board.

                     Make certain that load leads are attached

                                                                        to the correct

             load terminals. Make certain that connections are clean and tight.

Table 3-3. Troubleshooting, Continued.

Malfunction
  Test or Inspection
    Corrective Action

Step 4.   Inspect engine speed.

Adjust engine   o speed by turning manual speed control assembly until frequency meter on control panel
indicates 60 Hz or   400 Hz for AC units, or until rpm indicator indicates 3600 rpm for 28 volt DC units.

If engine speed   cannot be brought up so that generator operates at rated frequency, inspect external governor linkage for   binding and check steps under ENGINE STARTS BUT DOES NOT RUN SMOOTHLY.

If engine speed   o
governor linkage   cannot be brought down so that generator operates at rated frequency, inspect external
    for binding or notify higher echelon of maintenance.

o

Section V.OPERATOR MAINTENANCE PROCEDURES.

WARNING

Unless manual directs otherwise, do not attempt any of the following maintenance procedures when generator set is operating.Do not touch exposed electrical connection when
a source of power such as utility power or another generator set is connected to the load
terminals.   Severe electrical shock or death by electrocution can result.

3-11. GENERAL. This section contains information on the maintenance of the equipment that is the responsibility of the
operator.

3-12. FRAME.(See Figure 3-3)

   a. Inspect frame (7) for damage such as cracks, dents, rust or misalignment.
      Notify higher echelon of maintenance for repair or replacement of frame.

   b. Inspect for loose or missing hardware (items 1, 2, 3, 5, 6, 9, 10, 12, 13, 14,

      15, 16, 17, and 18). Notify higher echelon of replace hardware as neces- maintenance to tighten or
      sary.

3-13.   LIFTING EYE.   (See Figure 3-3)

Inspect lifting eye (8) for cracks, damage and secure echelon of maintenance to   mounting.Notify higher
tighten or replace mounting of spacer (11) or lifting eye (8).   bolts (9), or for replacement

3-14.   SKID BASE AND GROUND STUD.   (See Figure 3-3)

   a.   Inspect skid base (4) for damage such as cracks, dents, rust or misalignment.
      Notify higher echelon of maintenance for repair or replacement.

   b.   Inspect for loose or missing hardware (items 19, 20, 21, 22, 23, 25, 26, 29, 30, 31, 32, 33, 34, 35 and 36).Notify higher
      echelon of maintenance to tighten or replace hardware as necessary.

      Inspect rubber engine mounts (20) for damage, deterioration and secure mounting.Notify higher echelon of main-
   c.   tenance for replacement.

1. NUT
2. WASHER
3. WASHER
4. SKID BASE
5. WASHER
6. BOLT
7. FRAME
8. LIFTING EYE
9. BOLT
10. WASHER
11. SPACER
12. BOLT
13. WASHER
14. NUT
15. CLAMP
16. WASHER
17. WASHER
18. BOLT
19. BOLT
20. FOOT
21. WASHER
22. WASHER
23. NUT
24. STUD, GROUND
25. WASHER
26. NUT
27. BRACKET, REAR
28. BRACKET, FRONT
29. SCREW
30. WASHER
31. SCREW
32. WASHER
33. SCREW
34. LOCKNUT
35. SCREW
36. LOCKNUT
37. BRACKET
38. CLAMP
39. CLAMP

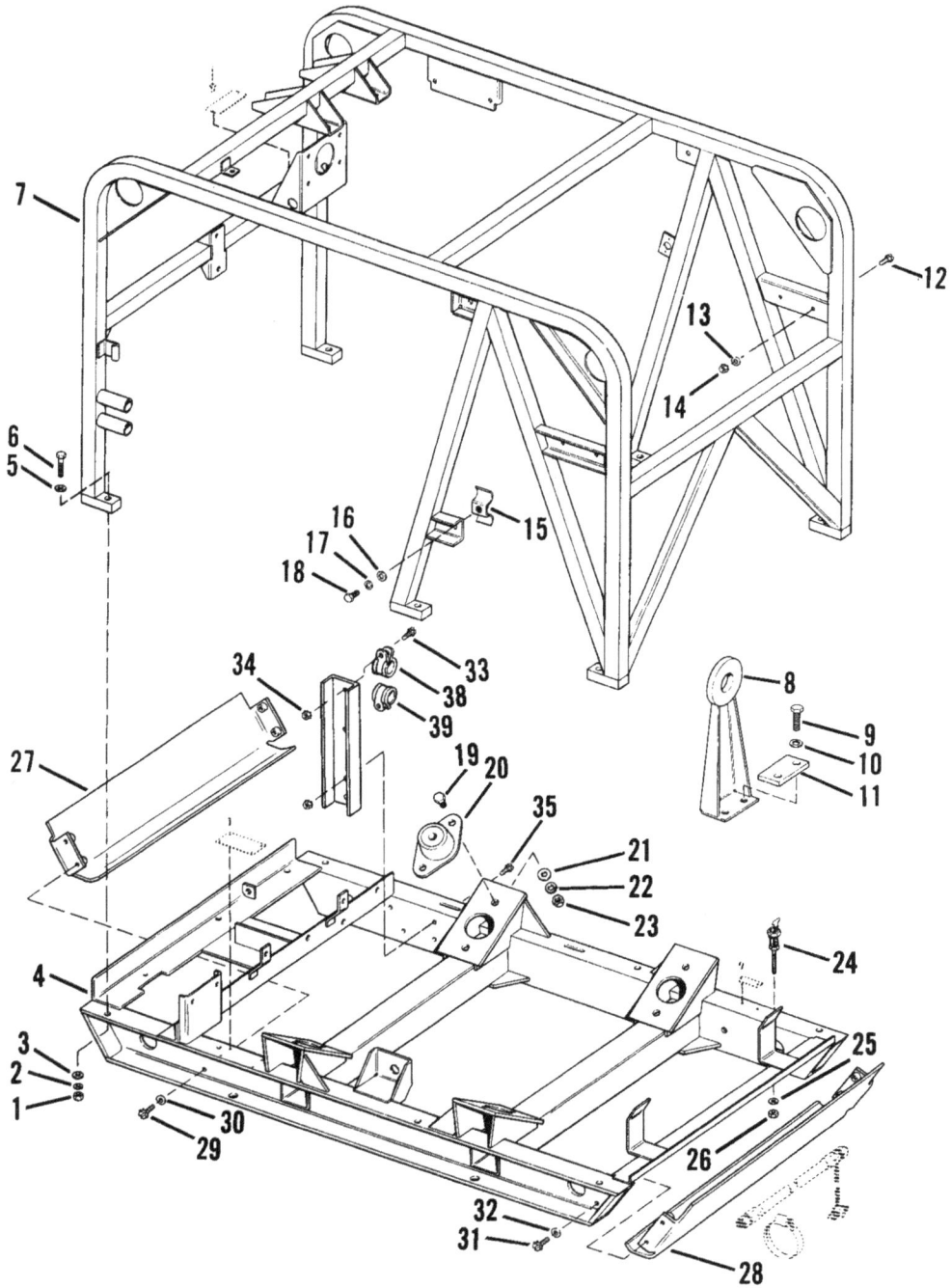

Figure 3-3.Frame and Skid Base.

d. Inspect ground stud (24) for damaged threads, corrosion or other damage. Check that ground stud is securely attached to the skid base. Check that connections to ground stud are clean and tight.

e. Inspect brackets (27, 28 and 37) for damage such as cracks, dents or misalignment. Notify higher echelon of maintenance for repair or replacement of brackets.

f. Inspect clamps (38 and 39) for wear or damage.Notify higher echelon of

maintenance for   replacement of clamps.

3-15.   BATTERY FRAME,TRAY AND BATTERY HOLDDOWN. (See Figure 3-4)

WARNING

Battery   electrolyte is an acid solution that gives off flammable fumes.   Do not smoke or use open flame when working around battery.   Doing so can cause an explosion that could result in serious personal injury. If skin is exposed to electrolyte, flush exposed area with water immediately.   If eyes are exposed to electrolyte, flush them with water and seek immediate medical attention.

a. Inspect battery top frame (9), bottom frame (6) and battery tray (5) for dents, cracks, corrosion or other damage.Clean all traces of battery electrolyte from components with fresh water or baking soda solution.Notify higher echelon of maintenance for replacement of components.

Inspect battery holddown bolts (23) for secure mounting, damage, corrosion or stripped threads.Tighten or replace damaged holddown bolts as necessary.

b. **BATTERY.** (See Figure 3-4)

3-16. **Inspect.**Inspect battery (8) for leaks or cracks.Make sure terminals are not damaged or loose. Make sure all vent covers are securely in place. Notify higher echelon of maintenance to replace battery.

a.

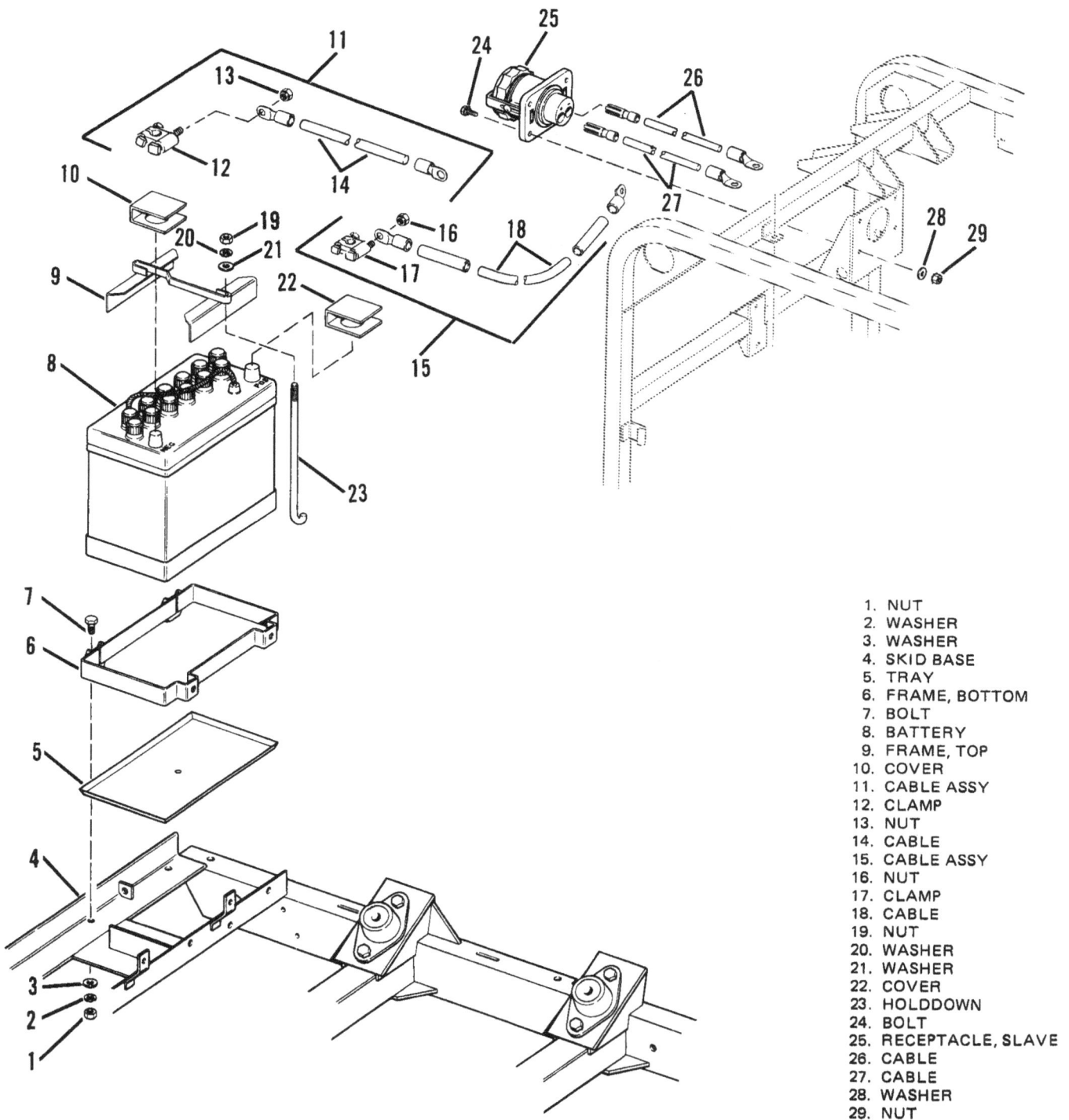

1. NUT
2. WASHER
3. WASHER
4. SKID BASE
5. TRAY
6. FRAME, BOTTOM
7. BOLT
8. BATTERY
9. FRAME, TOP
10. COVER
11. CABLE ASSY
12. CLAMP
13. NUT
14. CABLE
15. CABLE ASSY
16. NUT
17. CLAMP
18. CABLE
19. NUT
20. WASHER
21. WASHER
22. COVER
23. HOLDDOWN
24. BOLT
25. RECEPTACLE, SLAVE
26. CABLE
27. CABLE
28. WASHER
29. NUT

Figure 3-4. Battery, Battery Cables, Battery Tray, Frame and Holddown.

-------------------- WARNING --------------------

Battery electrolyte is an acid flammable fumes.
Do not smoke working around battery. Doing explosion that could result in

solution that gives off or use open flame
when so can cause an
serious personal injury.
If skin is exposed to electrolyte, flush exposed area with water immediately.
If eyes are exposed to electrolyte, flush them with water and seek immediate medical
attention.

Remove filler caps and check electrolyte (fluid) level in the battery. Electrolyte should cover the plates.   If electrolyte level is low,
add clean, distilled water as necessary. Tighten filler caps securely. Rinse top of battery with fresh water or
baking soda solution.

3-17. BATTERY CABLES AND SLAVE RECEPTACLE.   (See Figure 3-4)

a. <u>Inspect.</u>

(1) Inspect battery cable assemblies (11 and 15) and slave receptacle cables
   (26 and 27) for damage, deterioration and secure connections. Notify higher echelon of maintenance to replace
   battery cables.

(2) Inspect slave receptacle (25) for loose or missing hardware, or for damage.
   Notify higher echelon of maintenance to replace hardware or to replace a damaged slave receptacle.

b. _____ Service.   Disconnect battery cable assemblies (11 and 15) from battery
   (disconnect negative cable first). Use a wire brush to clean battery cable terminals (12 and 17) and battery posts. Connect positive cable to battery
   first when reconnecting.

3-18. VOLTAGE REGULATOR, BATTERY CHANGING.   (See Figure 3-5)

Inspect the voltage regulator for damage, secure mounting or damaged or frayed wires. Notify higher echelon of maintenance
for replacement.

3-19. STARTER RELAY. (See Figures 3-6 or 3-7)

Inspect the starter relay for damage, secure mounting, or damaged or frayed wires.
Notify higher echelon of maintenance for replacement.

MARINE CORPS TM 05926B/06509B-12
ARMY   TM 5-6115-615-12
NAVY   NAVFAC P-8-646-12

Figure 3-5. Voltage Regulator, Battery Charging.

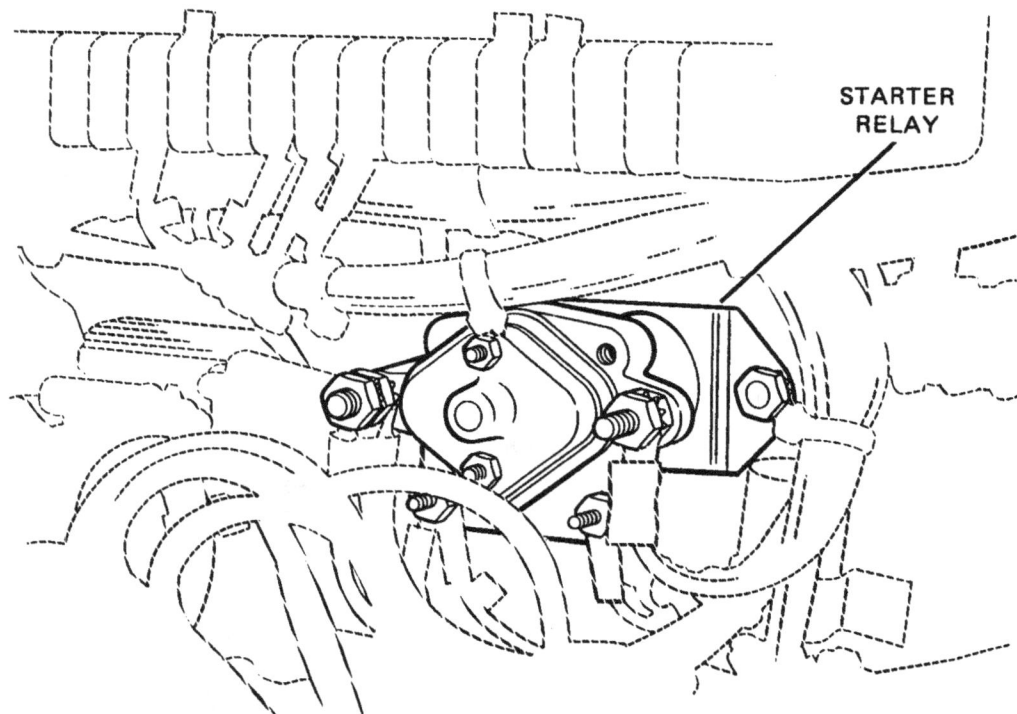

Figure 3-6.Control Box, Back Panel. Models HEP-

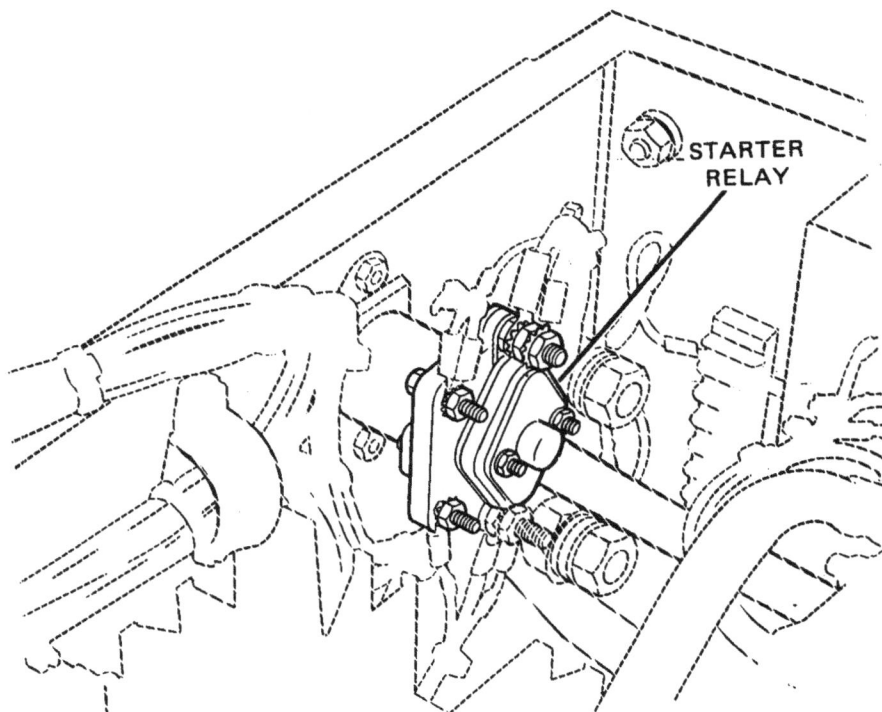

**Figure 3-7. Control Box, Back   Panel,
Model MEP-026B**

3-20.   CIRCUIT BREAKER, DC CONTROL.   (See Figure 3-8)

Inspect the DC control circuit breaker for secure mounting or damage.   Check to see that all wiring is securely connected to the circuit breaker.Notify higher echelon of maintenance to replace damaged circuit breaker. hardware, repair or replace wiring, or to replace a

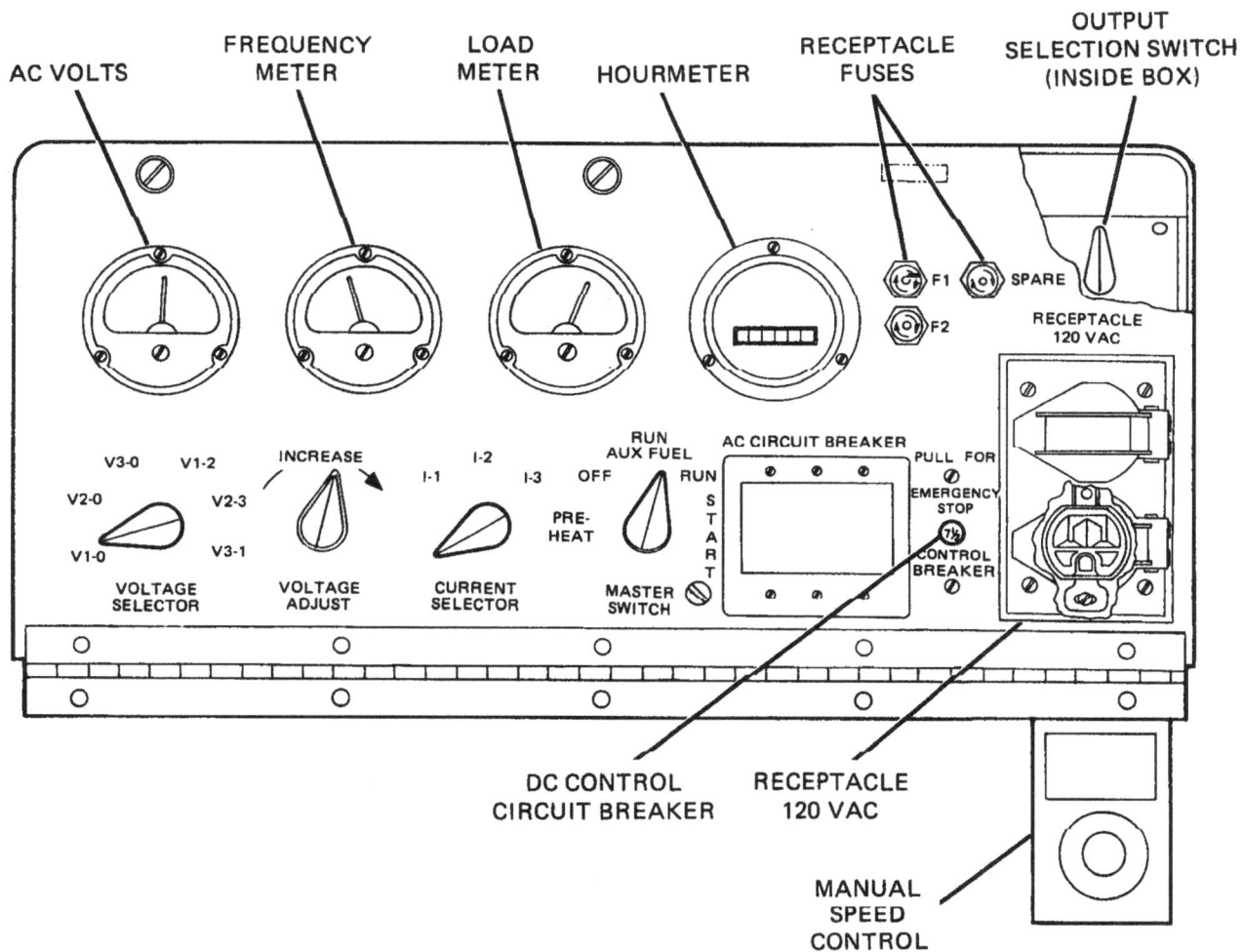

**Figure 3-8. DC   Control Circuit Breaker
(Model   MEP-016B Shown).**

3-21.   GROUNDING ASSEMBLY.   (See Figure 3-9)

   a.   Inspect ground rods (l), couplings (2), and driving bolt (3) for damage stripped threads.   If components or   or
         threads are damaged, they must be replaced.

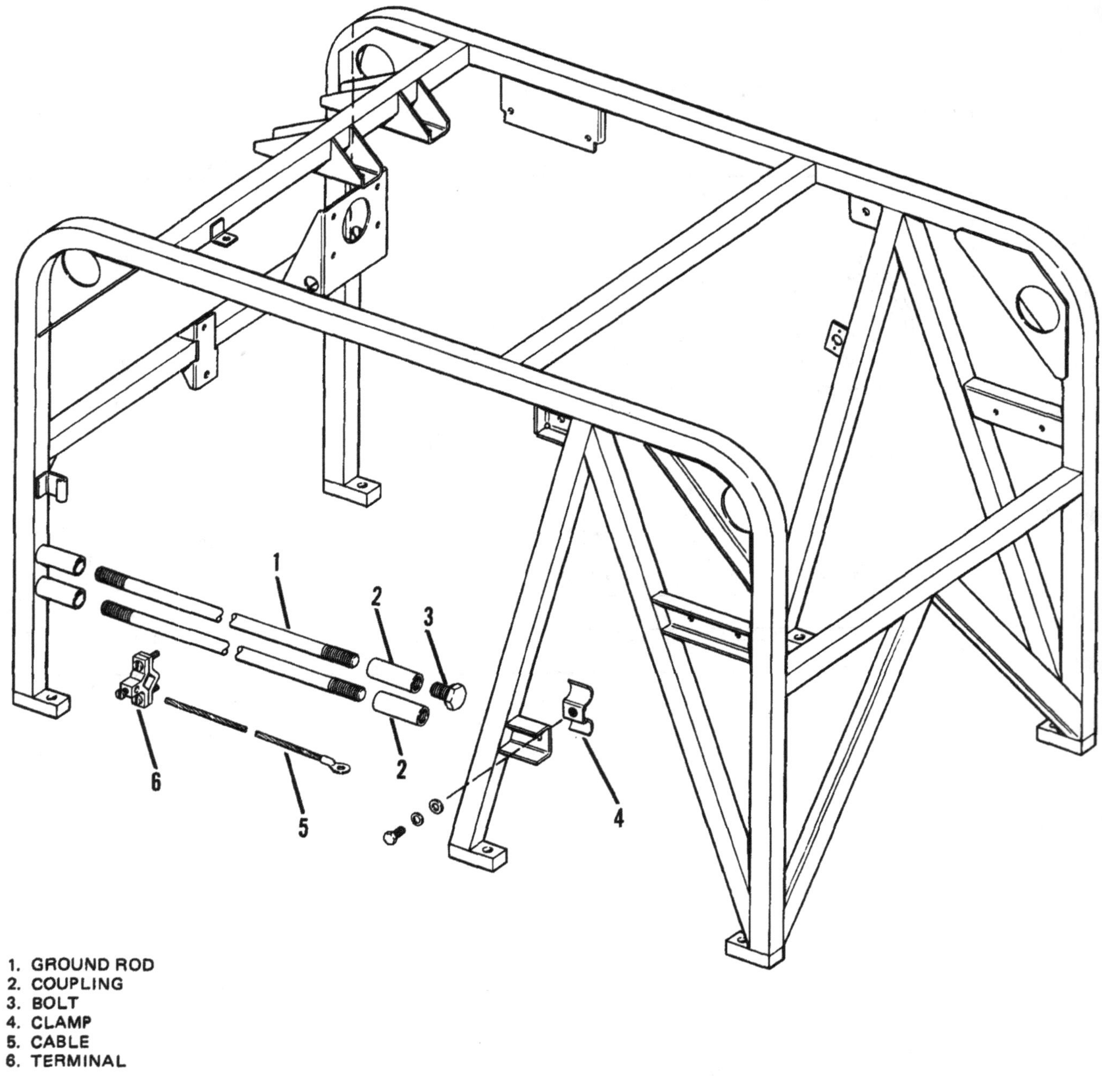

1. GROUND ROD
2. COUPLING
3. BOLT
4. CLAMP
5. CABLE
6. TERMINAL

Figure 3-9.Grounding Assembly.

b. Inspect ground stud clamp (4) for damage or missing mounting hardware. Replace parts as necessary.

NOTE

The ground cable and terminal are stored with the auxiliary fuel hose on the skid base beneath the generator.   The driving bolt and coupling are stored at the slave receptacle.

Inspect ground cable (8) for damage or frayed condition.   Replace a damaged cable.

c. Inspect ground cable terminal (9) for damage, or for missing hardware. Replace a damaged terminal.

d. Service ground rods and cables by cleaning dirty or corroded connections.

e. **LOAD TERMINAL BOARD.** (See Figure 3-10)

3-22. Inspect load terminal insulator plate (6) for damage and secure mounting. Inspect for loose or missing hardware.Notify higher echelon of maintenance to replace hardware, or to replace a damaged load terminal board.

a. Inspect load terminals (7) for damage and secure mounting.   Open control box and inspect to see that all internal wiring is securely connected to the load terminals.Notify higher echelon of maintenance to replace hardware, repair wiring, or to replace load terminals.

b.

c. Inspect cover (15), bracket (10) and hinge (11) mounting.   Notify   for damage and secure higher echelon of maintenance replace a damaged cover or hinge.   to replace hardware, or to

FUEL TRANSFER PUMP.   (See Figure 3-11)

3-23.

a. Inspect   fuel transfer pump for physical damage and secure mounting.

b. Inspect   for leaks around fittings and around base.

c. Inspect   to see that manual priming lever operates freely.

d. Notify higher echelon of maintenance to tighten hardware or replace components.

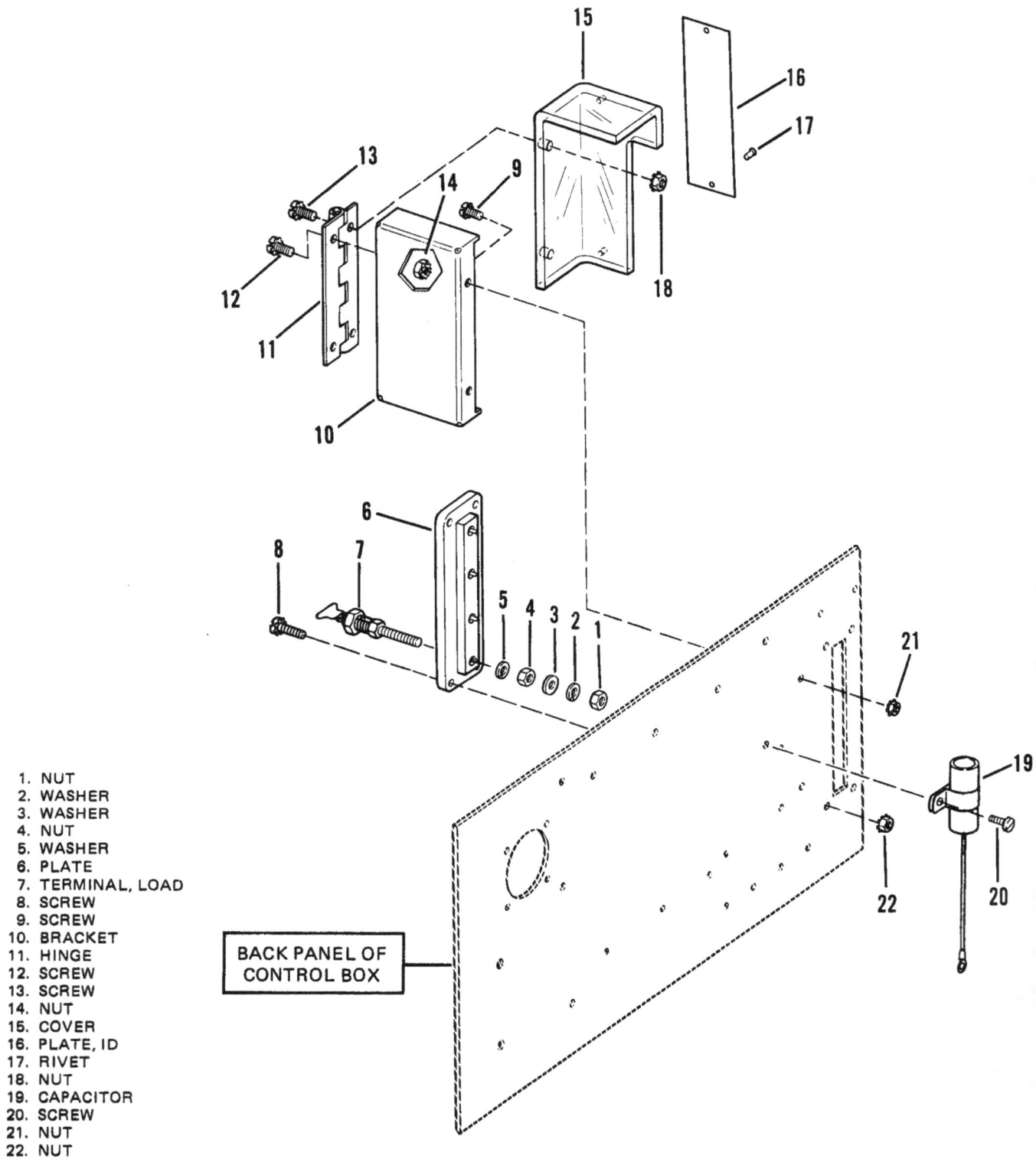

1. NUT
2. WASHER
3. WASHER
4. NUT
5. WASHER
6. PLATE
7. TERMINAL, LOAD
8. SCREW
9. SCREW
10. BRACKET
11. HINGE
12. SCREW
13. SCREW
14. NUT
15. COVER
16. PLATE, ID
17. RIVET
18. NUT
19. CAPACITOR
20. SCREW
21. NUT
22. NUT

BACK PANEL OF CONTROL BOX

Figure 3-10. Load Terminal Board (Model MEP-016B Shown).

FUEL
TRANSFER
PUMP

MANUAL
PRIMING
LEVER

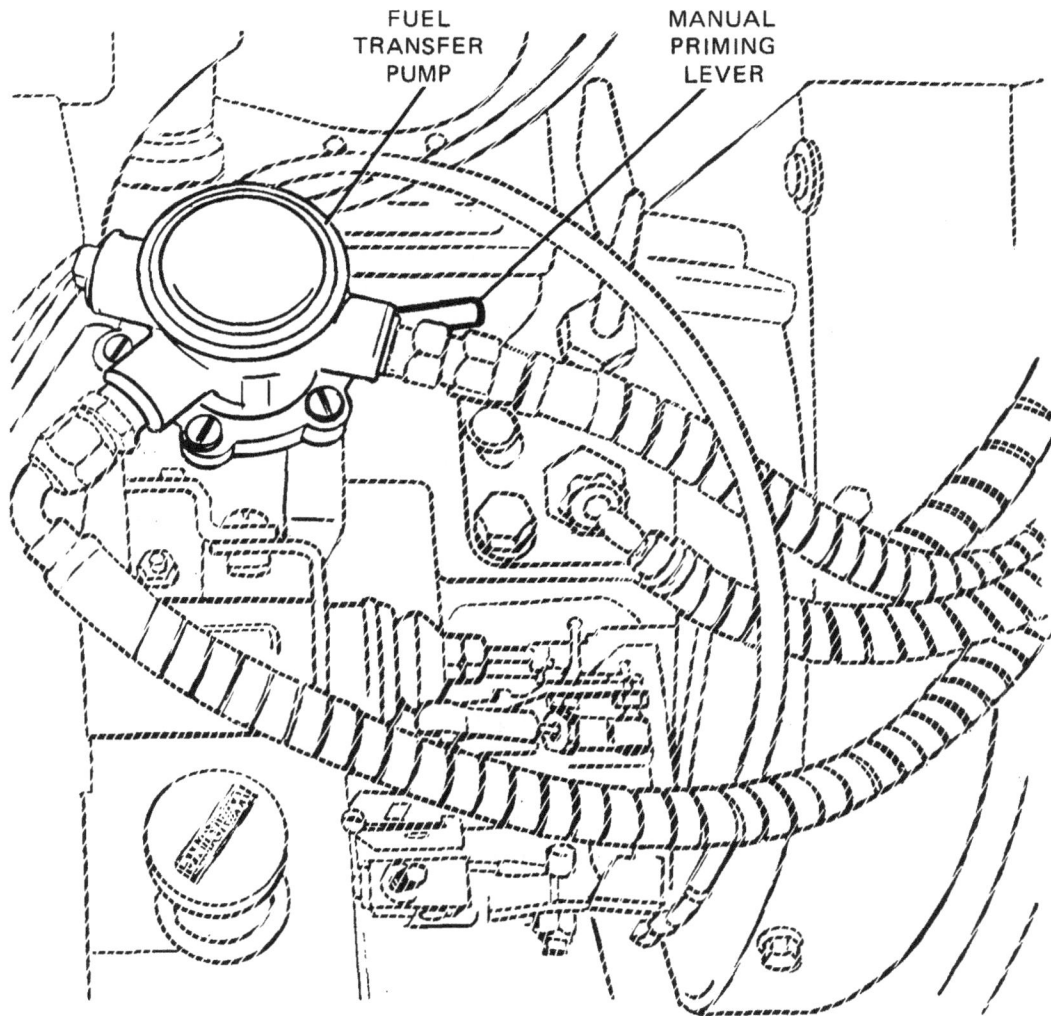

Figure 3-11. Fuel Transfer Pump.

3-24. AUXILLARY FUEL PUMP.   (See Figure 3-12)

    a. Inspect auxiliary fuel pump (2) for dents or other physical damage.Check to see that fuel pump is securely attached to the mounting bracket and that the bracket is securely attached to the skid base.

    b. Inspect for leaks around fuel line elbows (4, 5 and 7), and around fitting
      (6).

c. Inspect electrical leads to the fuel pump to see that wires are not broken, frayed or damaged. Check to see that connectors are securely connected.

d. Inspect to see that cap (9) on auxiliary fuel inlet is secured to fitting (8) when unit is not being run from an auxiliary fuel supply. Be sure chain for cap (9) is secured to skid base with screw (10) and nut (11).

e. Notify higher echelon of maintenance to tighten fittings and hardware, repair or replace electrical connections, or replace components.

1. SCREW
2. PUMP
3. HOSE
4. ELBOW
5. ELBOW
6. FITTING
7. ELBOW
8. FITTING
9. CAP
10. SCREW
11. NUT

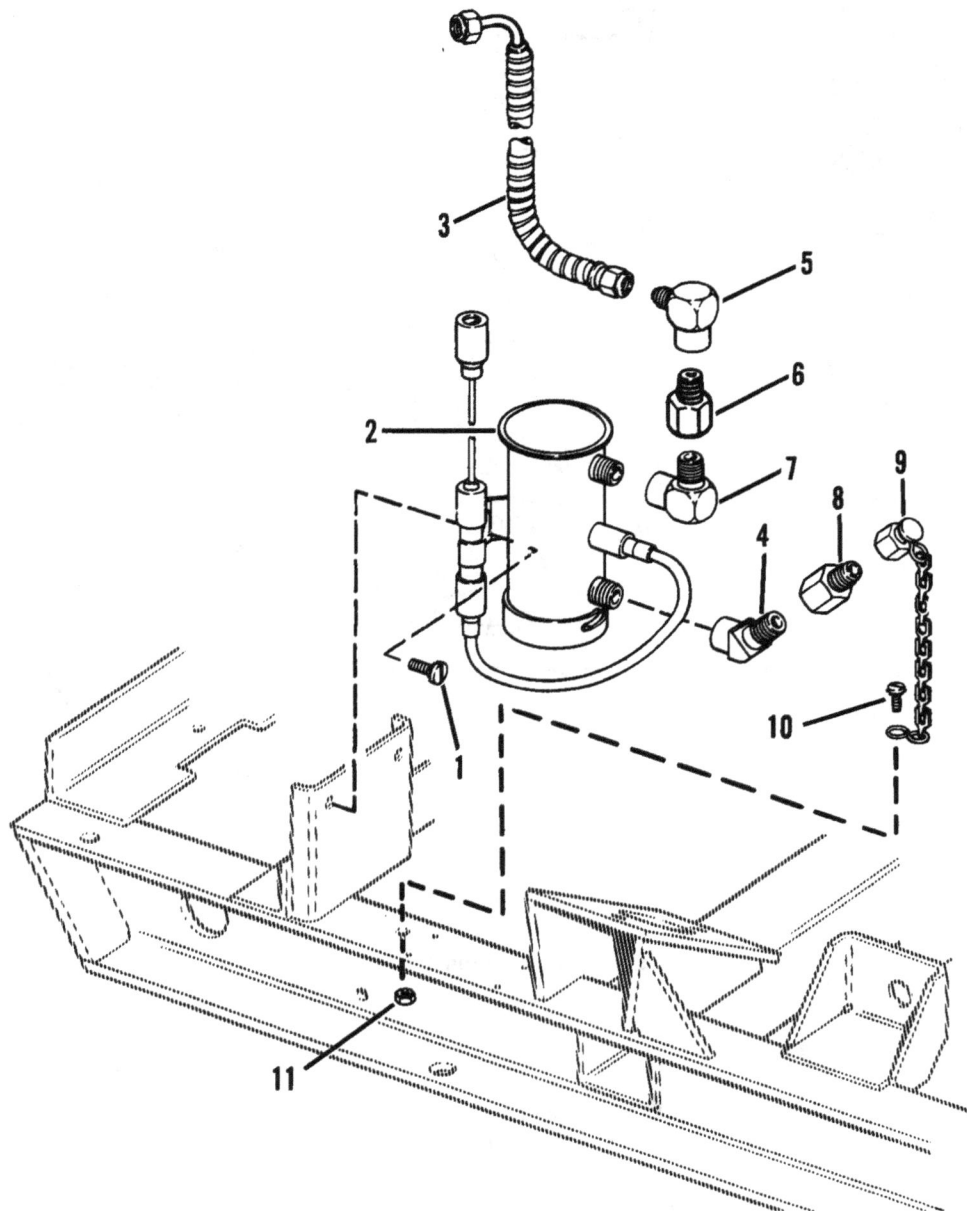

Figure 3-12. Auxiliary Fuel Pump.

3-25.   FUEL FILTER ASSEMBLY.   (See Figure 3-13)

a. Inspect filter assembly for dents or other physical damage.Check for leaks around drain valve (9), vent screw (2), and where cannister (8) joins cover

b.   Check   that fuel filter and that   assembly is securely attached to the mounting bracket is securely attached to the

c.   Notify higher echelon of maintenance to tighten fittings or hardware, or to replace components.

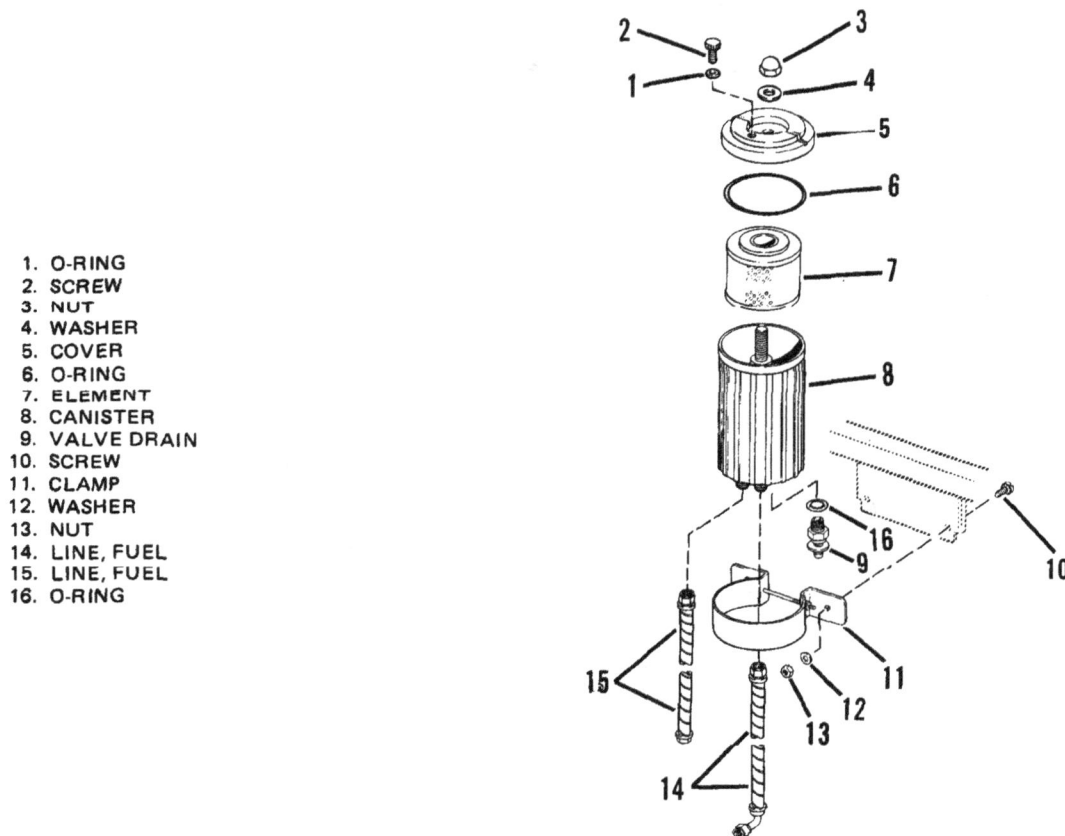

1. O-RING
2. SCREW
3. NUT
4. WASHER
5. COVER
6. O-RING
7. ELEMENT
8. CANISTER
9. VALVE DRAIN
10. SCREW
11. CLAMP
12. WASHER
13. NUT
14. LINE, FUEL
15. LINE, FUEL
16. O-RING

Figure 3-13. Fuel Filter Assembly, Exploded View.

3-26.   FUEL TANK.   (See Figure 3-14)

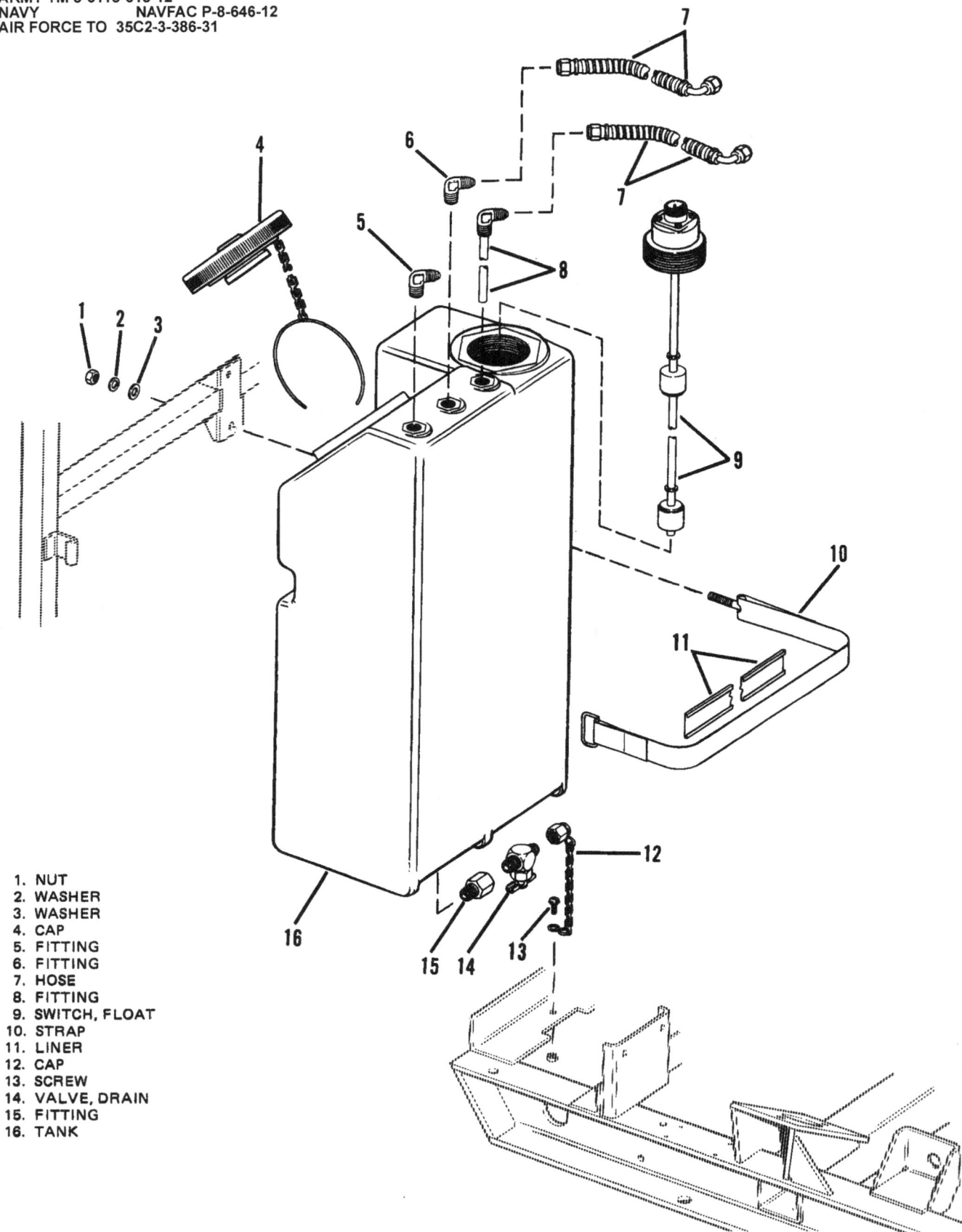

1. NUT
2. WASHER
3. WASHER
4. CAP
5. FITTING
6. FITTING
7. HOSE
8. FITTING
9. SWITCH, FLOAT
10. STRAP
11. LINER
12. CAP
13. SCREW
14. VALVE, DRAIN
15. FITTING
16. TANK

Figure 3-14. Fuel Tank and Fittings, Exploded View.

b. Inspect fittings threads or (5, 6 and 8), float switch (9) and hoses (7) for stripped damage.   Inspect wires to float switch
other condition.

c. Inspect underside of cap (4) to see that gasket is free of dam-   chain.
age.   vent is operational and rubber

d. Inspect to see that fuel drain fittings is secured to frame with

e. Make sure cap (4) fits securely on tank (16).

f. Be sure that strap (10) and liner (11) are not damaged, and that tank is held securely to frame.

g. Notify higher echelon of maintenance to tighten fittings and hardware, repair or replace damaged electrical connections, or replace damaged components.

FUEL LINES, VALVES AND FITTINGS. Inspect all fuel lines, valves and fittings for leaks or physical

3-27. damage.   Inspect to see that all fittings are tightened securely. Notify higher components.

echelon of maintenance to tighten fittings or replace

3-28.   ENGINE ASSEMBLY. Inspect engine assembly for leakage of fuel or lubricants, missing hardware. Notify higher
obvious damage, loose or repair or   echelon of maintenance for engine components.
replacement of

Figure 3-15)

3-29.   AIR FILTER.   (See

a. Inspect.

(8) and cover (4) for physical damage. Inspect to see that

housing is securely attached to frame.

(2) Inspect air filter indicator (9) for damage.

(3) Check hose (7) from air filter assembly to intake manifold for leaks and
secure mounting.

(4) Notify higher echelon of maintenance to tighten hardware or replace
components.

b. Service.

CAUTION

Before servicing air filter assembly, make sure all external surfaces of air filter
assembly and surrounding areas are clean.   Entry of dirt into engine can cause premature engine wear.

(1) Service air filter assembly per schedule in Table 3-2.

(2)   Loosen wingnut on cover (12), and remove cover from housing (8).

(3)   Remove and discard element (5).

NOTE

Under extreme conditions element can be cleaned with low pressure air.Use no more than four psi air pressure.

(4)   Be sure that all interior surfaces   of air filter assembly replace if   are clean.
Inspect gasket (10) for damage and dust valve (13).   Remove and clean

(5)   Install new element or   cleaned element (5) and install cover (4).

(6)   Reset filter indicator   (9) by pressingbutton.

1. SCREW
2. LOCKWASHER
3. WASHER
4. COVER
5. ELEMENT
6. CLAMP
7. HOSE
8. HOUSING
9. INDICATOR
10. GASKET
11. WASHER
12. WINGNUT
13. DUST VALVE

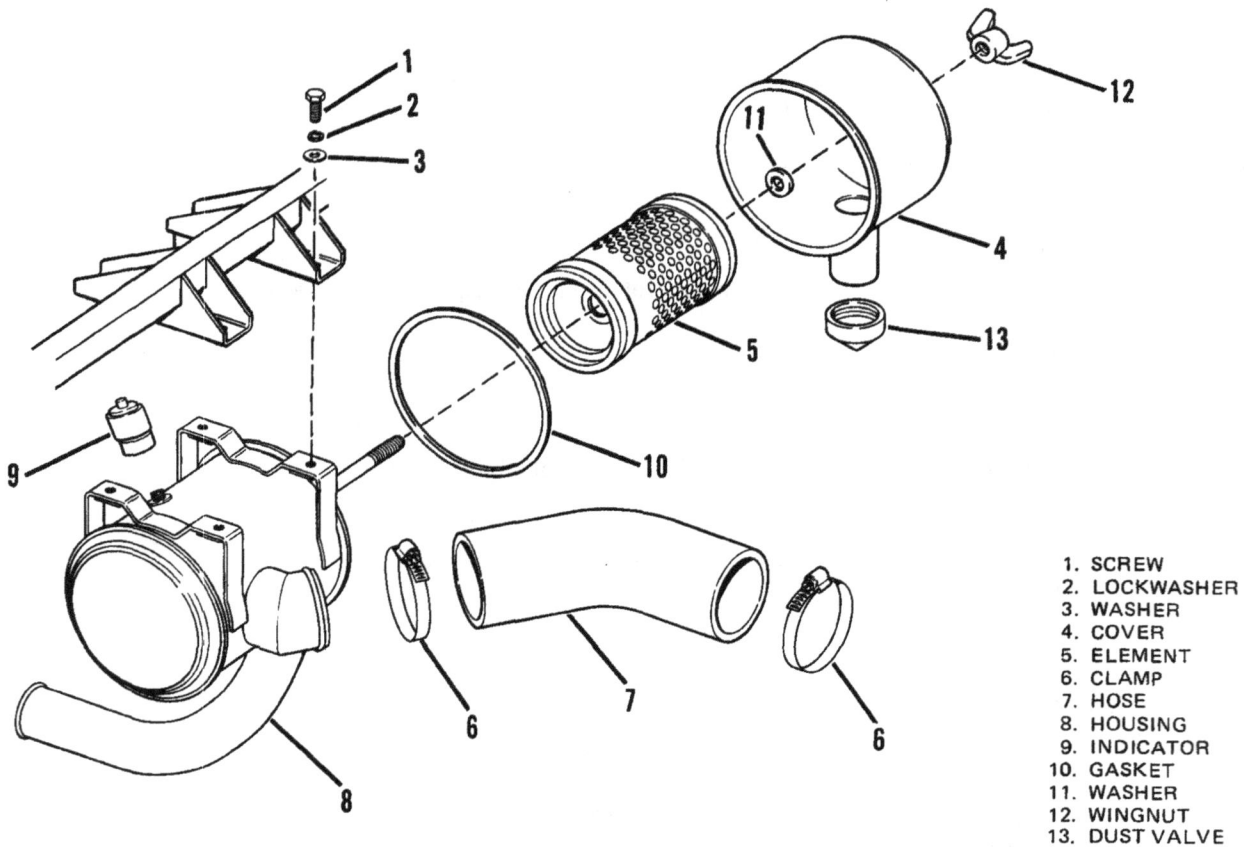

Figure 3-15. Air Filter Assembly, Exploded View.

3-30. **OIL PAN** **AND OIL DRAIN.**   (See Figure 3-16)

a. Inspect leaks.   oil pan, oil drain hose, fittings and drain valve for damage or Make sure all connections are secure.

b. Be sure   that oil drain valve does not leak.

c. Notify higher echelon of maintenance to tighten oil pan bolts, fittings or to replace components.

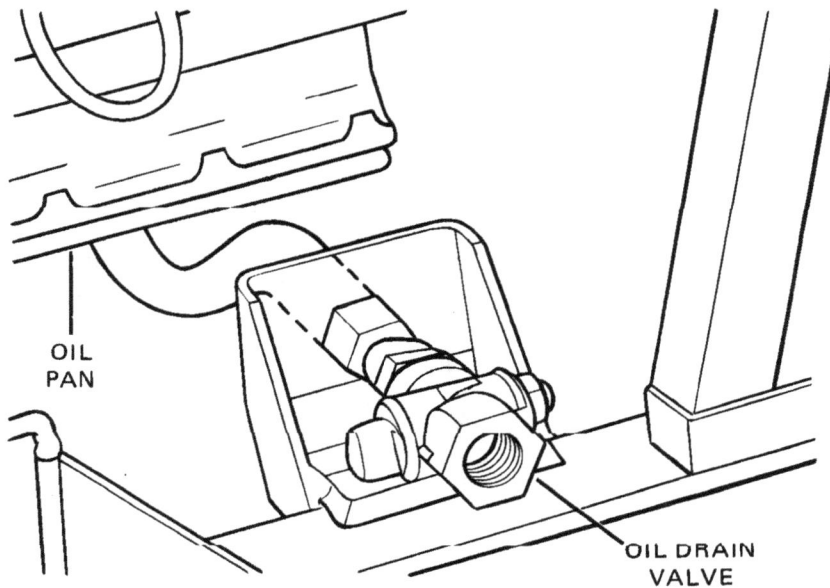

Figure 3-16.Oil Pan and Oil Drain.

3-31. **GOVERNOR LINKAGE.** (See Figure 3-17)

a. Inspect governor linkage (11) for damage and secure mounting.With engine off, manually operate linkage to see if linkage operates freely and without obstruction or binding.

b. Inspect manual speed control cable (30) for damage and secure mounting. Inspect to see that cable is not cut or kinked.Inspect to see that engine speed increases when the manual speed control (30) is turned in a clockwise direction, and that engine speed decreases when the speed control is turned in a counterclockwise direction.

Inspect to see that the engine fuel cutoff solenoid (15) is mounted securely and that the wires are not cracked or frayed.

c.

d. Inspect oil line (21) for damage or leaks. Check to see that all fittings
    are secure.

e. Notify higher echelon of maintenance to clean linkage, tighten hardware or
    fittings, repair or replace wiring, or replace components.

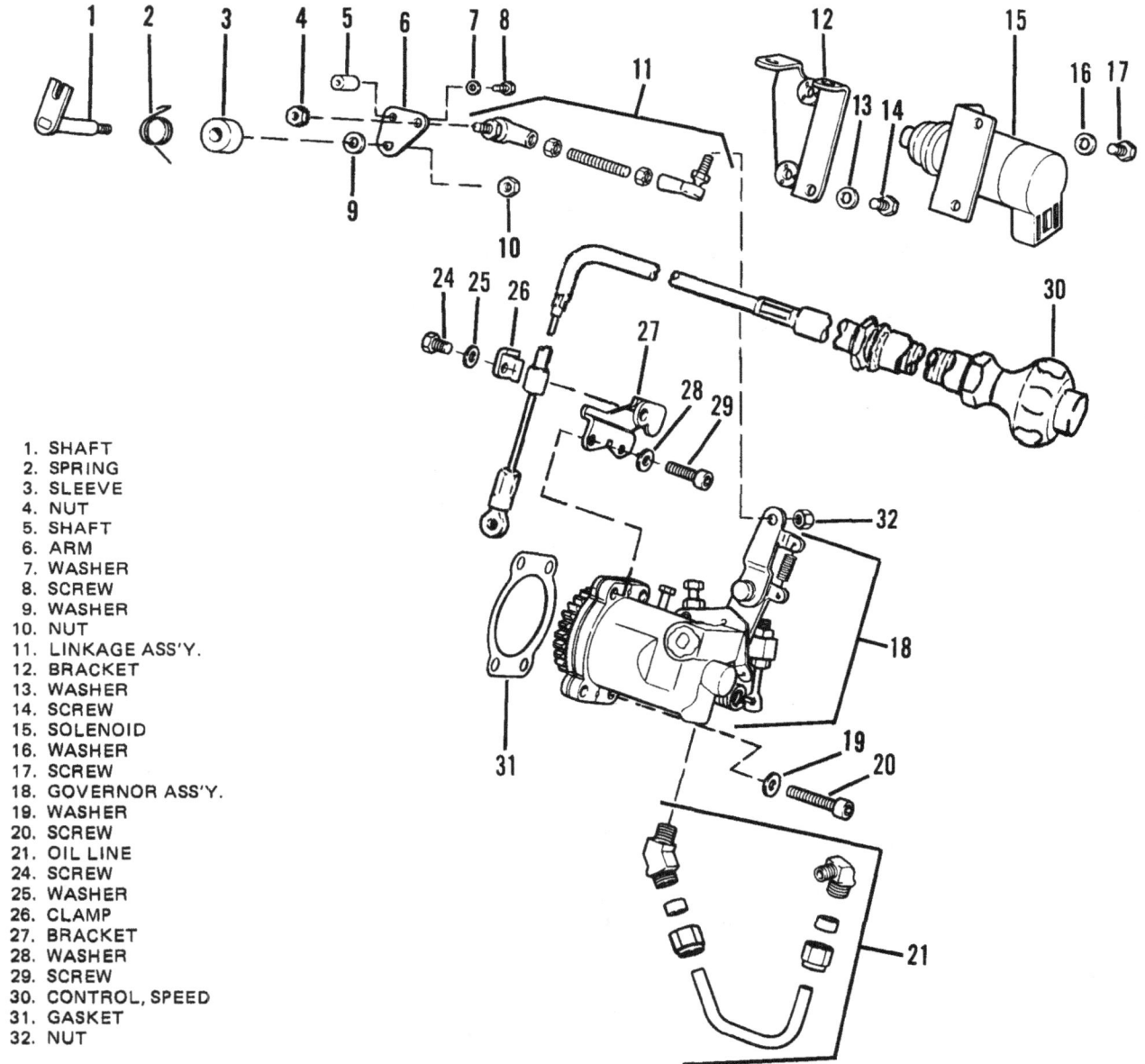

1. SHAFT
2. SPRING
3. SLEEVE
4. NUT
5. SHAFT
6. ARM
7. WASHER
8. SCREW
9. WASHER
10. NUT
11. LINKAGE ASS'Y.
12. BRACKET
13. WASHER
14. SCREW
15. SOLENOID
16. WASHER
17. SCREW
18. GOVERNOR ASS'Y.
19. WASHER
20. SCREW
21. OIL LINE
24. SCREW
25. WASHER
26. CLAMP
27. BRACKET
28. WASHER
29. SCREW
30. CONTROL, SPEED
31. GASKET
32. NUT

Figure 3-17. Governor Linkage, Exploded View.

3-32.   OIL FILTER.   (See Figure 3-18)

Inspect oil filter for leakage, damage and secure mounting.Notify higher echelon of maintenance to tighten or replace oil filter.

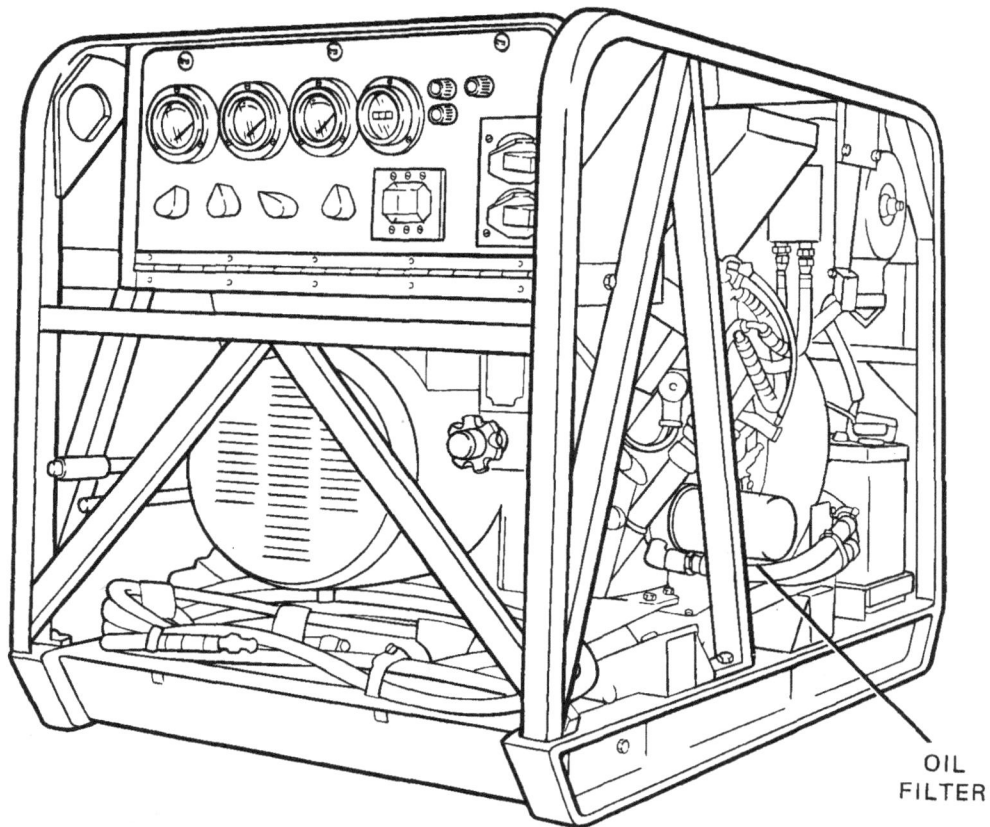

OIL
FILTER

Figure 3-18. Engine Oil Filter.

EXHAUST.   (See Figure 3-19)

NOTE

Perform steps a through c if engine exhaust is excessively noisy.

screws (l), nuts (2) and clamps (3) from heat shield (4).

a.   Remove

screws (5) and nuts (6).   Remove heat shield (4) from frame.

c. Inspect muffler (13) for dents, holes or corrosion.

d. Check to see that clamps (7, 8 and 16) are securely   attached and that
    they are not damaged.

e. Inspect exhaust pipe (19) for damage, obstructions and secure mounting.

f. Check mounting brackets (17) and (14) for damage and secure mounting.

9   Notify higher echelon of maintenance to tighten hardware or to replace components.

    Put heat shield (4) in position and install screws (5), clamps (3), nuts (2 and 6) and screws (1).
h.

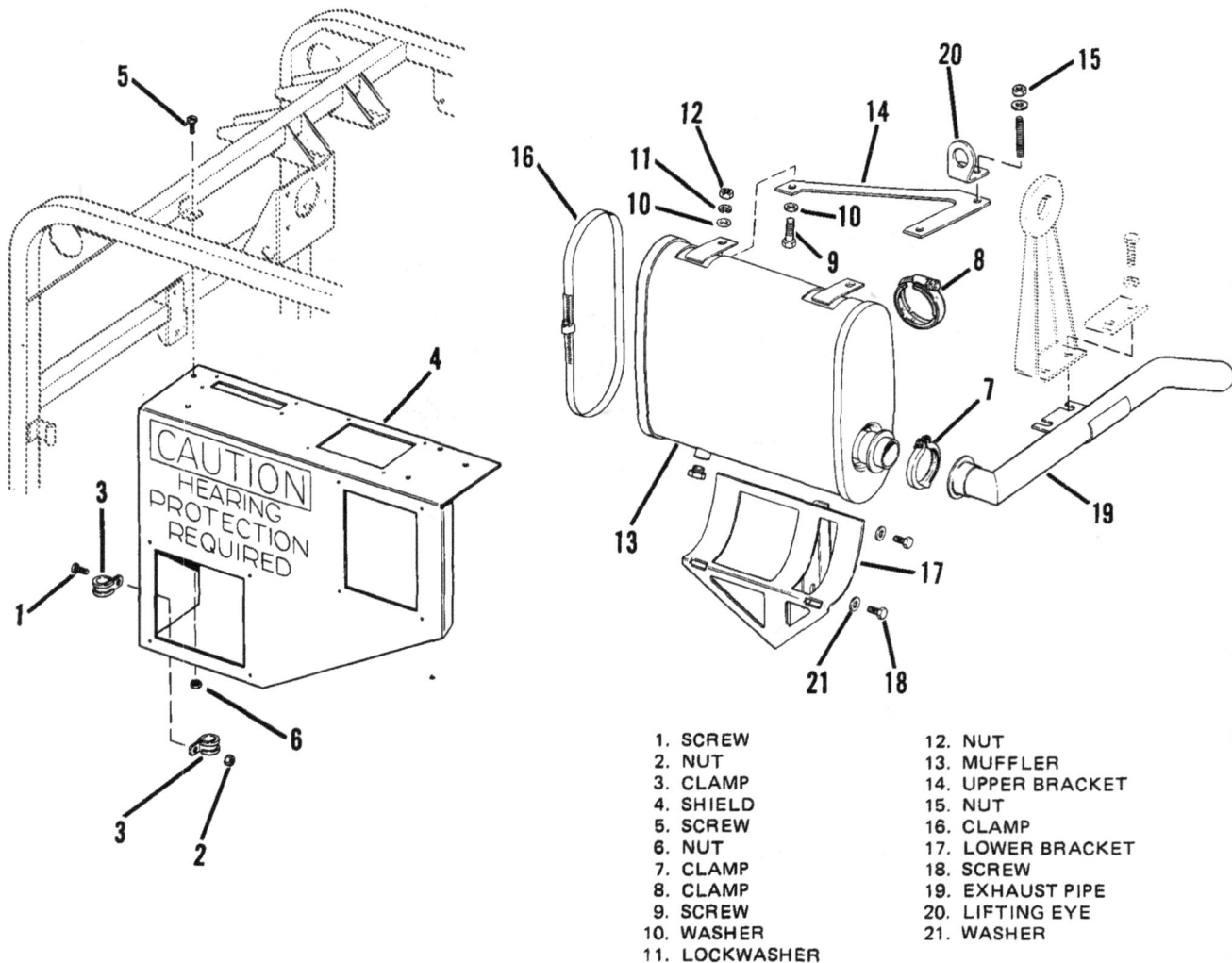

1. SCREW	12. NUT
2. NUT	13. MUFFLER
3. CLAMP	14. UPPER BRACKET
4. SHIELD	15. NUT
5. SCREW	16. CLAMP
6. NUT	17. LOWER BRACKET
7. CLAMP	18. SCREW
8. CLAMP	19. EXHAUST PIPE
9. SCREW	20. LIFTING EYE
10. WASHER	21. WASHER
11. LOCKWASHER	

Figure 3-19. Engine Exhaust, Exploded View.

3-34.   ENGINE CONTROLS AND INSTRUMENTS.   (See Figures 3-20, 3-21 and 3-22)

3-34.1. Panel Assembly, Engine Control.   Inspect the control panel assembly for cleanliness and damage.   Clean panel as required.   Inspect to see that all parts mounted to the panel are secure and that all hardware is in place. Loosen three fasteners and open control panel.   Inspect to see that hinge is secure and operates properly.   Inspect condition of retaining cord and be sure cord is mounted securely. Notify higher echelon of maintenance to tighten hardware or replace components.

3-34.2.Master Switch.Inspect master switch for physical damage.   Inspect electrical connections to see that they are clean and tight.   Inspect to see that when placed in the OFF, RUN AUX FUEL and RUN positions, the switch will remain in these positions.   Inspect to see that when placed in the PREHEAT position and released, the switch will return to the OFF position.With the DC Control Circuit Breaker pulled out (unit will not crank), inspect to see that when released from the START position the switch will automatically return to the RUN position.Notify higher echelon of maintenance to tighten hardware or electrical connections, or to replace a damaged switch.

3-34.3. Hourmeter. Inspect hourmeter for physical damage, and be sure that meter face is clean and readable.   Inspect electrical connections to see that they are clean and tight. Inspect to see that meter is mounted securely.Notify higher echelon of maintenance to tighten hardware or electrical connections or to replace damaged meter.

3-35.GENERATOR CONTROLS AND INSTRUMENTS. (See Figures 3-20,3-21, 3-22 and 3-23)

3-35.1. Inspect Control Box Assembly (all models).
 Clean box as necessary and inspect for damage and secure mounting.   Inspect condition of vibration isolators.

Loosen three fasteners and open control panel to permit access to inside of control box.   Inspect to see that all parts mounted to the box are secure and that all hardware is in place.   Inspect for broken, frayed or damaged wires. Inspect to see that all electrical connections are clean and tight.

a.   Inspect load connection terminals for damage and secure mounting.   Inspect to see that threads are in good condition and that nut retaining wire is in place.   Inspect to see that all electrical connections are clean and tight.

b.

Check that all instruments and controls operate normally when the unit is running.

Notify higher echelon of maintenance to tighten or replace hardware and vibration isolators, repair or replace electrical wiring, or to replace components as necessary.

c.

d.

e.

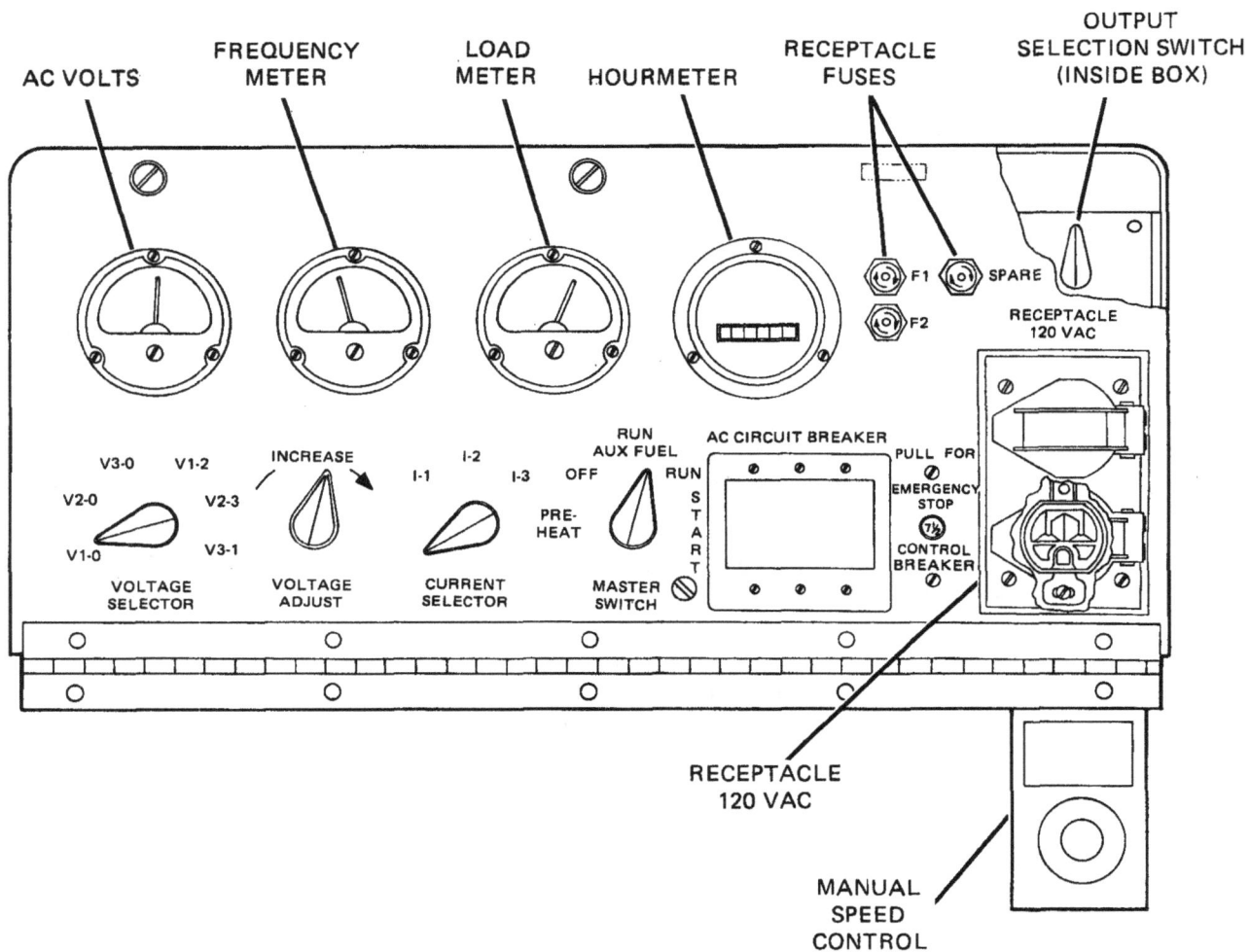

Figure 3-20. Controls and Instruments, Model MEP-016B.

MARINE   TM 05926B/06509B-12
CORPS   TM 5-6115-615-12
ARMY   NAVFAC P-8-646-12
            TO 35C2-3-386-31

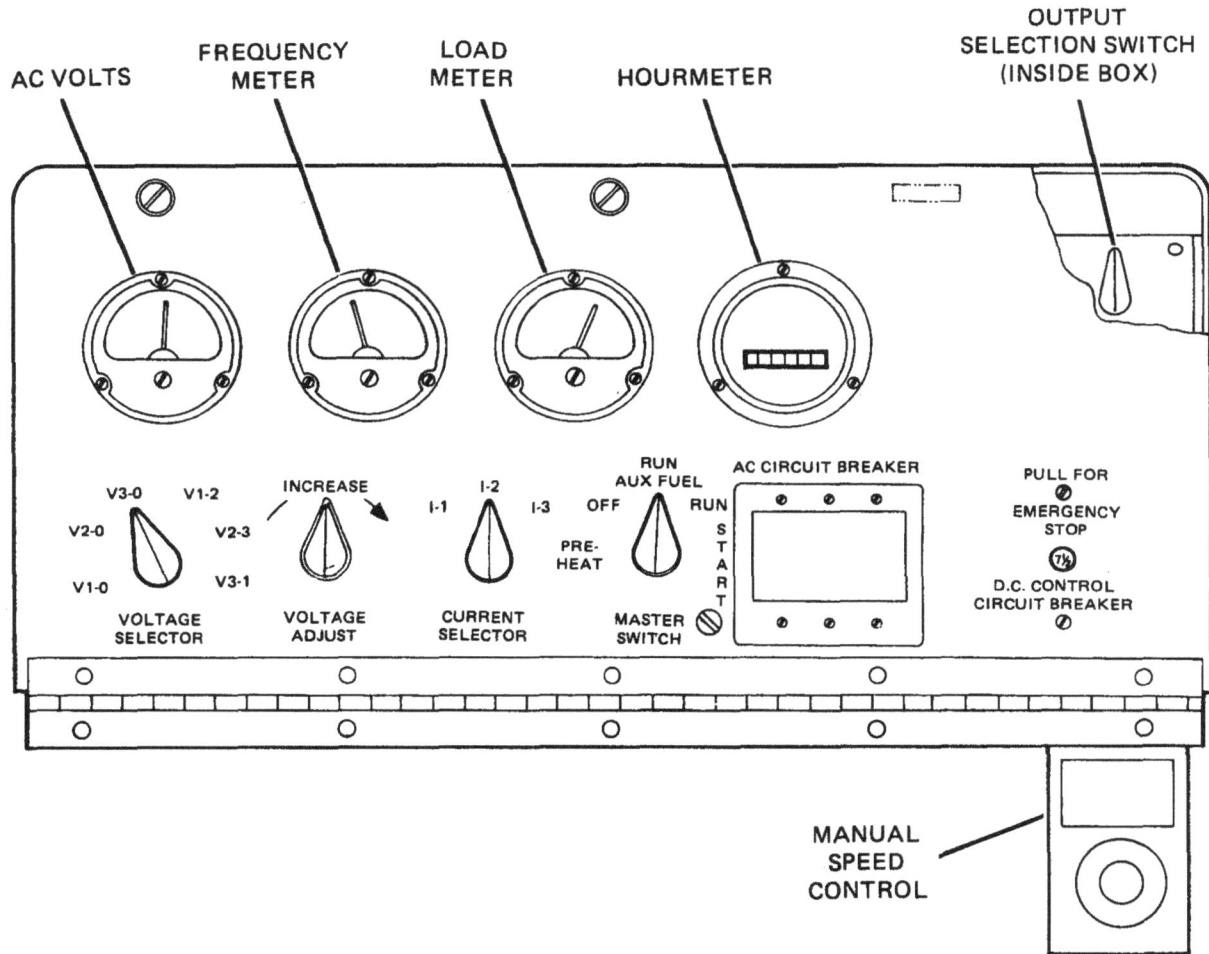

Figure 3-21. Controls end Instruments. Model MEP-021B.

DC VOLTS   LOAD METER   TACHOMETER   HOURMETER   CIRCUIT BREAKER

INCREASE

RUN AUX FUEL

OFF   RUN

PRE HEAT   START

VOLTAGE ADJUST   MASTER SWITCH

PULL • FOR EMERGENCY STOP

DC CONTROL CIRCUIT BREAKER

MANUAL SPEED CONTROL

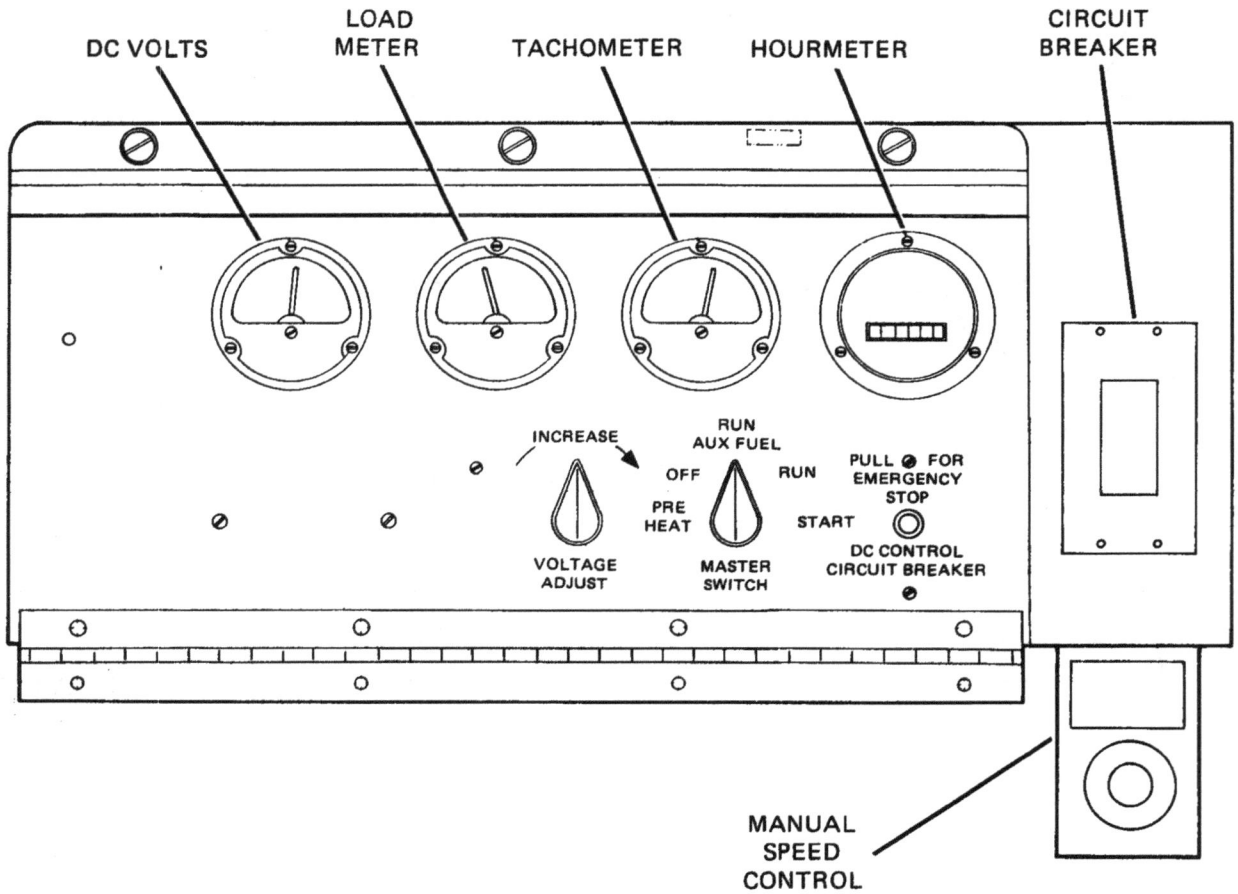

Figure 3-22. Controls and Instruments, Model MEP-026B.

MARINE CORPS TM 05926B/06509B-12
ARMY   TM 5-6115-615-12
NAVY   NAVFAC P-8-646-12

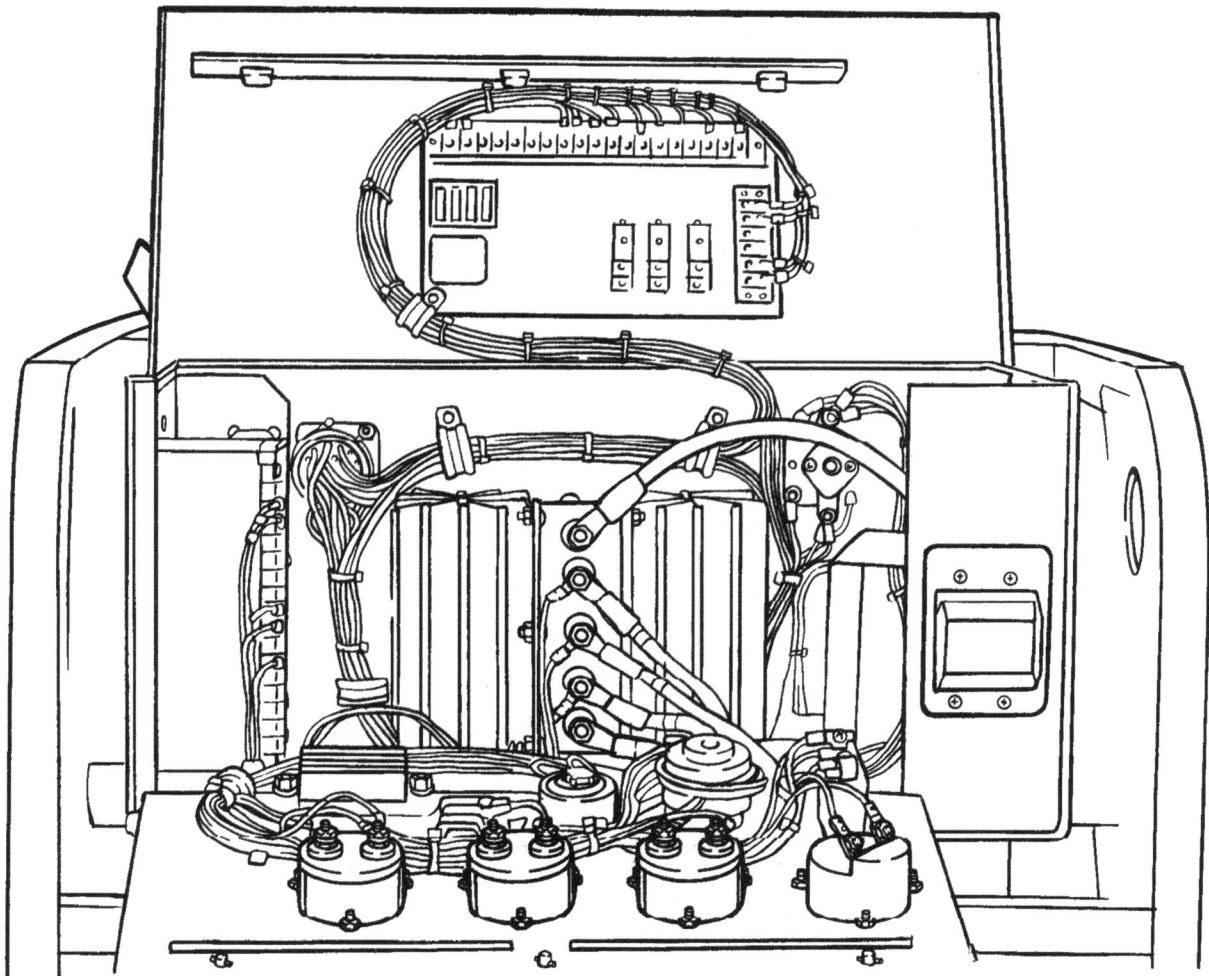

Figure 3-23.Inside of Control Box, Model MEP-026B Shown.

3-35.2.<u>Voltage Selector (Models MEP-O 16B and MEP-02lB)</u> .Inspect voltage selector switch for physical damage, and make sure that knob is secure on post. ^{Inspect to} see that switch remains in any position it is turned to. Inspect to see that all electrical connections are clean and tight.Notify higher echelon of maintenance to tighten mounting hardware or to replace a damaged voltage selector switch.

3-35.3.   Voltage Adjust (all models).   Inspect voltage adjust rheostat for physical damage, and make sure knob is secure on post.Mark rheostat position and inspect to see that rheostat turns smothly through its range.Return rheostat to its previously marked position.   Inspect to see that all electrical connections are clean and tight.Notify higher echelon of maintenance to tighten mounting hardware or to replace a damaged rheostat.

MARINE CORPS   TM 0926B/06509B-l2

ARMY

     TM -6115-615-12

NAVY

     NAVFAC P-8-646-12

AIR FORCE

     TO 35C2-3-386-31

3-35.4.Current Selector (Models MEP-016B and MEP-021B). Inspect current selector switch for phys damage and make sure that knob is secure on post. Inspect to see that switch remains in any position it is turned to. Inspect to see that all electrical connecti are clean and tight.Notify higher echelon of maintenance to tighten mounting hardware or to replace a damaged current selector sv

3-35.5.Main Circuit Breaker (all models). Inspect circuit breaker for damage. Inspect transparent cove for damage and secure mounting. Inspect to see that when the circuit breaker is switched to th ON or OFF positions, it will remain in these positions. Inspect to see that all electrical connections are clean and ti Notify higher echelon of maintenance to tighten mounting hardware or to replace a damaged circuit breaker.

3-35.6.Volt Meters (all models). Inspect meter to see that glass is not cracked or broken, and that meter face is readable. Inspect t that pointer is not bent. Inspect to see that all electrical connections are clean and tight. Notify higher echelon of maintenance to ti mounting hardware or to replace a damaged meter. Meters can be adjusted by turning the adjustment screw on the meter face.

3-35.7.Frequency Meter (Models MEP-016B and MEP-021B).Inspect meter to see that glass is not cracked or broken, and that r face is readable. Inspect to see that pointer is not bent. Inspect to see that all electr connections are clean and tight. Notify higher echelon of maintenance to tighten mounting hardware or to replace a damaged me Meter can be adjusted by turning the adjustment screw on the meter face.

3-35.8.Load Meter (all models). Inspect meter to see that glass is not cracked or broken, and that meter face i readable. Inspect to see that pointer is not bent. Inspect to see that all electrical connec are clean and tight.Notify higher echelon of maintenance to tighten mounting hardware or to replace a damaged meter. Meter can b adjusted by turning the adjustment screw on the meter face.

3-35.9. Hourmeter (all models). Inspect meter to see that glass is not cracked or broken, and that meter face is readable. Inspect to see that all electrical connections are clean and tight. Notify higher echelon of maintenance to tighten mou ing hardware or to replace a damaged meter.

3-35.10. DC Control Circuit Breaker (all models).Inspect circuit breaker for physical damage.Inspect to see that all electrical connections are clean and tight. Inspect to see that when circuit breaker is pu or pulled out it will remain in position. Notify higher echelon of maintenance to tighten mounting hardware or to replace a circuit breaker.

3-35.11. Fuses, Fl, F2, and Spare (Model MEP-016B only).Check fuseholders for damage and secure fit. Check condition c three fuses. Check to see that all electrical connections are clean and tight.Notify higher echelon of maintenance to replace a d aged fuse holder.

------- -- -- - - -
WARNING
------- -- -- --

The output receptacle system.   is connected to a floating ground
                                      be grounded as specified in
paragraph 2-5.1 to effectively ground the receptacle. Failure to do so may cause severe
injury or death.

3-35.12. <u>Receptacle (Model MEP-016B only).</u> Inspect receptacle and cover for damage and secure mounting. Check to see that spring loaded covers remain closed. Check to see that all electrical connections are clean and tight. Notify higher echelon of maintenance to tighten mounting hardware or to replace a damaged receptacle.

3-35.13.   <u>Output Selection Switch (Models MEP-016B and MEP-021B).</u>   (Switch is located inside of control box.)   Inspect switch for physical damage, and make sure that knob is secure on post.   Check to see that switch remains in any position it is turned to. Check to see that all electrical connections are clean and tight. Notify higher echelon of maintenance to tighten mounting hardware or to replace a damaged switch.

3-35.14.   <u>RPM Indicator (Model MEP-026B only).</u>   Inspect meter to see that glass is not cracked or broken, and that meter face is readable. Check to see that pointer is not bent.   Check to see that all electrical connections are clean and tight. Notify higher echelon of maintenance to tighten mounting hardware or to replace a damaged meter. RPM indicator can be adjusted by turning the adjustment screw on the meter face.

CHAPTER 4

OGANIZATIONAL MAINTENANCE

Section I.

4-1. INSPECTING AND SERVICING   THE EQUIPMENT.

4-1.1. Unloading theEquipment.   The dry weight of the generator set is 440 lbs (199 kg). The crated generator set may be lifted by forklift,crane or similar lifting device.   If slings are used, take care in placing them around the crate to insure proper balance of the load and to minimize the danger of it slipping.The crate must be kept in the UP position at all times.   Do not use a lifting device with a capacity of less than 600 lbs (272 kg).   Do not allow the crated generator set to swing while it is suspended.Failure to observe this warning

WARNING

may result in serious injury or death to personnel.

4-1.2. Unpacking.

a. Before unpacking, move the generator set as near as possible to the location where it will be operated.

b. Remove the top and sides of the crate. To avoid damaging the equipment, be careful When using bars, hammers or other tools to uncrate the generator set.

CAUTION

c. Remove the unit from the wooden skid base.

4-1.3. Depreservation.   Prepare the generator set for inspection and operation as outlined on DA Form 2258, attached on or near the operational controls.

4-1.4. Inspection.

a. Inspect the generator set for damage or missing parts and accessories.

b. Perform the Preventive Maintenance Checks and Services as outlined in Table 4-1.

4-1.5. Servicing.Service the generator set in accordance with Tables 3-2, 4-1, and Lubrication Instruction/Lubrication Order LI 05926B/06509B-12/5 (See Figure 3-1).

4-2.1. General. The generator set should be with ample ventilation, and level within +15

4-2.2. Outdoor Installation. When preparing installed on a site clear of obstacles, degrees maximum.

for a permanent installation, be sure the base is solid enough to support the weight of the unit. See Figure 1-4 for dimensions of the base.   Select a site where there will be sufficient space on all sides for operation and servicing of the set.When preparing a temporary installation, move the generator set as close to the worksite as practical.   Use suit-able planks, logs or other material for a base in an area where the ground is soft .

4-2.3. Indoor Installation. Keep the area well ventilated at all times so that the generator set receives a maximum supply of air. If a free supply of fresh air is not available, provide ductwork which will assure at least 600 cubic feet of air per minute.   If louvers are used at the entrance, increase the size of the duct work by 25 to 50 percent.   Install a gas tight metal pipe from the exhaust outlet of the unit to the outside of the installation.The termination of the exhaust pipe shall be such that hot gases or sparks will be discharged harmlessly, and will not be directed against combustible materials or into an area con-taining flammable gases or vapors.   Use as few bends in the pipe and as short a pipe as possible.   The exhaust pipe should include a low point with suitable means for draining condensation. Provide metal shields, 12 inches larger in diameter than the exhaust pipe, where the line passes through flammable walls.

WARN-

Do not operate the generator set in an enclosed area unless the exhaust gases are piped to the outside. Inhalation of exhaust fumes will result in serious illness or death.

4-2.4. Leveling The generator set is a portable unit and is designed to operate satisfactorily up to 15 degrees out-of-level in all directions.   Set up the unit as level as possible, and keep it as level as possible during operation.

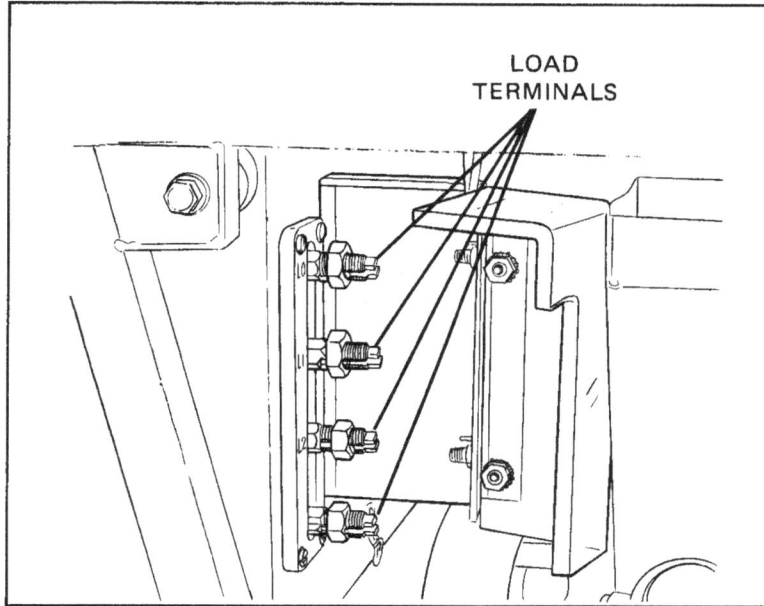

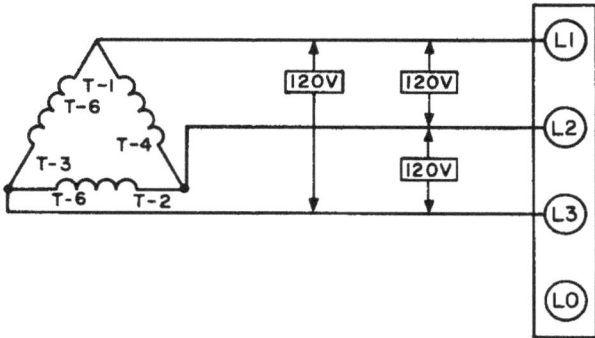

120 VOLT 3 PHASE CONNECTION

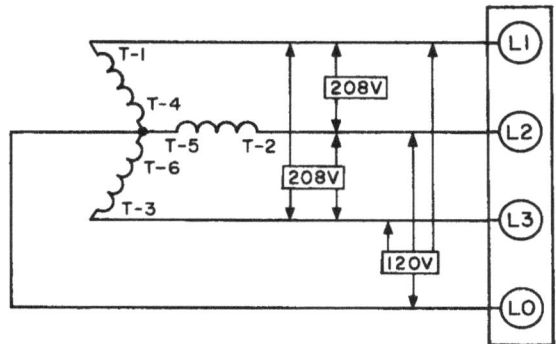

120/208 VOLTS 3 PHASE CONNECTION

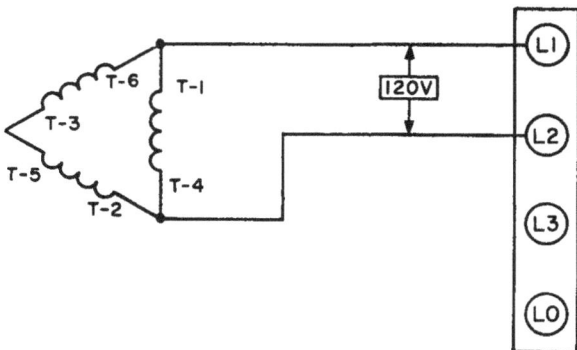

120 VOLT 1 PHASE CONNECTION

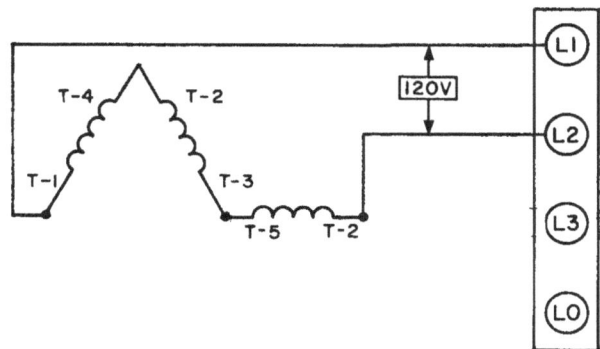

240 VOLT 1 PHASE CONNECTION

**Figure 4-1. Load Terminals and Load Connection Diagrams, Models
MEP-016B and MEP-021B.**

WARNING

Prior to connection of load cables, be certain all switches and circuit breakers are in the OFF or OPEN position and that the generator set is not running. Failure to do so can result in death from electrocution.

a. Models MEP-016B and MEP-021B.Refer to Figure 4-1 and connect the load cables to the generator set as described below.   Be sure that the output selection switch (located inside the control box) corresponds to the load terminal connections.

(1) For 120 volts,   single phase; connect cables to terminals L1 and L2.

(2) For 240 volts,   single phase; connect cables to terminals L1 and L2.

(3) For 120 volts,   three phase; connect cables to terminals L1, L2, and L3.

(4) For 120/208 volts, three phase, four wire; connect   cables to terminals LO, L1, L2, and L3.

b. Model MEP-026B.Refer to Figure 4-2 and connect the   negative (-) cable from the load to negative (-) terminal on the back of   the control box.

Connect the positive (+) cable from the load to the positive (+) terminal on the back of the control box.

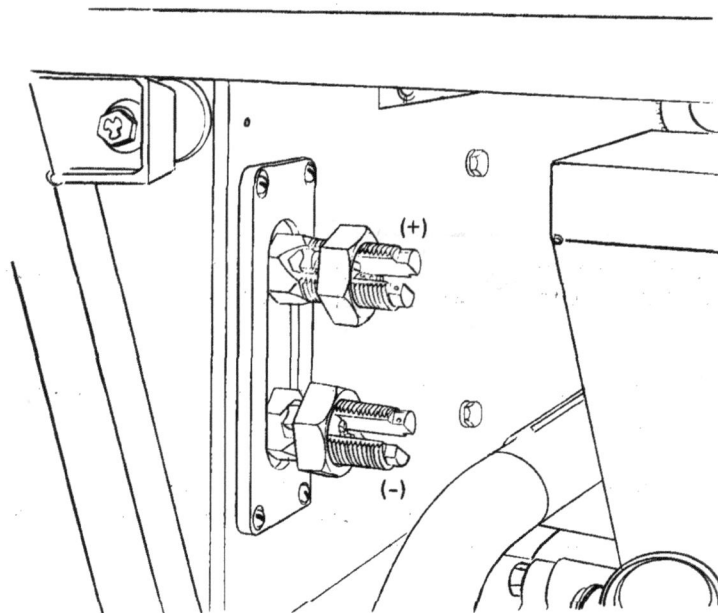

Figure 4-2.Load Terminals, Model MEP-026B.

MARINE CORPS TM 05926B/06509B-12
ARMY   TM 5-6115-615-12
NAVY   NAVFAC P-8-646-12

Section II. MOVEMENT TO   A NEW WORKSITE.

4-3.DISMANTLING GENERATOR SET FOR MOVEMENT.

4-3.1.   Preparation for Movement.

a.   Stop operation of the generator set (see paragraph 2-7).

b.   Disconnect the load cables.

c.   Remove exhaust pipe extension if used.

d.   Close vent on underside of fuel filler cap.

e.   Disconnect the ground lead from the ground stud.

f.   Pull up, disassemble, and store ground rod.

g.   Disconnect the negative battery cable from the battery and secure cable to prevent contact with battery post during movement of set.

h.   Disconnect the auxiliary fuel line from the auxiliary fuel pump (if used) and secure protective cap to fitting on pump.

4-3.2.   Movement. Preferred means of movement is to transport on a suitable vehicle. for short distances is permissible if the terrain is suitable.
Towing

4-4. REINSTALLATION AFTER MOVEMENT.

Refer to paragraph 4-2 for instructions on worksite.   installation after movement to a new

Section III. REPAIR PARTS;

**AND DIAGNOSTIC EQUIPMB-
NT   SPECIAL TOOLS; SPECIAL TEST, MEASUREMENT (TMDE);   AND SPECIAL SUPPORT EQUIPMENT.**

4-5. SPECIAL TOOLs AND EQUIPMENT.
needed to replace glow plug.   A 3/8 in. 6 point, deep socket (1/4 in. drive) is

4-6. MAINTENANCE REPAIR PARTS.
Repair parts and equipment are listed and illustrated in the Repair Parts and Special Tools List SL-4-05926B/06509B/ TM 5-6115-615-24P/NAVFAC P-8-646-24P/TO 35C2-3-386-34.

Section IV. LUBRICATION INSTRUCTIONS.

4-7. LUBRICATION.

4-7.1. _Lubrication Order._ Lubrication instructions contained in LI   for the generator set are
05926B/06509B-12/5/LO 5-6115-615-12   (see Figure 3-1).

4-7.2. _Oil Change._ (See Figure 4-3.)

WARNING

Do not remove oil filler cap when engine is running.
Hot oil can splash up and cause burns.

a.   Run set for a minimum then shut of 5 minutes to bring oil to operating temperature,

WARNING

Keep feet clear when tilting and working around tilted generator set. Failure to do so
can result in personal injury.

CAUTION

Do not apply air pressure to the crankcase to speed the oil drain process. Air pressure can
force the oil seals out of the crankcase.

Using two people, tilt and block side of generator set opposite drain cock
b.   (1).

Open drain cock (1) on skid base and allow oil to drain into a suitable
c.   container. The capacity of the engine lubrication system is 3.0 qts (2.8 1).

Remove blocks and lower generator set to level position.
d.

Remove oil filter (2) from engine and allow oil to drain.
e.

Install a new oil filter. Be sure to lubricate the filter gasket with clean oil prior to installation. Turn filter on by hand until
f.   it contacts filter base, then tighten 3/4 turn more.

Close drain cock (1) on skid base.

g.   Add oil through filler tube (3).   Refer to Table 3-1 for oil specifications.

h.

i. Start engine and check   for leaks.   Check oil level and add oil as necessary. with oil.
    Do not overfill engine

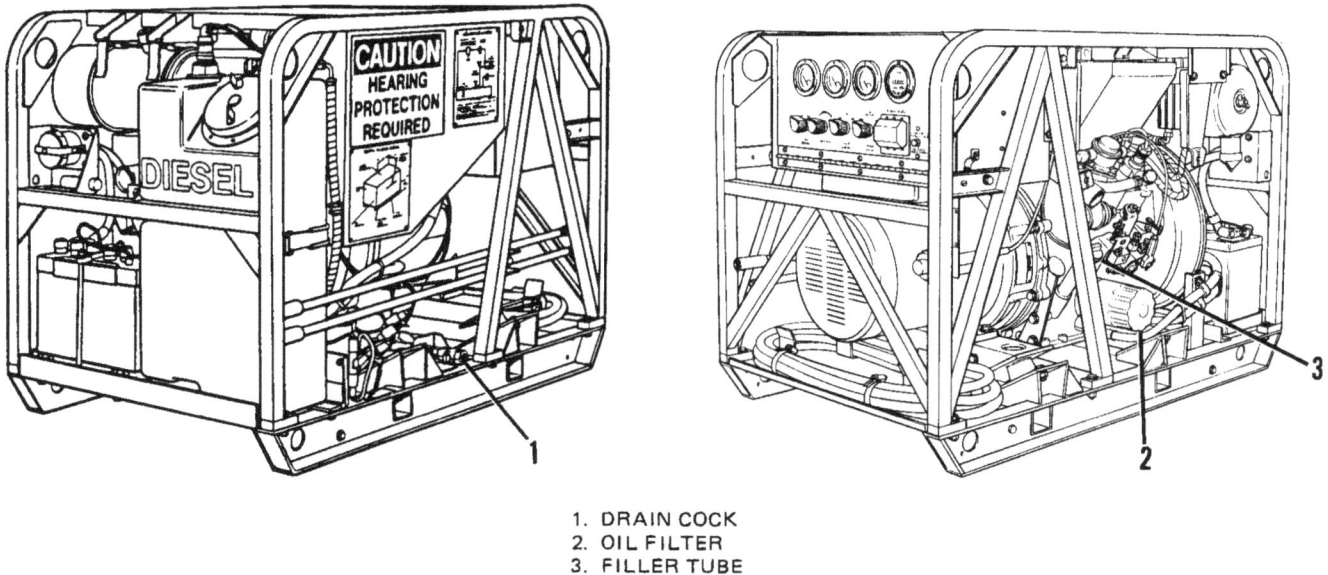

1. DRAIN COCK
2. OIL FILTER
3. FILLER TUBE

Figure 4-3. Engine Oil Change.

Section V. PREVENTIVE MAINTENANCE CHECKS AND SERVICES.

4-8. PMCS, GENERAL. To ensure that the generator set is ready for operation at all times, it must be inspected systematically so defects may be discovered and corrected before they result in serious damage, failure, or present a safety hazard. The necessary Preventive Maintenance Checks and Services that are to be performed by organizational personnel are listed and described in the following paragraphs. Defects discovered during operation will be noted for future correction.   stop operation immediately if a deficiency is noted which would damage the equipment or present a safety hazard.   All faults will be recorded together with the corrective action taken on the applicable form.

(MC) Marine Corps users should refer to current issue of TM 4700-15/1.

4-9. DETERMINING PMCS INTERVALS. Organizational PMCS on this unit should be performed on a "per hours of operation" basis.   The hourmeter on the control panel should be used to determine the generator set operating time.

4-10. PMCS FOR UNITS IN CONTINUOUS OPERATION. For PMCS performed on a "per hours of operation" basis, perform PMCS as close as possible to the time intervals indicated. For units in continuous operation, perform PMCS before starting operation if continuous operation will extend past the service interval indicated.

4-11. (MC, A, N) PREVENTIVE MAINTENANCE CHECKS AND tabulated listing of PMCS which shall be performed item numbers are listed consecutively and indicate requirements.

SERVICES. Table 4-1 contains a by Organizationa personnel. The the sequence of minimum

Table 4-1. Preventive Maintenance Checks and Services.

Interval W = Weekly (40 Hrs) M=Monthly (100 Hrs)
S= Semi Annual (500 Hrs)   Total M/H: 1.9
Organization   Work Items To Se Inspected

WMS	Time Inspection Procedures	M/H
	GENERAL	
1	Make a visual inspection of the entire generator set for any	0.3

obvious faults such as loose or missing bolts, nuts, end pins, and check for bent, cracked or broken parts.   Inspect all wires
and terminals for damage and/or loose connections.

	BNGINE	
2	Inspect engine for loose connections, leaks in oil and fuel systems, and free action of all moving parts.	0.3
3	Drain oil from crankcase and oil filter, and change lubricating oil every 125 operating hours as outlined in paragraph 4-7.	0.2
4	Replace oil filter element every 125 operating hours as outlined in paragraph 4-7.	0.2

Avoid prolonged contact and inhalation of fumes of dry cleaning solvent.Use dry clean-

WARNING

ing solvent only in a well ventilated area.

5	Clean governor linkage with dry cleaning solvent (Table 3-1, item 9).	0.2
6	Inspect muffler. Refer to paragraph 4-54.	0.2
7	Clean fins on oil cooler - Refer to paragraph 4-52.	0.2

Table 4-1. Preventive Maintenance Checks and Services, Continued.

Interval W = Weekly (40 Hrs)
M= Monthly (100 Hrs)
S = Semi Annual (500 Hrs) Total M/H: 1.9

Organization	Work Item to be Inspected	
	Time	
WMS	Inspection Procedures	M/H

FUEL SYSTEM

8   Change the filter element as outlined in paragraph 4-36.   0.1
Change fuel filters in fuel pumps as outlined in paragraph 4-34 and 4-35.

ELECTRICAL SYS-

Section VI. TROUBLESHOOTING.

4-12. GENERAL. This section contains information for locating and correcting operating troubles which may develop in the generator set.Each malfunction for an individual component, unit or system is followed by a list of tests or inspections which will help you to determine probable causes and corrective actions to take. You should perform the tests/inspections and corrective actions in the order listed.

4-13. MALFUNCTIONS NOT CORRECTED BY THE USE OF THE TROUBLESHOOTING TABLE. This manual cannot list all of the malfunctions that may occur, nor all the tests and inspections and corrective actions.   If a malfunction is not listed, or cannot be corrected by the listed corrective actions, notify your supervisor.

Table 4-2. Troubleshooting.

Malfunction
    Test Or Inspection
       Corrective Action

------

1. ENGINE FAILS TO   CRANK WHEN MASTER SWITCH IS HELD IN THE START POSITION.

    Step 1.   Test battery (paragraph 4-22).

       Replace defective battery.

Malfunction

    Test Or Inspection

        Corrective Action

---

1. ENGINE FAILS TO CRANK WHEN MASTER SWITCH IS HELD IN THE START POSITION (CONT'D).

    Step 2.   Inspect battery and starter cables for corrosion and loose connections.

        Clean or tighten battery cables.

    Step 3.   Test START position of master switch (paragraph 4-59).

        Replace defective master switch.

    Step 4.   Inspect and test starter relay ground (paragraph 4-26).

        Replace defective solenoid.

2. STARTER TURNS BUT WILL NOT ENGAGE.

    Step 1.   Inspect starter drive assembly (paragraph 4-48).

        Clean drive assembly if it is sticking.

    Step 2.   Inspect starter drive gear assembly for worn or broken teeth (paragraph 4-48).

        Replace starter.

3. ENGINE CRANKS NORMALLY BUT FAILS TO START.

    Step 1. Test fuel cutoff solenoid for proper operation (paragraph 4-37).

        Replace defective solenoid.

    Step 2.   Inspect fuel filter element (paragraph 4-36).

    Step 3.   Check

Clean strainer element.

Replace fuel filter element.

for clogged fuel line to

injector pump (paragraph 4-39).

Clean or replace clogged

fuel line.

    Step 4.   Test for worn or damaged fuel

transfer pump (paragraph 4-34).

        Replace defective pump.

Table 4-2. Troubleshooting, Continued.

Malfunction

    Test Or Inspection

        Corrective Action

Step 5.   Inspect and test glow plug (paragraph 4-50).

Replace defective glow plug.

4. ENGINE STARTS BUT DOES NOT RUN SMOOTHLY (MISFIRES, KNOCKS, OR MAKES UNUSUAL NOISES).

Check Steps 1, 2, 3, 4 and 5 under ENGINE CRANKS NORMALLY BUT FAILS TO START above.

Perform corrective action as necessary.

5. LOW ENGINE POWER, ESPECIALLY UNDER HEAVY LOAD.

Step 1. Check Steps 1, 2, 3,4 and 5 under ENGINE CRANKS NORMALLY BUT FAILS TO START above.

Perform corrective action as necessary.

Step 2.   Check air filter assembly for obstructions (paragraph 3-29).

Remove obstructions.

6. ENGINE   HAS EXCESSIVE OIL CONSUMPTION.

Check   air filter element.

Clean or replace air cleaner element as necessary (paragraph 3-29).

7. ENGINE   VIBRATES.

Check   for loose or worn engine mounts.

Tighten mounting hardware or replace as necessary (paragraph 4-17)

If vibration continues, notify higher echelon of maintenance.

Table 4-2. Troubleshooting, Continued.

Malfunction
   Test Or Inspection
         Corrective Action

8. MAIN CIRCUIT BREAKER WILL NOT CLOSE.

   Test for short circuit in load: Shut down set and disconnect the load.   Connect a multimeter (set to read ohms) across the load lines to check for a short circuit. A low reading (0 ohms) indicates a short circuit.

   Correct short circuit before reconnecting load.

ENGINE 0VERSPEEDS.

9.   Step 1.   Inspect for correct oil level (Figure 3-1).

      Step 2.   Inspect for free operation of governor linkage (paragraph 4-44).

      Step 3.Inspect air filter element (paragraph 3-29).
                Section VII. RADIO INTERFERENCE SUPPRESSION.

4-14. GENERAL METHODS FOR PROPER SUPPRESSION. For proper suppression of radio interference, a low resistance path to ground must be provided for stray currents. The methods used include shielding the high frequency wires, grounding the frame with bonding straps, and using filtering systems.

4-15. RADIO INTERFERENCE SUPPRESSION COMPONENTS.
4-15.1. Primary Suppression Components.The primary suppression components are those whose primary function it is to suppress radio interference. The primary suppression components are shown in Figure 4-4.
4-15.1.1.Model MEP-016B.The primary radio interference suppression components are four capacitors; C1, C2, C3 and

C4.All of these four capacitors are 0.10 microfarad, 5 percent tolerance 1000V.They are recognizable by their tie-down straps. The silkscreened numbers "C1," etc. are clearly visible.

4-15.1.2.Model MEP-021B.The primary radio interference suppression components are four capacitors; C1, C2, C3 and C4.All of these four capacitors are 0.018 microfarad, 5 percent tolerance, 1000V. They are recognizable by their tie-down straps. The silkscreened numbers "C1," etc. are clearly visible.

4-15.1.3.Model MBP-026B.The primary radio interference suppression components are three capacitors: C1, C2, and C3. Two of these capacitors, C1 and C3, are 0.47 microfarad, 200V.These two are mounted to the side wall connected to the terminal block which has wires running to the output terminals.The third one, C2, is 50 microfarad, 50V, and is supported by a cable clamp.

4-15.2. Secondary Suppression Components.These components have radio interference suppression functions which are incidental or secondary to their primary function. They are internal-external tooth lockwashers on the fuel pump and battery charging regulator-rectifier.

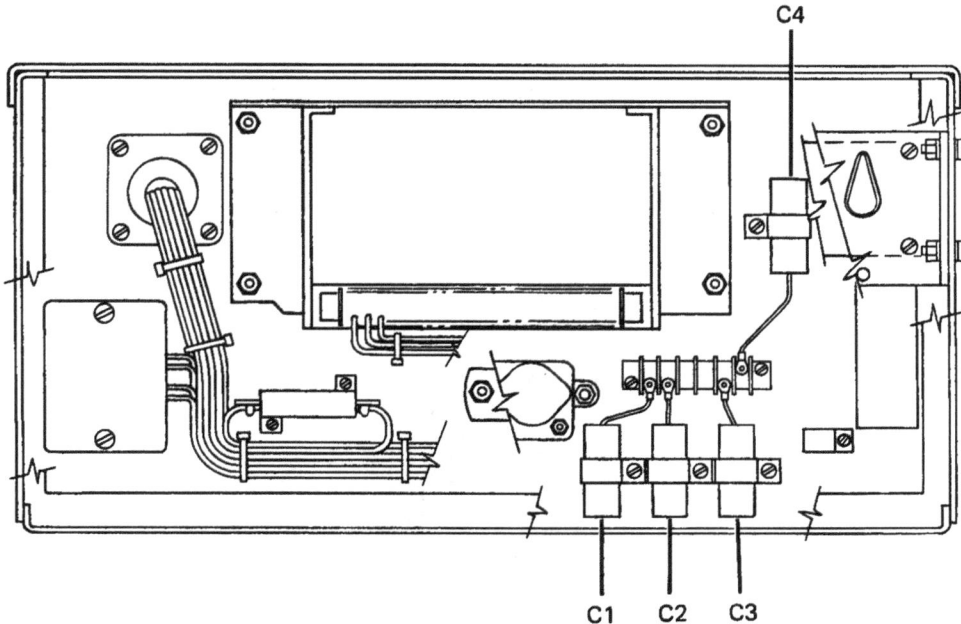

Figure 4-4. Primary   **Radio Interference Suppression Components.**
Models   **MBP-016B and MBP-021B Shown.**

Section VIII.   **MAINTENANCE OF FRAME AND SKID BASE.**

WARNING

Unless manual directs otherwise, do not attempt any of the following maintenance procedures when generator set is operating.   Do not touch exposed electrical connections when a source of power such as utility power or another generator set is connected to the load terminals.   Severe electrical shock or death from electrocution may result.

4-16. GENERAL. The frame and skid base consists of the frame suppression panel and muf- (including the sound eye, battery fler shield), the skid base, lifting battery holddown, and the related hardware.

4-17.   **ENGINE MOUNTING BRACKETS.**   (See Figure 4-5)

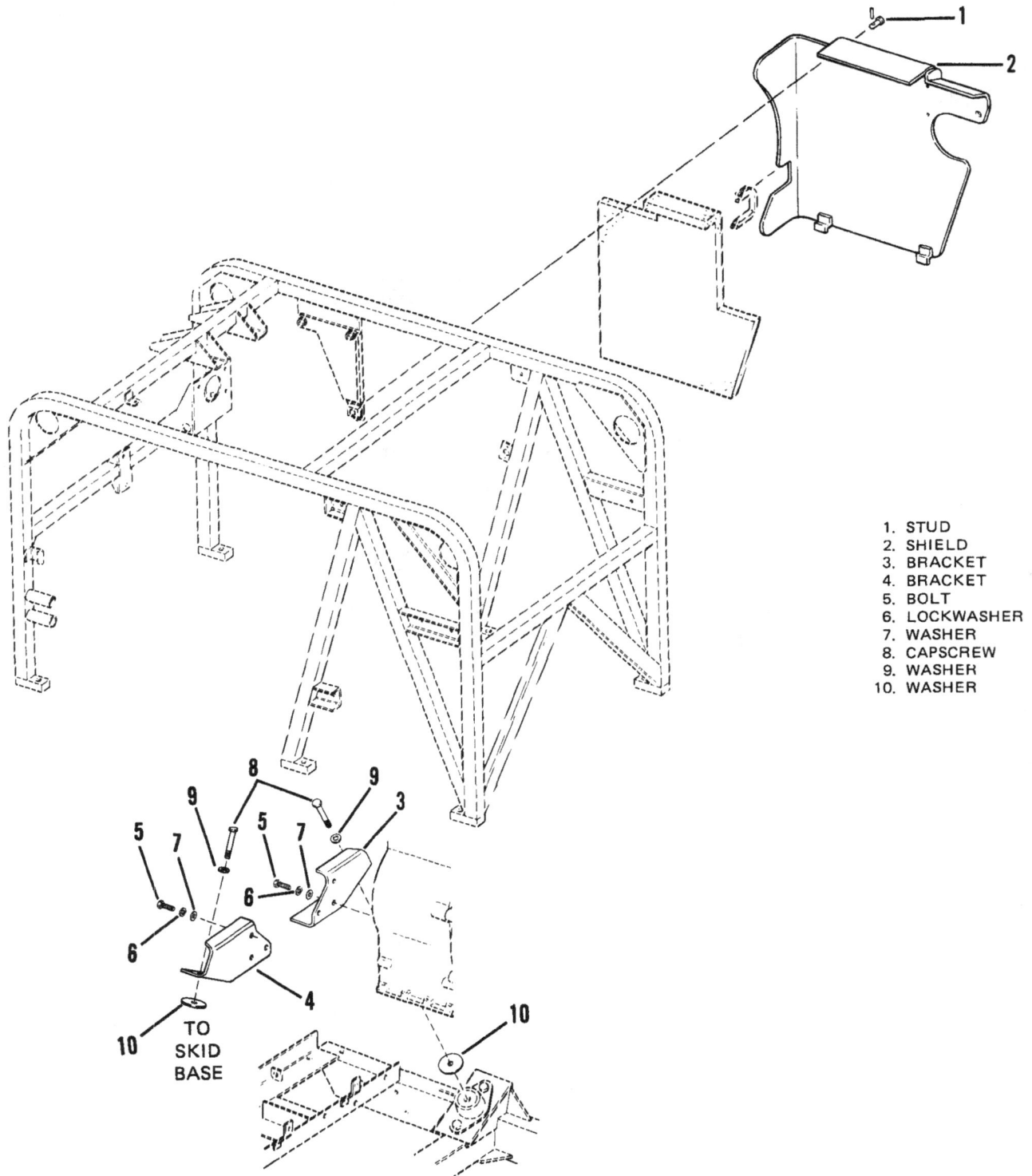

1. STUD
2. SHIELD
3. BRACKET
4. BRACKET
5. BOLT
6. LOCKWASHER
7. WASHER
8. CAPSCREW
9. WASHER
10. WASHER

Figure 4-5.Engine Mounting Brackets.

MARINE
CORPS
ARMY

TM 05926B/06509B-12
TM 5-6115-615-12
NAVFAC P-8-646-12
T0 35C2-3-386-31

It is necessary to loosen studs (1) and remove noise suppression shield (2) to gain access to one of the engine mounting brackets.

a. Inspection. Inspect engine mounting brackets (3 and 4) to see that they are tightly secured and free of cracks, bends or other damage.

b. Removal.

(1) Place a suitable block under end of engine (opposite generator) to support engine after mounting brackets are removed.

(2) Remove bolts (5), lockwashers (6) and washers (7).

(3) Remove capscrews (8), washers (9), brackets (3 and 4) and washers (10).

c. Installation.

(1) Install capscrews (8), washers (9), brackets (3 and 4) and washers (10).

(2) Install bolts (5), lockwashers (6) and washers (7).

(3) Remove block from under engine.

(4) Install noise suppression shield (2) and secure with studs (1).

4-18. LIFTING EYE. (See Figure 4-6)

a. Removal. Remove the two bolts (1), washers (2), and spacer (3) that secure lifting eye (4) to the top of the generator housing and remove the lifting eye.

1. BOLT
2. WASHER
3. SPACER
4. LIFTING EYE

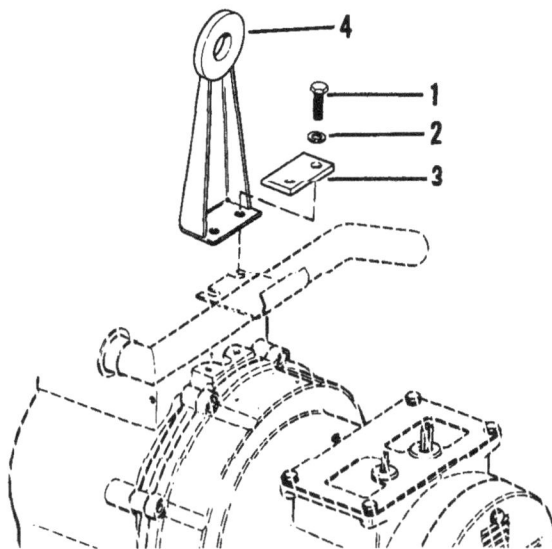

Figure 4-6. Lifting Eye.

MARINE CORPS TM 05926B/06509B-12
ARMY   TM 5-6115-615-12
NAVY   NAVFAC P-8-646-12
AIR FORCE   TO 35C2-3-386-31

b.Installation.   Put lifting eye (4) in position on top of the generator housing and install spacer (3), two bolts (1) and washers (2) that hold it.

4-19. BATTERY FRAME AND TRAY, BATTERY HOLDDOWN.   (See Figure 4-7)

WARNING

Battery electrolyte is an acid solution that gives off flammable fumes.   Do not smoke or use open flame when working around battery.Doing so can cause an explosion that could result in serious personal injury. If skin is exposed to electrolyte, flush exposed area with water immediately. If eyes are exposed to electrolyte, flush them with water and seek immediate medical attention.

NOTE

Disconnect negative reconnect (-) cable from battery first, and if last.

(1)   Disconnect battery cable 22) from   assemblies (11 and 15) and rubber covers (10 and

(2)   Remove nuts (19), lockwashers (20) and washers (21).   Remove battery top frame

(3)   Remove   four nuts (1), battery   lockwashers (2), washers (3) and bolts (7).
        Remove   bottom   frame (6) and battery tray (5) from skid base (4).

(4)   Remove   (23) from skid base (4).

b. Inspection and Repair.

(1) Inspect battery tray (5), battery bottom frame (6), battery top frame (9), and hook bolts (23) for dents, rust, or bent condition. Repair by straightening and/or welding.Prime paint all items.

(2) Inspect all bolts and screws for stripped threads or damaged heads. Replace hardware as necessary.

c.Installation.

(1) Install hook bolt (23) in rear position on skid base (4). Put battery tray (5) and battery bottom frame (6) in position on skid base (4). Install four bolts (7), washers (3), lockwashers (2) and nuts (1).

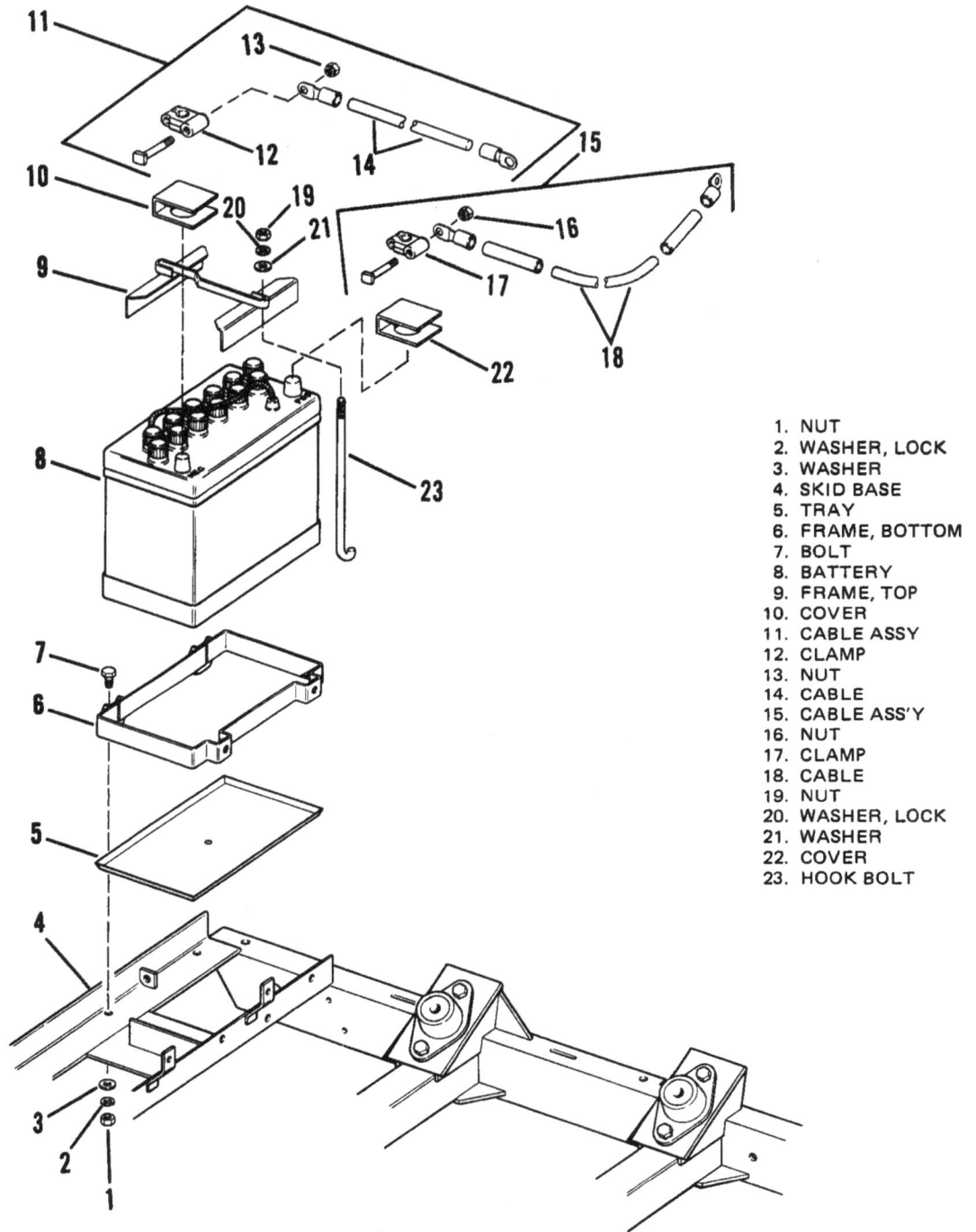

1. NUT
2. WASHER, LOCK
3. WASHER
4. SKID BASE
5. TRAY
6. FRAME, BOTTOM
7. BOLT
8. BATTERY
9. FRAME, TOP
10. COVER
11. CABLE ASSY
12. CLAMP
13. NUT
14. CABLE
15. CABLE ASS'Y
16. NUT
17. CLAMP
18. CABLE
19. NUT
20. WASHER, LOCK
21. WASHER
22. COVER
23. HOOK BOLT

Figure 4-7. Battery, Battery Cables, Battery Tray and Holddown.

<hr>
WARNING
<hr>

Battery electrolyte is an acid solution that gives off flammable fumes.Do not smoke or use open flame when working around battery.Doing so can cause an explosion that could result in serious personal injury. If skin is exposed to electrolyte, flush exposed area with water immediately.   If eyes are exposed to electrolyte, flush them with water and seek immediate medical attention.

Put battery (8) in position.   Install hook bolt (23) on skid base. Put battery top frame (9) in position and install washers (21), lockwashers (20) and nuts (19) on hook bolts (23).

(2)

NOTE

Connect positive (+) cable to battery first, and negative (-) cable last.

Install rubber cable covers (10 and 22) and connect battery cables (11) and (15) to battery.Tighten clamping nut on cable clamp.

(3)

4-20.GROUND ROD- REPLACEMENT. (See Figure 4-8)

1. BOLT
2. LOCKWASHER
3. WASHER
4. CLAMP
5. GROUND ROD ASSY

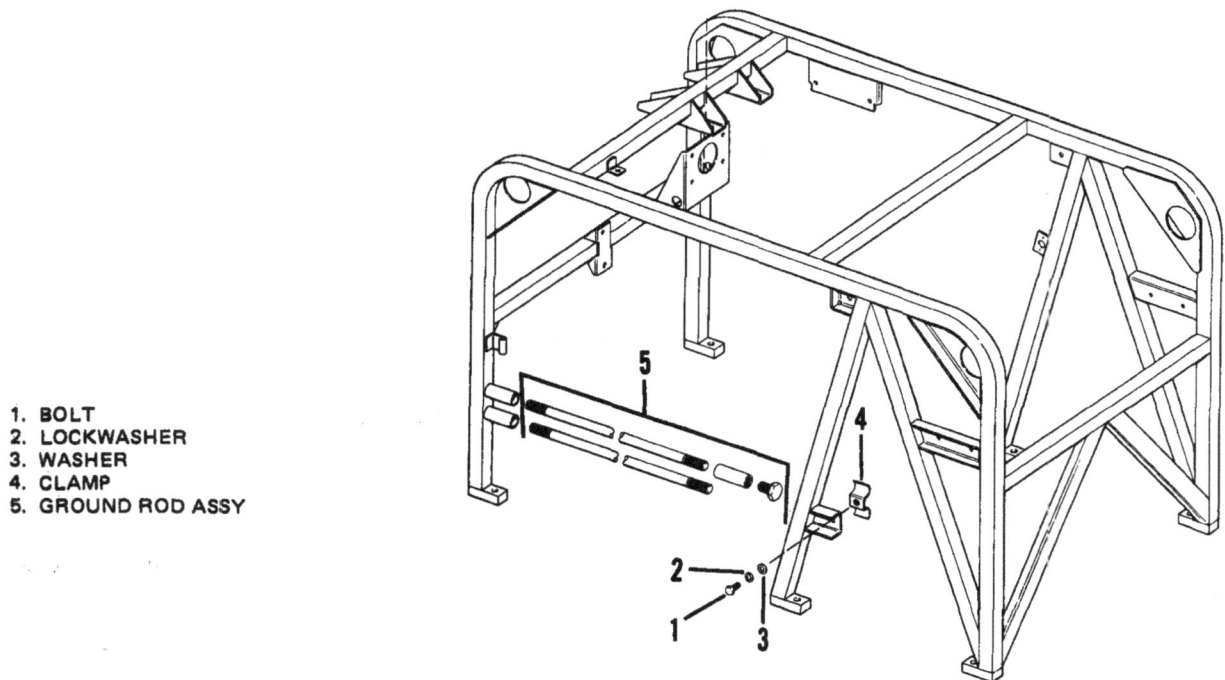

Figure 4-8. Ground Rod.

CORPS

MARINE   TM 05926B/06509B-12
             TM 5-6115-615-12
ARMY   NAVFAC P-646-12
NAVY   T0 35C2-3-386-31
AIR FORCE

Remove ground

a.   Remove bolt (1), lockwasher (2), washer (3), and clamp (4).
    rod (5).

Section IX. MAINTENANCE OF DC ELECTRICAL SYSTEM.

WARNING

Unless manual directs otherwise, do not attempt any of the following procedures when generator set is operating.   Do not touch exposed electrical connections when a source of power such as utility power or another generator set is connected to the load terminals. Severe electrical shock or death from electrocution may result.

4-21. PURPOSE, CONSTRUCTION, FUNCTION, AND OPERATION. The DC electrical control system provides starting power for the generator set and it charges the battery while the set is running.   It consists of the battery, battery cables and slave receptacle, the starter relay, the voltage regulator the battery charger, the master switch, the DC control circuit breaker and the wiring harness.

4-22. BATTERY.   (See Figure 4-7)
    Battery electrolyte can cause severe burns to the skin.
    Always flush exposed parts of the skin with clear water as soon as possible after
      contact with electrolyte.

WARNING

Batteries generate explosive gas during charging.
Utilize extreme caution.   Do not smoke or use open flame in the
vicinity of the generator set when servicing batteries.

    a. Test. Charge battery long enough for the battery to be fully charged (1 to 2 hours). Test each cell of battery separately using a hydrometer.   Remove cap from one cell of battery and draw fluid from that cell into hydrometer. Hydrometer must register a specific gravity of 1.260 to 1.280.   If specific gravity is below 1.260, charge battery. Replace fluid into cell and replace cap.   If one or more cells in a battery will not take a charge, replace the battery.

b.   <u>Removal.</u>_____(See Figure

WARNING

Disconnect battery cables before servicing generator components.The high current output of the DC electrical system can cause arcing and/or burns if a short circuit occurs.

NOTE

Disconnect negative reconnect (-) cable from battery first and

(1)   Disconnect battery clamps (12 and 17) and covers (10 and 22) from battery (8).

(2)   Remove nuts (19), lock washers (20) and washers (21).   Remove battery top frame (9) from battery (8). Remove battery.

c.   <u>Installation.</u>

WARNING

Disconnect battery cables before servicing generator components.The high current output of the DC electrical system can cause arcing and/or burns if a short circuit occurs.

(1) Put battery (8) in position. Put install washers (21),   battery top frame   (9) in position and on hook
                                                             (20) and nuts (19)

NOTE

Connect positive (+) cable to battery first.

(2) Install rubber cable covers (10 and 22) and connect battery cables (11 and 15) to battery.Tighten clamping nuts on cable clamps (12 and 17).

4-23.BATTERY CABLES.   (See Figure 4-9)

a.   **Removal.**

_____

WARNING

_____

Battery electrolyte is an acid solution that gives off flammable fumes.   Do not smoke or use open flame when working around battery. Doing so can cause an explosion that could result in serious personal injury. If skin is exposed to electrolyte, flush exposed area with water immediately. If eyes are exposed to electrolyte, flush them with water and seek immediate medical attention.

NOTE

Always disconnect the negative (-) reattach it last.   cable first and

(1)   Disconnect negative (-) cable assembly (1) from battery. Disconnect positive (+) cable assembly (2) from battery.   Remove rubber covers (9 and 10).

(2)   Remove nut (3) and remove battery cable (5) and slave receptacle cable (not shown) from terminal (4). Remove nut (7) and remove battery cable (6) and slave receptacle cable (not shown) from terminal (8).

(3)   Tag and disconnect cables (5 and 6) from solenoid.   the starter case and starter

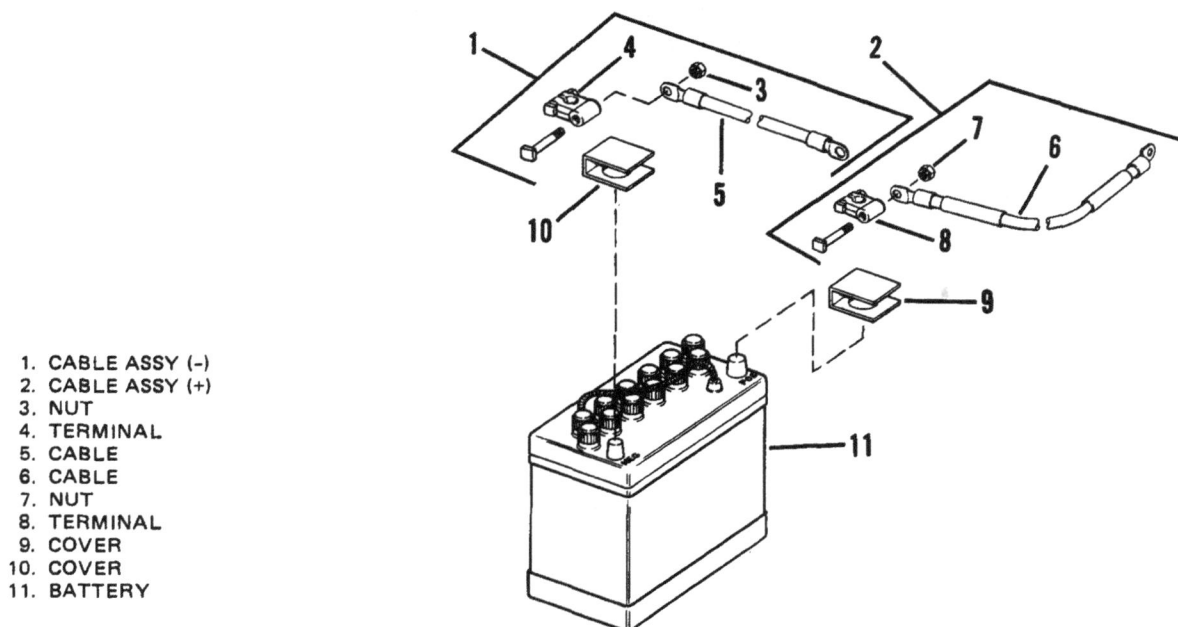

1. CABLE ASSY (-)
2. CABLE ASSY (+)
3. NUT
4. TERMINAL
5. CABLE
6. CABLE
7. NUT
8. TERMINAL
9. COVER
10. COVER
11. BATTERY

Figure 4-9. Battery Cables and Slave Receptacle.

---

(1) Inspect battery cables for damaged terminals or insulation. Replace damaged cables.

(2) Inspect terminals and hardware for damage or stripped threads. Replace damaged terminals.

(3) Inspect battery post covers for torn condition. Replace damaged covers.

c. Installation.

(1) Attach the positive (+) battery cable (6) to the stud on the starter motor solenoid.Attach the negative (-) battery cable (5) to the starter case.

(2) Put positive (+) battery cable (6) and slave receptacle cable (not shown) in position on terminal (8) and install nut (7). Put negative (-) battery cable (1) and slave receptacle cable (10) in position on terminal (4) and install nut (3).

NOTE

Always disconnect the negative (-) cable first and reattach it last.

(3) Put cover (9) in position and connect positive (+) cable assembly (2) to the positive (+) battery post.Put cover (10) in position and connect negative (-) cable assembly (1) to the negative (-) post on the battery. Tighten terminal connectors securely.

4-24.   SLAVE RECEPTACLE.   (See Figure 4-10)

a.   Test.   Using a voltmeter check the battery voltage at the battery terminals, and then check the battery voltage at the slave receptacle. If the voltage at the slave receptacle is not the same as the battery voltage, check for continuity between the battery and slave receptacle. Replace the defective component.

b.   Removal.

NOTE

Always disconnect the negative (-) cable first.
Disconnect the negative (-) cable assembly from the battery.Disconnect the positive (+) cacle assembly from the battery.

Remove nut (1) and remove slave receptacle cable (2) from negative terminal

(1)

(2)

(4).   Remove nut (5) and terninal remove slave receptacle cable (3) from positive
(7).

(3)   Remove nuts (8), washers from the   (9) and screws (10).   Remove slave receptacle (11)

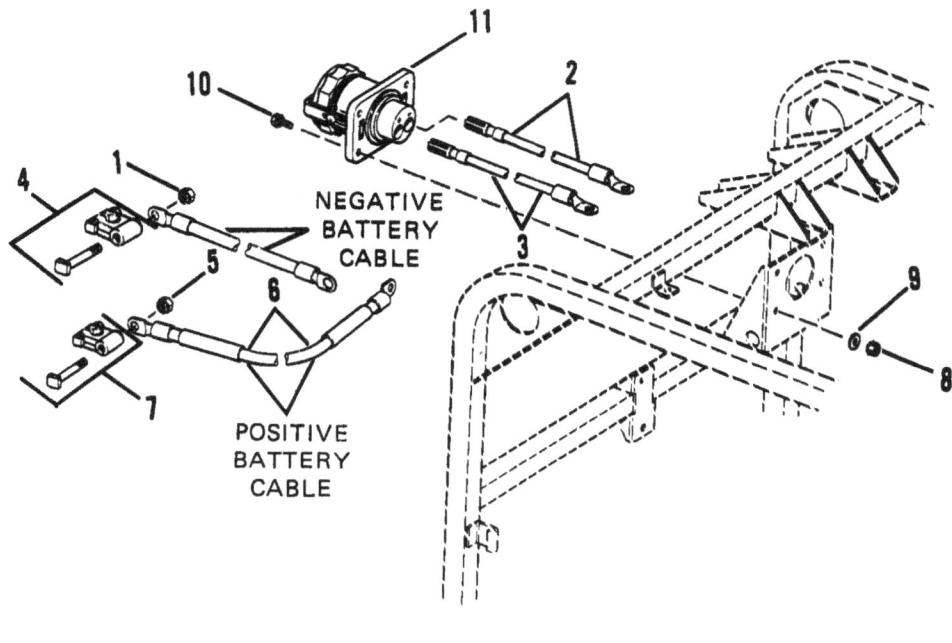

1. NUT
2. CABLE (-)
3. CABLE (+)
4. TERMINAL
5. NUT
6. CABLE
7. TERMINAL
8. NUT
9. WASHER

Figure 4-10. Slave Receptacle.

c.   Installation.

(1)   Put slave receptacle (11, Figure 4-10) in position on the frame and install screws (10), washers (9) and nuts (8).

(2)   put slave receptacle cable (3) in position on positive terminal (7) and install nut (5). Put slave receptacle cable (3) in position on negative terminal (4) and install nut (1).

NOTE

Always connect the positive (-) cable first.

Connect the positive (+) and negative (-) cable assemblies to the battery.
4-25. VOLTAGE REGULAR BATTERY CHARGING.

(3)

a.   General. The battery charging voltage regulator converts the AC current from the engine alternator to approximately 28 volts DC for charging the battery whenever the unit is running.

b.   Test. Using a DC Voltmeter, test voltage regulator (see Figure 4-11 for test points).   Check the voltage with the unit running. Output voltage from the regulator must be 28.5 $\pm$ .7 volts DC at 77°F (25°C). If voltage is not within specifications, the voltage regulator must be replaced.

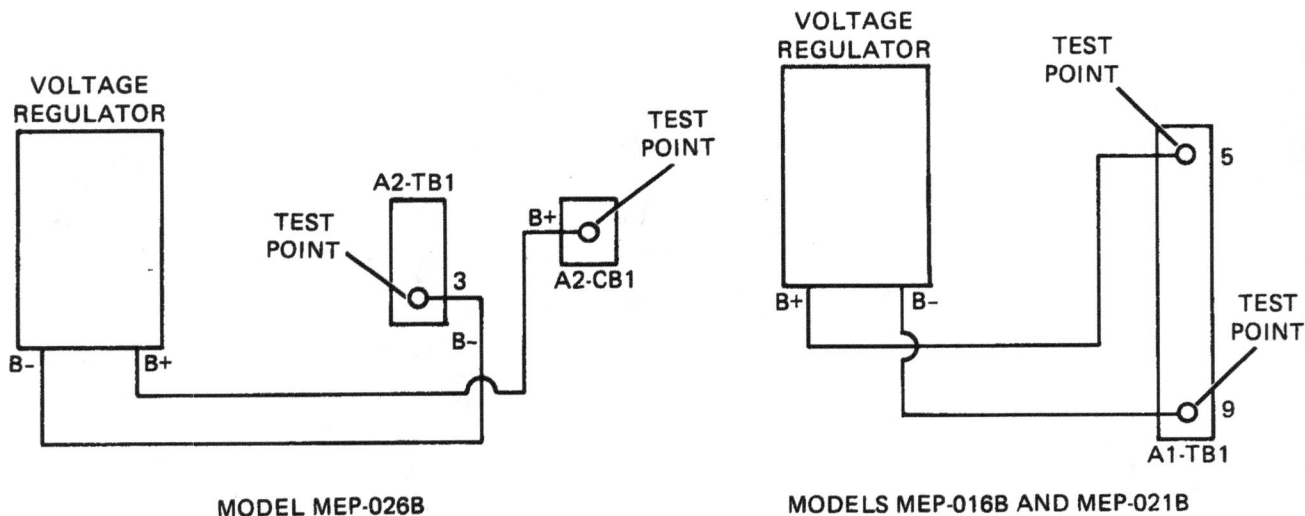

Figure 4-11.Voltage Regulator Test Points.

c. Removal.(see Figure 4-12)

(1) Follow four wires from voltage regulator to connection points. Tag wires and note locations of terminals. Disconnect four wires. Cut plastic tie wraps as necessary.

(2) Remove nuts (l), and screws (2).   Remove the voltage regulator (3).

d. Installation.
Put voltage regulator (3) in position and install screws (2), and nuts (1).

(1)   Using tags for identification, connect four wires.   Replace plastic tie wraps (4)
(Table 3-1,item 18) as necessary.
(2)   **4-26.STARTER RELAY.**   (See Figure 4-13)

a. Inspect.

(1) Inspect the starter relay for damage, secure mounting, and check to see that all wires are connected securely.

(2) Tighten or replace hardware as necessary. Repair or replace damaged or loose wiring. Replace a damaged starter relay.

b. Test.

(1) Tag and disconnect the wires from the starter relay.

(2) Connect the leads of an ohmmeter to terminals (A1 and A2) of the starter relay. The resistance between (A1 and A2) should be high (infinite) when no voltage is applied to terminals (x1 and x2).

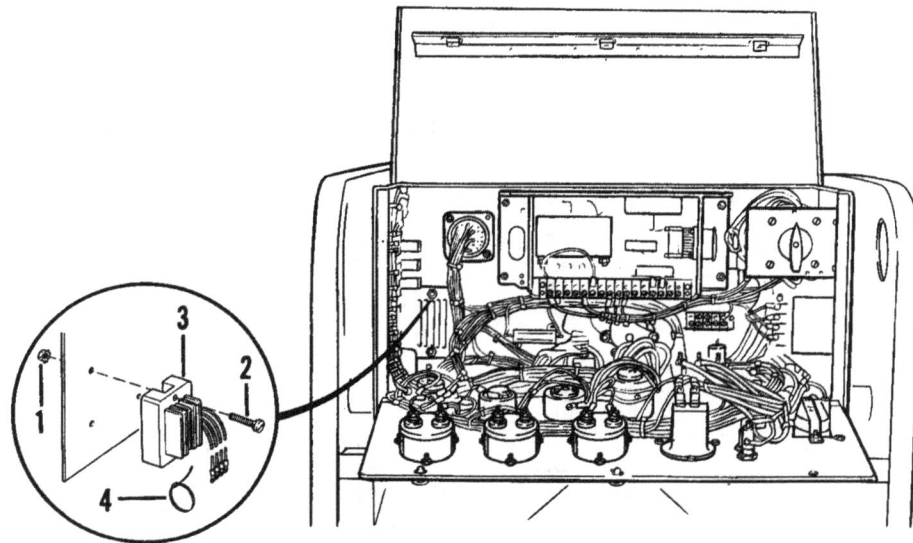

MODELS MEP-016B AND MEP-021B

1. NUT
2. SCREW
3. REGULATOR, VOLTAGE
4. PLASTIC TIE WRAP

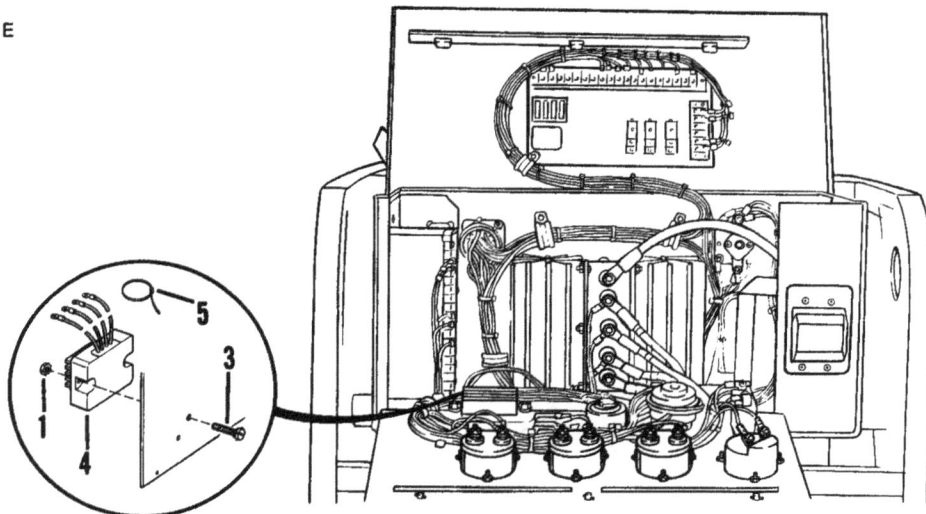

Figure 4-12. Voltage Regulator, Battery Charging.

(3) With the ohmmeter still connected to terminals (A1 and A2) ,apply battery voltage to terminals (X1 and X2).   The resistance between terminals (A1 and A2) should drop to zero. If the resistance does not drop with battery voltage applied, replace the relay.

c. <u>Removal.</u>

(1) Tag and disconnect all wires from the starter relay (3).

(2) Remove screws (1) and nuts (2). Remove relay (3).

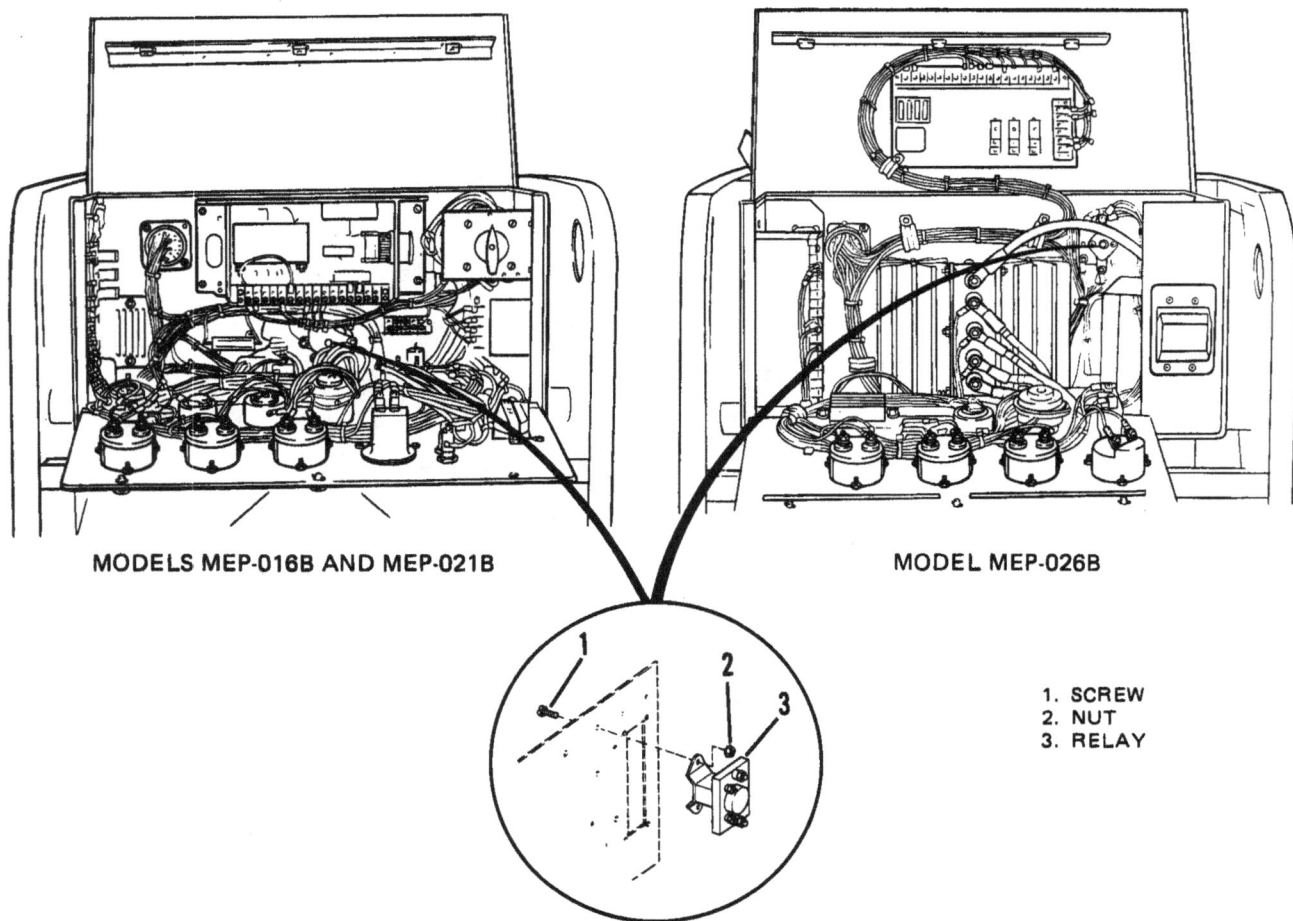

MODELS MEP-016B AND MEP-021B

MODEL MEP-026B

1. SCREW
2. NUT
3. RELAY

Figure 4-13. Starter Relay.

d. Installation.

(1) Put the starter relay (3) in position and install screws (1) and nuts (2).

(2) Using the tags for identification, connect the wires to the relay (3).

4-27.   DC CONTROL CIRCUIT BREAKER. (Emergency Stop) (see Figure 4-14)

a.   Test.

(1) Tag and disconnect wiring from circuit breaker.

(2) With circuit breaker off, use an ohmmeter to check for resistance between terminals.

(3) Ohmmeter should indicate no continuity (high ohms) between terminals.

(4) Flip circuit breaker to the ON position. Recheck resistance between terminals.

(5) Ohmmeter should now indicate continuity (low ohms) between terminals.

b. Removal.

NOTE

Procedures for all three sets MBP-016B, MEP-021B, and MEP-026B are the same. Figure 4-14 illustrates the MEP-021B (400 Hz) set.

(1) Tag and disconnect wiring from circuit breaker (1).

(2) Remove screws (2) and circuit breaker (1).

Figure 4-14. DC Control Circuit Breaker (Model MEP-021B Shown).

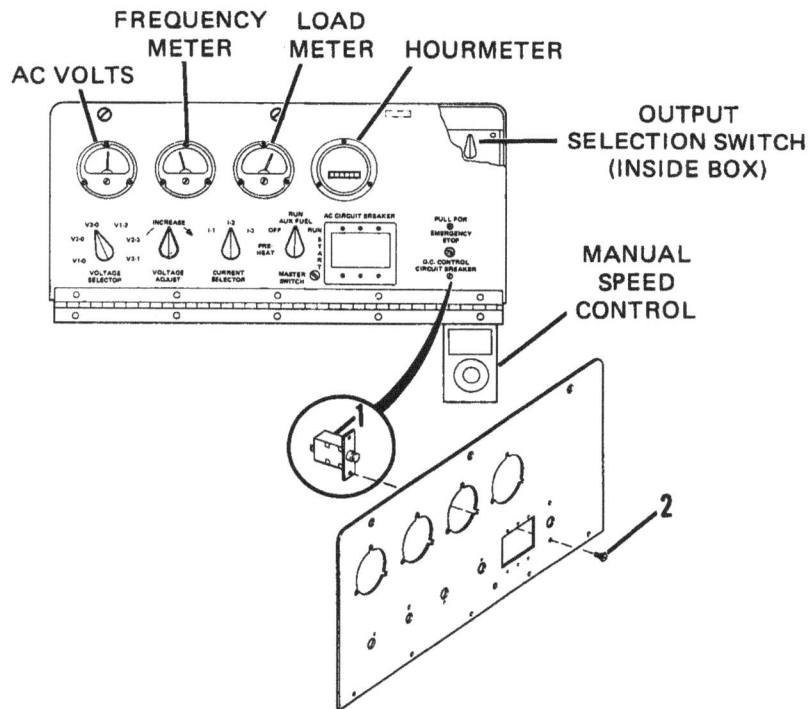

1. CIRCUIT BREAKER
2. SCREW

c. Installation.

(1) Secure circuit breaker (1) to front panel with screws (2).

(2) Using tags for identification, connect wiring to circuit breaker.

Section X. MAINTENANCE OF THE POWER GERNERATION SYSTEM.

Unless manual directs otherwise, do not attempt any of the following maintenance procedures when generator set is operating. Do not touch exposed electrical connections when a source of power such as utility power or another generator set is connected to the load terminals.   Severe electrical shock or death from electrocution may result.

GENERATOR ASSEMBLY.   (See Figure 4-15)

4-28.   Inspect generator housing for cracks, distortion or other visible damage.

a.

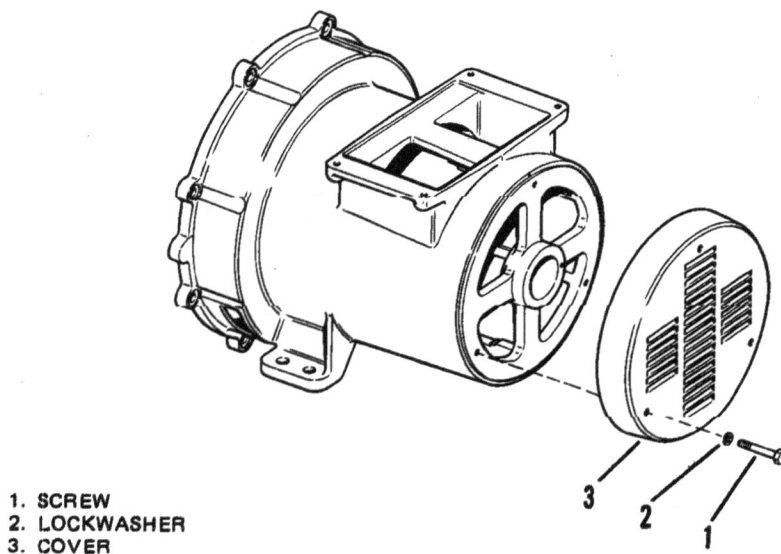

1. SCREW
2. LOCKWASHER
3. COVER

Figure 4-15.   Generator (Model MEP-02113 Shown).

b.   Remove three screws (1),   lockwashers (2) and cover (3).

c.   Visually inspect physical generator windings for signs of   burning, shorting or other damage.

d.   with screws (1) and lockwashers   (2).

MARINE CORPS TM 05926B/06509B-12
ARMY   TM 5-6115-615-12
NAVY   NAVFAC P-8-646-12
AIR FORCE TO 35C2-3-386-31

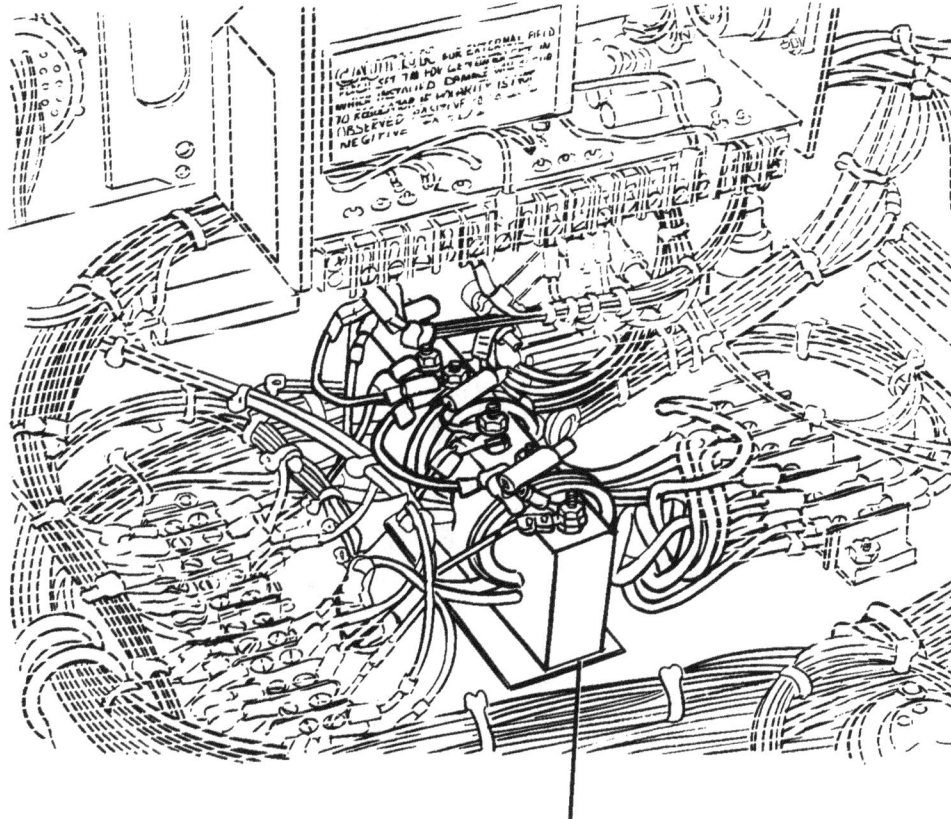

CURRENT TRANSFORMER

Figure 4-16. Current Transformer, Models MEP-016B and MEP-021B Only.

a.   Inspect that current transformer is mounted securely.

b.   Check that all wiring is securely connected to terminals.

c.   Check that resistors   are not damaged or loose.

4-30.   LOAD TERMINAL BOARD.   (See Figures 4-17 and 4-18)

NOTE

Procedures for Models MEP-016B, MEP-021B and MEP-026B are similar. For Models MEP-016B and MEP-021B, refer to Figure 4-16. For Model MEP-026B, refer to Figure 4-17.

Removal.

a.

(1) If removal of cover is necessary remove screws (1) and nuts (2) to remove cover and bracket assembly from back of control box.

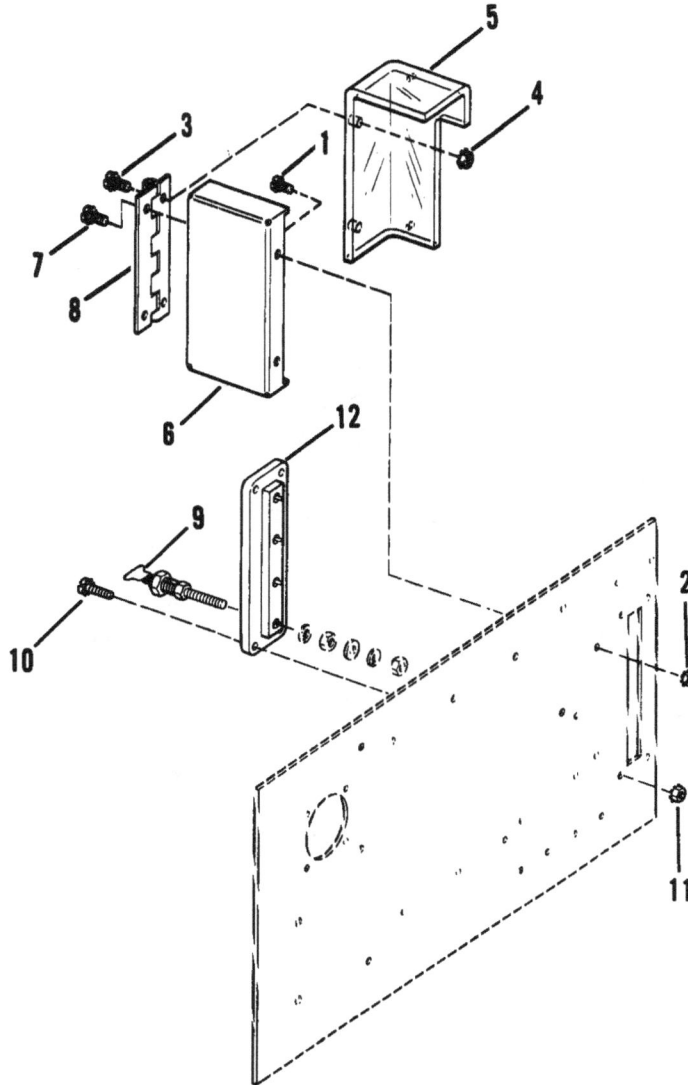

Figure 4-17.Load Terminal Board, Models MEP-016B and MEP-021B.

1. SCREW
2. NUT
3. SCREW
4. NUT
5. COVER
6. BRACKET
7. SCREW
8. HINGE
9. TERMINAL, LOAD
10. SCREW
11. NUT
12. BOARD

(2) If necessary to remove cover (5), bracket (6). remove screws (3) and nuts (4) from

(3) Remove screws (7) to remove hinge (8) from bracket (6).

(4) Tag and disconnect wiring from load terminals (9)

(5) Remove screws (10) and nuts (11) to remove terminal board (12).

(6) Load terminals (9) damaged. are not removed from terminal board (12) unless visibly

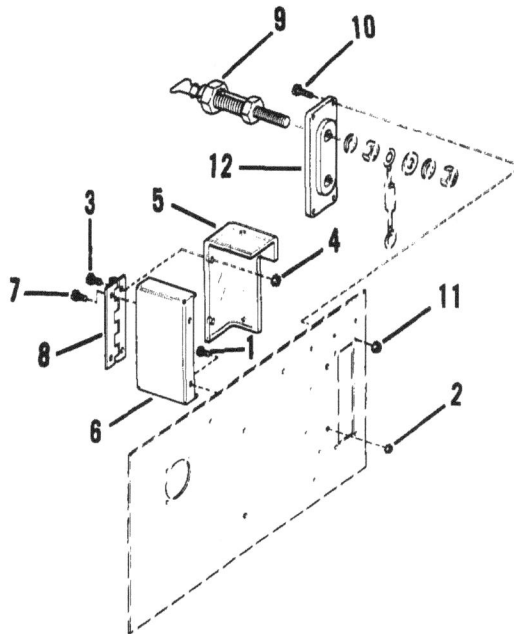

Figure 4-18. Load Terminal Board, Model MEP-026B.

b. Installation.

    (1)   If removed, secure load terminals (9) to terminal board (12).

    (2)   Secure terminal board (12) to control box with screws (10) and nuts (11).

    (3)   Using tags for identification, attach leads to terminals.

    (4)   Install bracket (6) on back of box with screws (1) and nuts (2)

    (5)   Install hinge (8) to bracket (6) with screws (7).

    (5)   Put cover (5) in position on hinge (8) and install screws (3)and nuts (4).

4-31.   CONVENIENCE OUTLET. (Model MEP-016B Only)   (See Figure 4-19)

a.   Inspect.

    (1) Inspect convenience outlet and cover for damage and secure mounting.
        Open control box and check that wires are securely attached to the outlet terminals.

    (2) Tighten or replace hardware as necessary.   Repair or replace damaged
        wiring.   Replace a damaged outlet or cover.

b.   Test.

    (1) Start and run generator set.

MARINE CORPS TM 05926B/06509B-12
ARMY   TM 5-6115-615-12
NAVFAC P-8-646-12
AIR FORCE   TO 35C2-3-386-31

(2)   Plug a 120 volt test lamp into the convenience outlet.

(3)   If test lamp does not illuminate, stop generator and recheck connections and convenience outlet fuses. Refer to paragraph 4-32.

(4)   Test continuity of wires.   Replace wiring as necessary.

(5)   Test convenience outlet again.   If test lamp still fails to illuminate, remove convenience outlet.

Figure 4-19.Convenience Outlet (Model MEP-016B Only).

c. Removal.

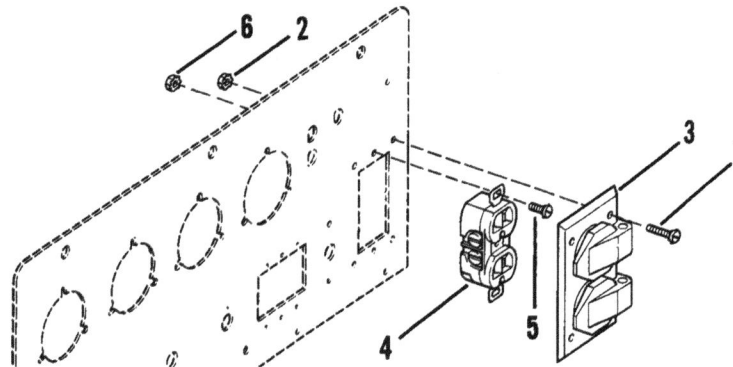

Remove screws (1), nuts (2) and outlet cover (3).

Tag and disconnect wiring from outlet (4).

(1)

(2)

(3)   Remove screws   (5), nuts (6) and outlet (4).

d. Installation.

(1)   Secure outlet   (4) to control box with screws (5) and nuts (6).

(2)   Using tags for identification, connect wiring to outlet (4).

(3)   Secure outlet cover (3) with screws (1) and nuts (2).

4-32.   CONVENIENCE OUTLET FUSES.   (Model MEP-016B only) (See Figure 4-20)

a.   Removal.

(1) Remove cap (1) from fuseholder (2) and remove fuse (3).

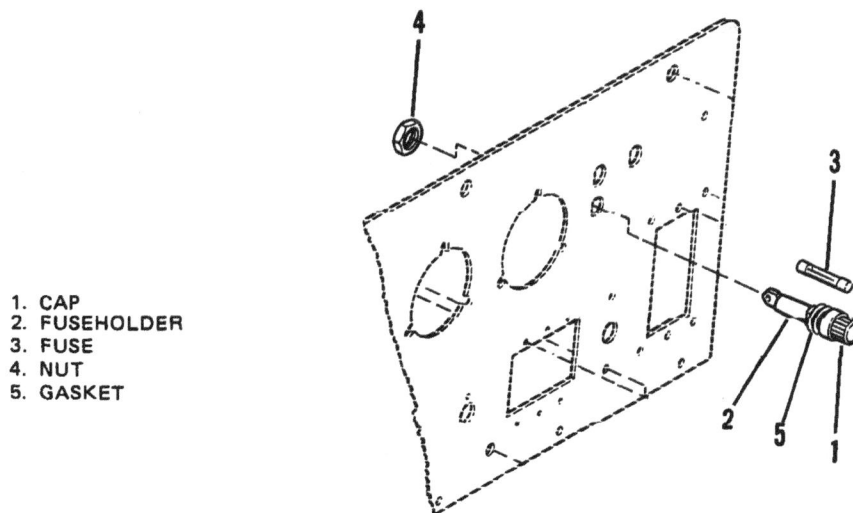

1. CAP
2. FUSEHOLDER
3. FUSE
4. NUT
5. GASKET

Figure 4-20.Convenience OutletFuses (Model MEP-016B Only).

(2) Inspect that fuse filament is not broken.   If unsure, test for continuity with a multimeter or test lamp. Inspect fuseholder for damage and secure mounting. Check that wires are securely connected to the fuse holder.

(3) If fuseholder is damaged, disconnect wires and remove nut (4) to remove fuseholder.

c. Installation.

(1) Put gasket (5) and fuseholder (2) in position and install nut (4). Connect wires to fuseholder.

(2) Install fuse (3) and securely tighten cap (1).

Section XI. MAINTENANCE OF FUEL SYSTEM.

WARNING

Unless manual directs otherwise, do not attempt any of the following maintenance procedures when generator set is operating.   Do not touch exposed electrical connections when a source of power such as utility power or another generator set is connected to the load terminals.   Severe electrical shock or death from electrocution may result.

4-33. DESCRIPTION AND FUNCTION. The fuel system consists of the fuel tank, fuel filter, transfer pump, fuel injection pump, fuel injection nozzle, auxiliary fuel pump and a fuel shutoff solenoid. Refer to paragraph 1-9.1.1. for a functional description of the fuel system.

4-34. PUEL TRANSFER PUMP.   (See Figures 4-21 and 4-22)

   a. Test.

Connect a fuel pressure gage to outlet port of transfer fuel pump with a tee fitting.

Start and run generator set.

   (1)   Fuel pressure gage should indicate a pressure of 4 to 7 psi (27 to 48 kPa) at engine idle.

      b.   Service.   Remove and clean the transfer pump filter after every 500 hours of operation (See Figure 4-21).

   (2)

        (1) Remove filter element (3) and gasket (4).

   (3)

     (2) Clean filter element (3) and reinstall with gasket (4).

   c. Removal.

   (1)   Drain the fuel filter (see paragraph 4-36).

   (2)   Remove hose (11, Figure 4-21).

   (3)   Disconnect hose (2) from the pump.

   (4)   Remove screws (5) and washers (6).

   (5)   Remove fuel cutoff solenoid brackets (refer to paragraph 4-37).

   (6)   Remove fuel transfer pump (7) and gasket (8).

   (7)   If necessary, remove fittings (1 and 10).

(8)Remove pump push rod (9) from crankcase bore.Use a magnetic retriever to raise rod out of bore.

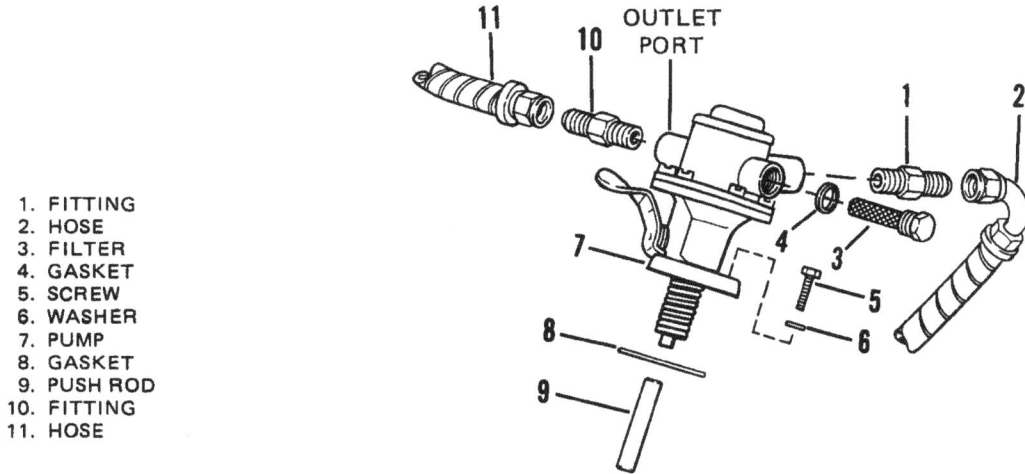

1. FITTING
2. HOSE
3. FILTER
4. GASKET
5. SCREW
6. WASHER
7. PUMP
8. GASKET
9. PUSH ROD
10. FITTING
11. HOSE

Figure 4-21.Fuel Transfer Pump.

d. Disassembly.

(1) Notch pump cover reas- and body with a file for location purposes when pump is

(2) Remove filter (3, Figure 4-21) and gasket (4).

CAUTION

Do not pry cover (2, Figure 4-22) off of pump. Damage to the diaphragm may result.

(3) Remove six cover screws (1, Figure 4-22). Tap pump cover (2) with a soft plastic hammer to separate the two parts.

(4) Compress push rod spring (3) against pump body and remove retaining ring (4), spring plate (5) and spring (3).

(5) Remove diaphragm (6) from pump body.

(1) Wipe parts with a with clean lint-free cloth that has been slightly dampened solvent.

(2) Inspect parts for damaged threads, cracks, deformation, or other visible damage.

NOTE

Service kit contains gaskets, diaphragm, and filter.
If any other components are damaged or worn, replace the pump assembly.

f. Assembly.

(1)   Lubricate diaphragm shaft (7, Figure 4-22) with clean diesel fuel and insert carefully through oil seal into pump body.

(2)   Place spring (3) and spring plate (5) onto diaphragm shaft (7). Compress spring and secure with retaining ring (4).

(3)   Push diaphragm shaft upward against spring force.Assemble cover (2) to body with notch marks aligned. Install screws (1) but do not tighten.

(4)   Release tension of diaphragm shaft and uniformly tighten screws (1) to 17 to 26 in lbs (2 to 3 N.m).

     Install fuel filter (3, Figure 4-21) and gasket (4).

(5)

1. SCREW
2. COVER
3. SPRING
4. RETAINING RING
5. SPRING PLATE
6. DIAPHRAGM
7. DIAPHRAGM SHAFT

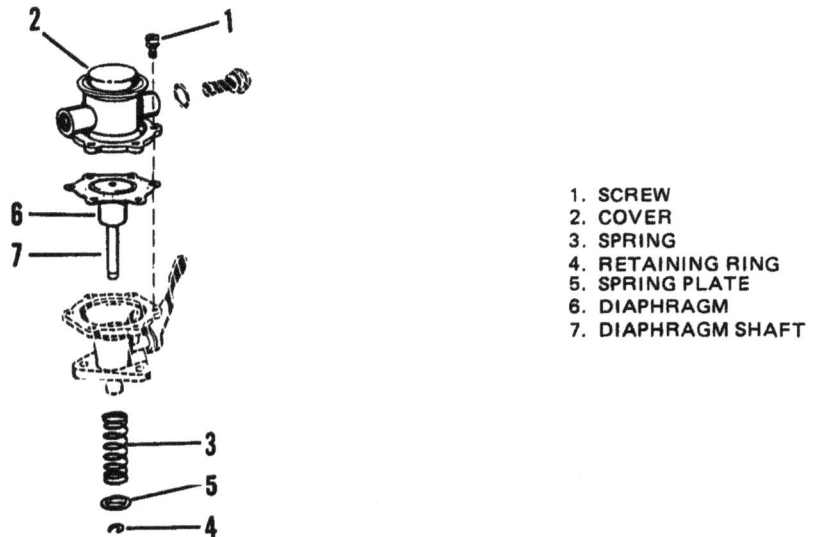

Figure 4-22.Fuel Transfer Pump, Exploded View.

g.   Installation.

(1) Lubricate pump push rod (9, Figure 4-21) with clean engine oil and push in bore until camshaft is contacted.

(2)   Install fittings (1 and 10) on pump.

(3)   Place gasket (8) on crankcase.

(4)   Place pump (7) on crankcase.

(5)   Install fuel cutoff solenoid (see paragraph 4-37).

NOTE

Fuel shutdown solenoid bracket is held down by two fuel pump mounting screws.

(6)   Apply loctite (Table 3-1, item 10) to screws (5) and install screws with washers (6).Do not tighten screws at this time.

CAUTION

Fuel priming lever must be held in the up position while tightening pump mounting screws.   Failure to do this can result in damage to the pump plunger rod.

(7)   Tighten screws (5) a few turns at a time alternately while lifting up on fuel priming lever.This prevents pump plunger from jumping off transfer rod.

(8)   Tighten screws (5) securely.

(9)   Install hose (11).   Connect hose (2) to pump.

(lo)   Test fuel transfer pump for correct fuel pressure.

(11)   Check engine shutdown solenoid adjustment (refer to paragraph 4-37).

4-35.   AUXILIARY FUEL PUMP.   (See Figures 4-23 and 4-24)

a.   Test.   Perform operational test for auxiliary fuel pump.

(1)   Connect an accurate fuel pressure gage to output port of auxiliary fuel pump. Turn master switch to RUN AUX FUEL position.Pump should provide a fuel pressure of 6 to 7 psi.   Replace

(2)   defective fuel pump.

b. Service.Every 6 months (or 500 operating hours), clean or replace the internal filter (4) of the auxiliary fuel pump. To do this, follow the removal procedures below, and steps (a and b) of the disassembly procedures.

c.Removal.

(1) Remove ground rods to gain access to fuel pump (refer to paragraph 4-20).

(2) Disconnect fuel pump electrical connector (3).

(3) Disconnect fuel line (1, Figure 4-23) from fuel pump (2).

(4) Remove cap (4).

1. FUEL LINE
2. PUMP, AUX. FUEL
3. CONNECTOR
4. CAP
5. SCREW

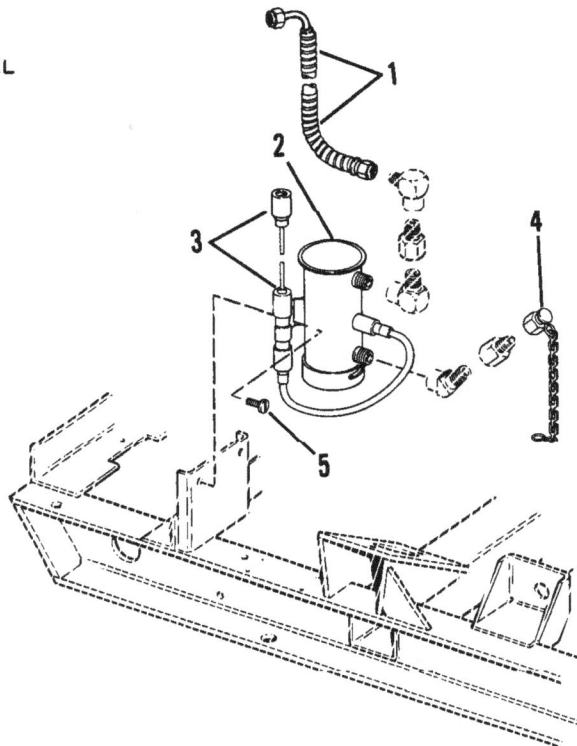

Figure 4-23.Auxiliary Fuel Pump.

d. Repair.

(1) Disassembly.

(a) With a wrench,release bottom cover (1, Figure 4-24) from bayonet
    fittings.Twist cover by hand to remove from pump body.

(b) Remove filter (4) magnet (3) and cover gasket (2). Wash filter in cleaning solvent and blow out dirt and
    cleaning solvent with air
    pressure. Check cover gasket and replace if deteriorated. Clean cover.

(c)   Remove retainer spring (5) from plunger tube (11) using thin nose pliers to spread and remove ends of retainer from tube.Then remove washer (6), O-ring seal (7), cup valve (8), plunger spring (9) and plunger (10) from tube (11).   Do not disassemble plunger (10).

(d)   Wash parts in cleaning solvent and blow out with air pressure. If plunger does not wash clean or if there are any rough spots, gently clean surface with crocus cloth. Wash the pump assembly in cleaning solvent. Blow out the tube with air pressure.   Swab the inside of the tube with a clean, dry cloth.

Figure 4-24. Auxiliary Fuel Pump, Exploded View.

(2) Assembly.

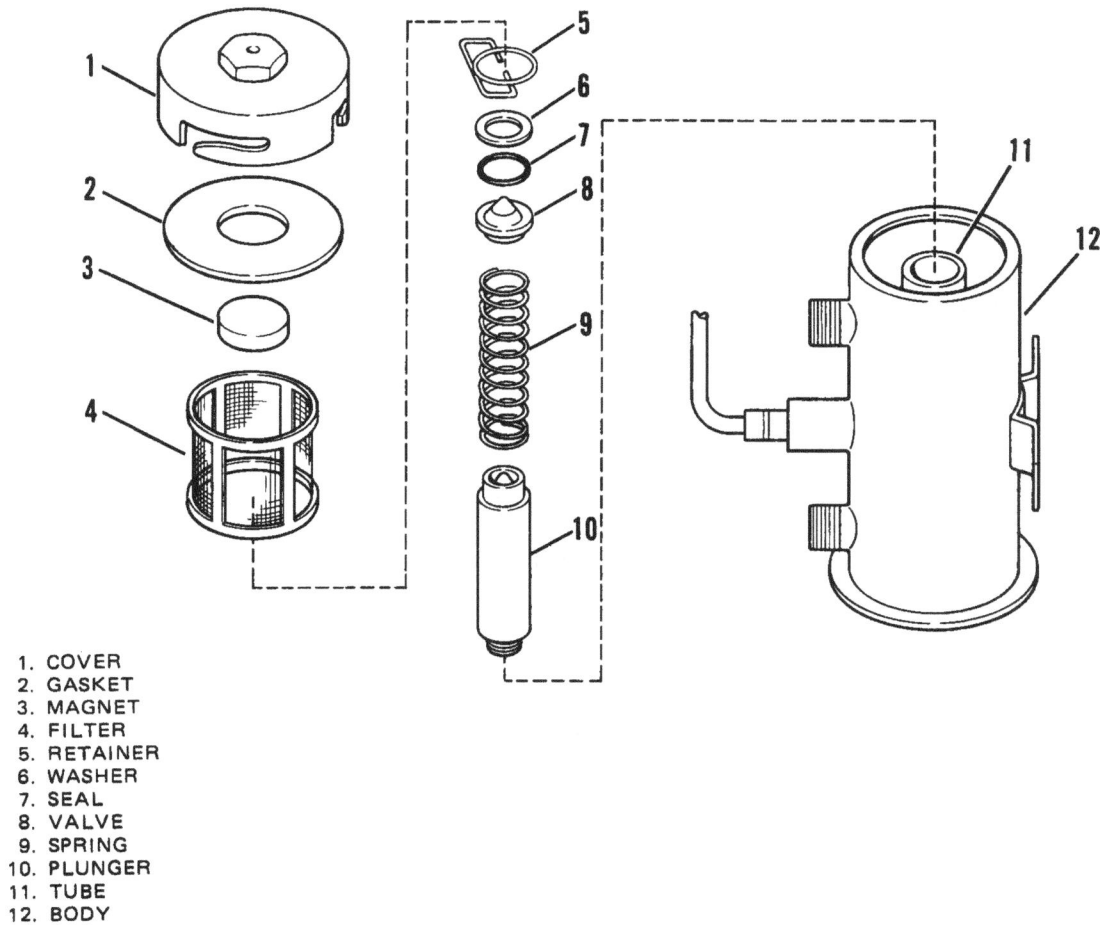

1. COVER
2. GASKET
3. MAGNET
4. FILTER
5. RETAINER
6. WASHER
7. SEAL
8. VALVE
9. SPRING
10. PLUNGER
11. TUBE
12. BODY

(a) Moisten the plunger assembly and tube with engine oil.   Insert the plunger assembly (10) in the tube with the buffer spring end first. Check fit by slowly raising and lowering the plunger in the tube. It should move fully without any tendency to stick.   If a click cannot be heard, the interrupter assembly is not functioning properly in which case the pump should be replaced.

(b) To complete the assembly, install the plunger spring (9), cup valve (8), O-ring seal (7) and washer (6) as shown. Compress spring (9) and assemble retainer (5) with ends of retainer in side holes of tube (11).

(c) Place cover gasket (2) and magnet (3) in bottom cover (1) and assemble filter (4) and cover assembly. Twist cover by hand to hold in position on pump housing. Tighten bottom cover with a wrench.

e. Installation.

(1) Secure fuel pump (2, Figure 4-23) to frame with screws (5).

(2) Install cap (4).

(3) Reconnect fuel line (1) to fuel pump (2).

(4) Connect fuel pump electrical connector (3).

(5) Reinstall ground rods (refer to paragraph 4-20).

4-36. FUEL FILTER.   (See Figure 4-25)

a. Service. Daily (before operation) loosen bleed screw (3) and push up on drain valve (11) to drain any water or sediment from the fuel filter. Tighten bleed screw (3).

b.   Replace.   (Changing filter element)

(1) Remove   nut (4) and O-ring (5).

(2) Remove   cover (6) and O-ring (7). Remove and discard filter (8).

(3) Install new filter (8) in filter cannister (9).

(4) Install O-ring (7) and secure cover (6) with nut (4,   and O-ring (5).

c.   Removal.

(1) Open bleed screw (3) and drain all fuel from filter canister (9) by pushing up on the drain valve (11).

(2) Disconnect fuel lines (13) and (17).

(3) Remove nuts (16), washers (15) and screws (10).

(4) Remove filter assembly (1) with bracket (14).

(5) If necessary, remove drain valve (11).

d.   Installation.

(1) If removed, install drain valve (11).

1. FILTER ASSY
2. O-RING
3. SCREW, BLEED
4. NUT
5. O-RING
6. COVER
7. O-RING
8. FILTER
9. CANISTER
10. SCREW
11. VALVE, DRAIN
12. O-RING
13. FUEL LINE
14. BRACKET
15. WASHER
16. NUT
17. FUEL LINE

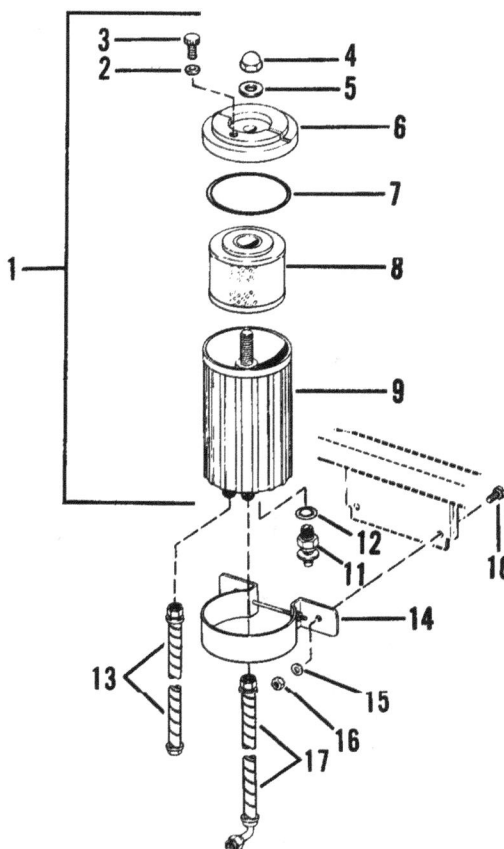

Figure 4-25. Fuel Filter.

(2)   Secure filter assembly (1) with bracket (14) to frame with screws (10), washers (15) and nuts (16).

(3)   Connect fuel lines (13) and (17).

(4)   After installation, make sure that bleed screw (3) on filter is closed and operate manual priming lever on fuel transfer pump to prime fuel system before starting the engine.

4-37.   SOLENOID, FUEL CUTOFF.   (See Figure 4-26 and 4-27)

a. Inspect.   Inspect the fuel cutoff solenoid for damage and secure mounting. Check the electrical connections and the wires for damage. Check that retainer clip on plug engages with slot on solenoid. Check that governor and pump linkage are clean and move freely.

b. Test.

NOTE

A minimum of 1 gallon of fuel is required in the fuel tank to energize the solenoid.

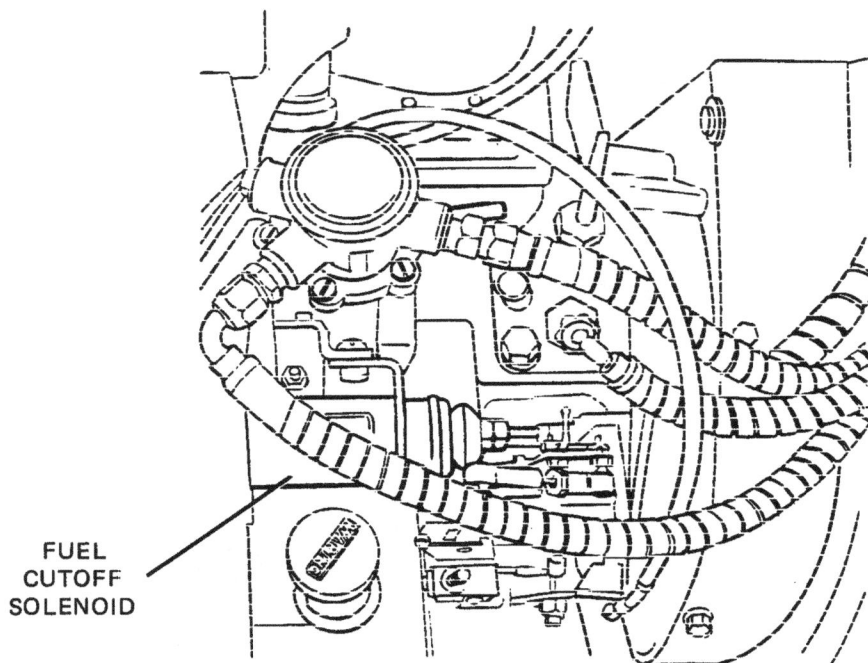

Figure 4-26.Location of Fuel Cutoff Solenoid.

(1)   With master switch in the RUN position, voltage across the solenoid connections should be 24 volts.   If no or low voltage is read, check the solenoid circuit.

(2)   Disconnect negative (-) battery cable from battery.

(3)   Remove the connector from the solenoid.

(4)   Use a multimeter to measure the resistance across the solenoid terminals as follows:

(a)   With the solenoid plunger in its de-energized (extended) position, the resistance should be 1.82 ohms + 10 percent.

(b)   Manually push the plunger all the way in.The resistance should be 83.5 ohms + 10 percent.

c.   Removal.

(1) Disconnectnegative (-) battery cable from battery.

CAUTION

Retainer clip on connector must be depressed to disconnect connector from sole-noid. Failure to do so could damage wires.

(2) Disconnect connector release the   retainer   (1).   Push in to

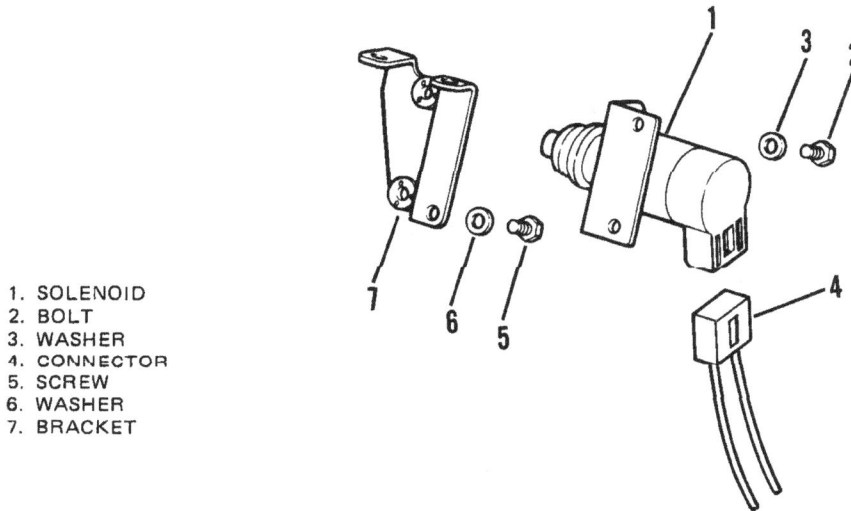

1. SOLENOID
2. BOLT
3. WASHER
4. CONNECTOR
5. SCREW
6. WASHER
7. BRACKET

Figure 4-27.Fuel   Cutoff Solenoid.

(3) Remove bolts (2) and washers (3),   and remove solenoid (1).

(4) If necessary, remove screws (5), washers (6) and bracket (7).

d. Installation. _____

(1)   If necessary, put bracket (7) in position and install screws (5) and washers (6).

         Put solenoid (1) in position and install bolts (2) and washers (3).
(2)
         Connect connector (4) to solenoid (1).   Be sure retainer clip on connector engages with slot
(3)   in solenoid.

         Reconnect negative (-) battery cable.
(4)
         The solenoid plunger should be adjusted so it fully stops fuel injection when in the de-energized position:
(5)
             Disconnect governor linkage at injection pump end and energize solenoid.
         With the injection pump external arm in full counterclockwise position, adjust solenoid plunger screw for 0.01 to
     (a) 0.02 in. (0.25 to 0.50 mm) clearance with stop pin on external arm.

     (b) Tighten locking nut on plunger shaft to lock shaft position.

         Reconnect governor linkage at injection pump end.

     (c)

     (d)

4-38. FUEL LEVEL FLOAT SWITCH.   See Figure 4-28).

a. Inspect.

(1) Inspect to see that fuel   level float switch is securely mounted to fuel

(2) Inspect to see that fuel level float switch connectors are secure.

b.   Test.

(1) Unscrew electrical connector from top of fuel tank.

(2) Drain fuel tank.

(3) Use a multimeter to check for continuity across switch pins (1). The pins are labeled A, B, C, D, and E.

(4) A multimeter across ohms).   pins A and B should indicate an open circuit (high

(5) A multimeter across   pins D and E should indicate a closed circuit (0 ohms).

(6) Fill fuel tank with   diesel fuel.

(7) Use a multimeter to   check for continuity across the switch pins (1).

(8) A Multimeter across ohms).   pins A and B should now indicate a closed circuit (0

A multimeter across ohms).

(9)   pins D and E should indicate an open circuit (high

(10) Replace switch if continuity requirements are not met.

c. Removal.

(1) Unscrew electrical connector from switch connector pins (1).

(2) Use a wrench to remove float switch (2) from fuel tank (3).

d. Installation.

(1) Coat the threads of float switch (2) with sealing compound (Table 3-1, item 13) and install switch (2) on tank (3).

(2) Connect electrical connector to switch connector pins (1).

4-39. FUEL LINES, FLEXIBLE. Refer to Figure 4-29 and replace fuel hoses (1), elbows (2), valves (3) and fittings (4)

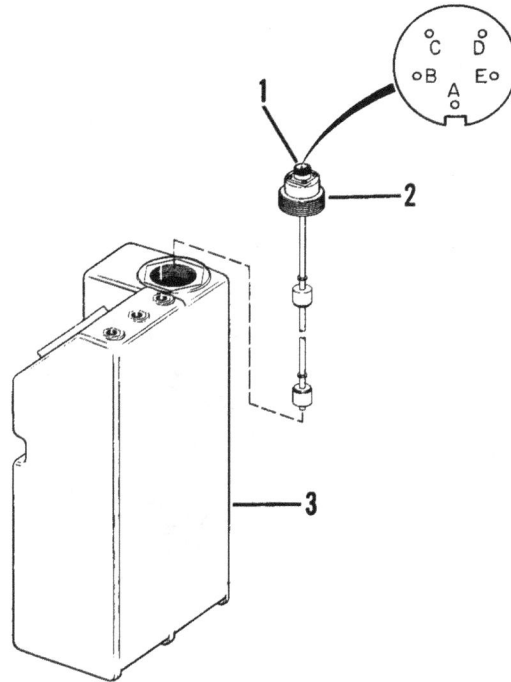

1. PINS
2. SWITCH, FLOAT
3. TANK, FUEL

Figure 4-28. Fuel Level Float Switch.

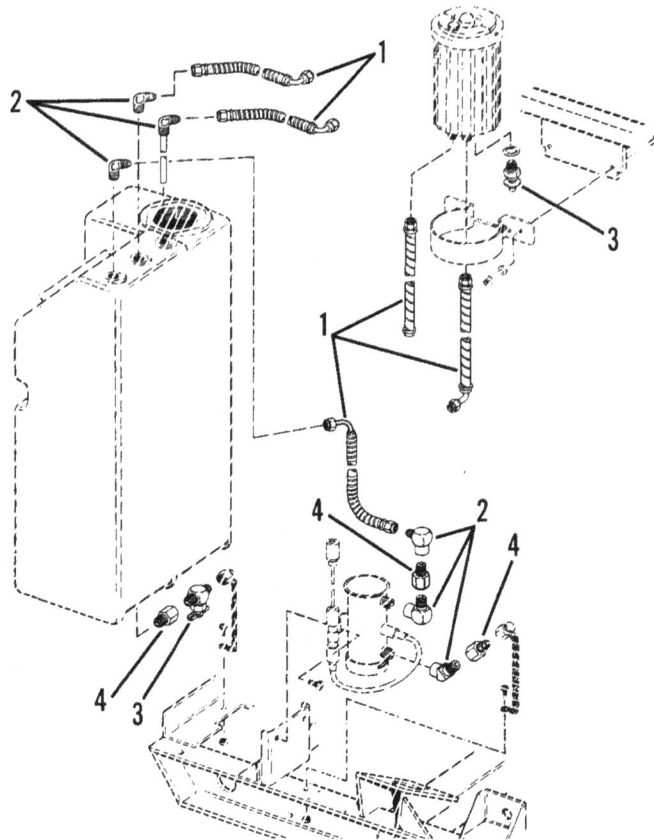

1. HOSE
2. ELBOW
3. VALVE
4. FITTING

Figure 4-29. Flexible Fuel Lines.

4-40. FUEL INJECTION PUMP. Refer to Figure 4-30 and inspect the pump for physical damage or fuel injection deficient leaks around gaskets or fittings. If higher echelon of maintenance.

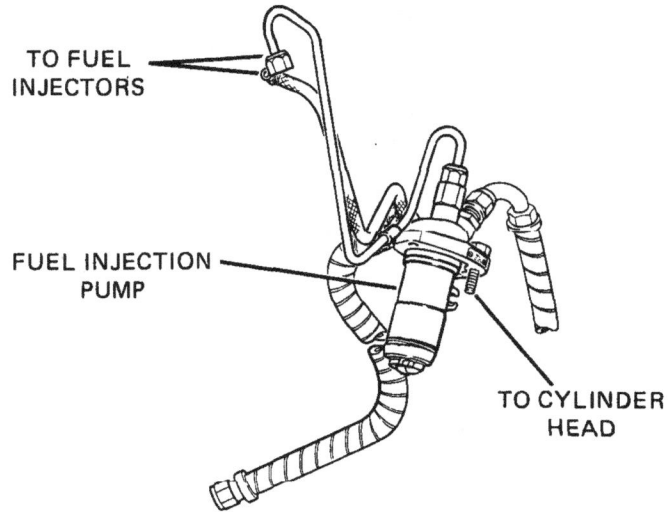

Figure 4-30.Fuel Injection Pump.

4-41.FUEL INJECTOR. Refer to Figure 4-31 and inspect the fuel injector for physical damage or leaks.   If deficient notify higher echelon of maintenance.

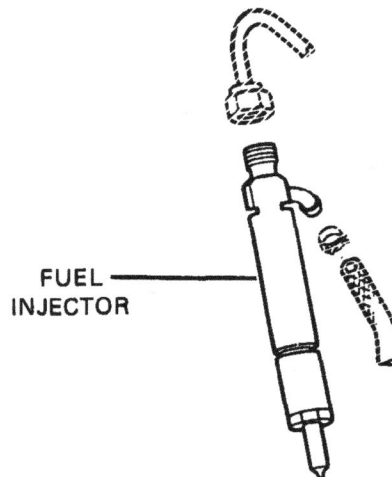

Figure 4-31. Fuel Injector.

Section XII. MAINTENANCE OF THE ENGINE.

.

WARNING

Unless manual directs otherwise, do not attempt any of the following maintenance procedures when generator set is operating. Do not touch exposed electrical connections when a source of power such as utility
power or another generator set is connected to the load terminals.   Severe electrical shock or death from electrocution can result.

4-42.   AIR CLEANER.   (See Figure 4-32)

a.Removal.

CAUTION

Before removing air cleaner, clean all external surfaces of air cleaner assembly and surrounding areas. Entry of dirt will cause premature engine wear.

(1) Remove wingnut (1), cover (2), washer (3), filter element (4) and gasket (5).

(2) Loosen clamps (6).

(3) Remove bolts (7), lockwashers (8) and washers (9). Remove air cleaner housing (10) and hose (11) from frame.

(4) Remove air filter indicator (12) from air cleaner housing (10).

(5) If necessary, remove dust valve (13) from cover (2). b.Installation.
(1) If removed, install dust valve (13) on cover (2).

(2) Install air filter indicator (12) on air cleaner housing (10).

(3) Put   air cleaner housing (10) in position and install bolts (7), lockwashers and washers (9).
     (8)
(4) Put   hose (11) in position and tighten clamps (6).

(5) Install gasket (5), filter element (4), washer   (3), cover (2)   and wingnut
     (1).

1. WINGNUT
2. COVER
3. WASHER
4. FILTER ELEMENT
5. GASKET
6. CLAMP
7. BOLT
8. LOCKWASHER
9. WASHER
10. HOUSING
11. HOSE
12. INDICATOR
13. DUST VALVE

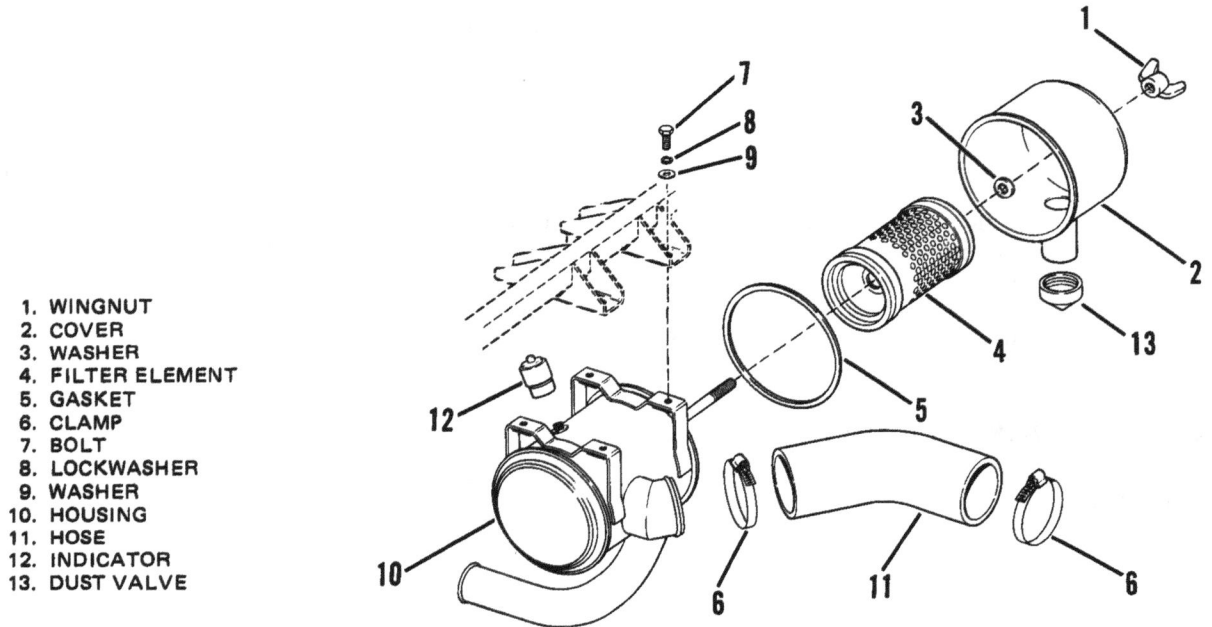

Figure 4-32. Engine Air Cleaner

4-43. Rocker Arms. (See Figure 4-33)

a. Inspection.

1. SCREW
2. WASHER
3. LIFTING EYE
4. CLAMP
5. EXIT DUCT
6. CLAMP
7. HOSE, BREATHER
8. TUBE, BREATHER
9. BOLT
10. WASHER
11. COVER
12. GASKET

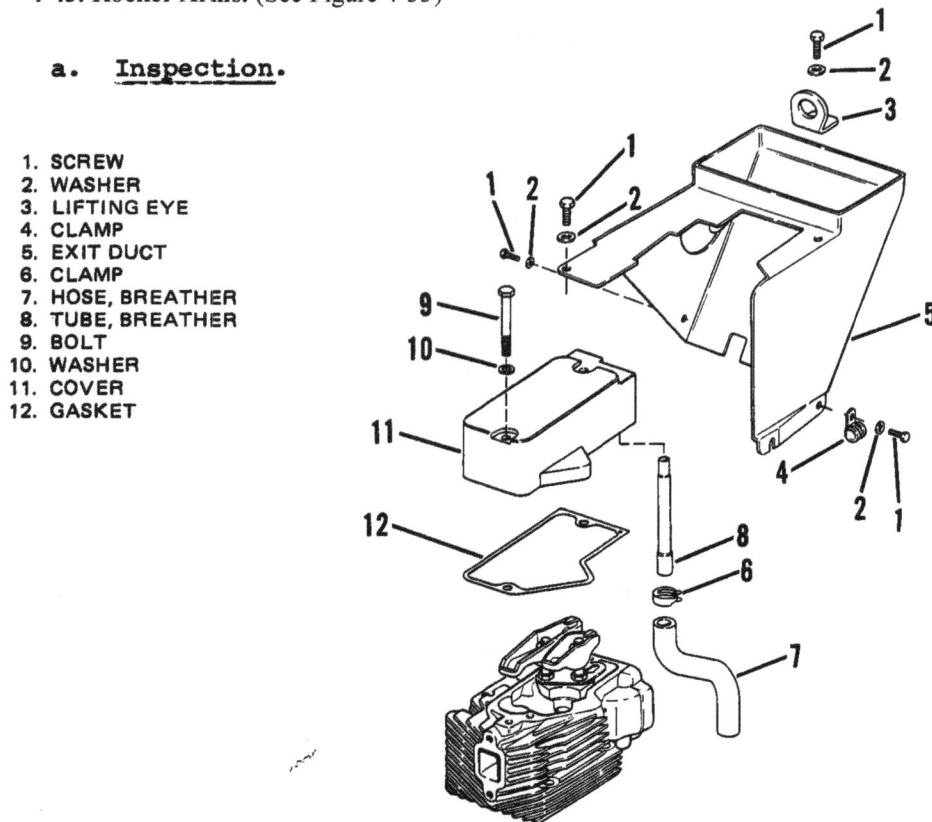

Figure 4-33. Rocker Arm Inspection.

(1)   Remove air cleaner (refer to paragraph 4-42) and remove air intake hose from intake manifold.   Cover all openings to prevent entry of dirt.

(2)

(3)   Remove five screws (1), five washers (2), lifting eye exit duct (5).   (3), clamp (4) and

        Loosen clamp (6) and disconnect breather hose (7) from

(4)   breather tube (8).

(5)   Remove two screws (9), washers (10), and rocker arm cover (11) with gasket (12).   Discard gasket (12).

(6)   Inspect rocker arms for cracks or other visible damage.

(7)   Install rocker arm cover with new gasket.   Be sure that small breather hole in rocker cover is alined with hole in rocker cover gasket.

(8)   Install breather hose (7) on breather tube (8) and tighten clamp (6).

(9)   Install exit duct (5), clamp (4) and lifting eye (3) with five screws (1) and five washers (2).

        Connect air intake hose to intake manifold and install air cleaner assembly
(10)   (refer to paragraph 4-42).

4-44. GOVERNOR AND LINKAGE.   (See Figures 4-34, 35 and 36)

a.   Cleaning and Inspection.

(1)   Clean governor and all linkage components.   Inspect governor for secure mounting.

        Ensure that governor linkage operates smoothly and that all springs are attached.
(2)
        Check that the throttle cable is securely attached.

(3)   Tighten mounting hardware as necessary.   Replace missing or damaged components.

(4)

(1)   Remove   oil filter (see paragraph 4-46).

(2)   Rotate   engine support down.

(3)   Remove   governor oil line (1, Figure 4-33).

(4)   Remove   screw (2), washer (3), and clip (4).

(5)   Disconnect throttle cable (5), washer (6) and pin (7).

(6)   Remove governor linkage assembly (8) by removing locknuts (9) and (26).

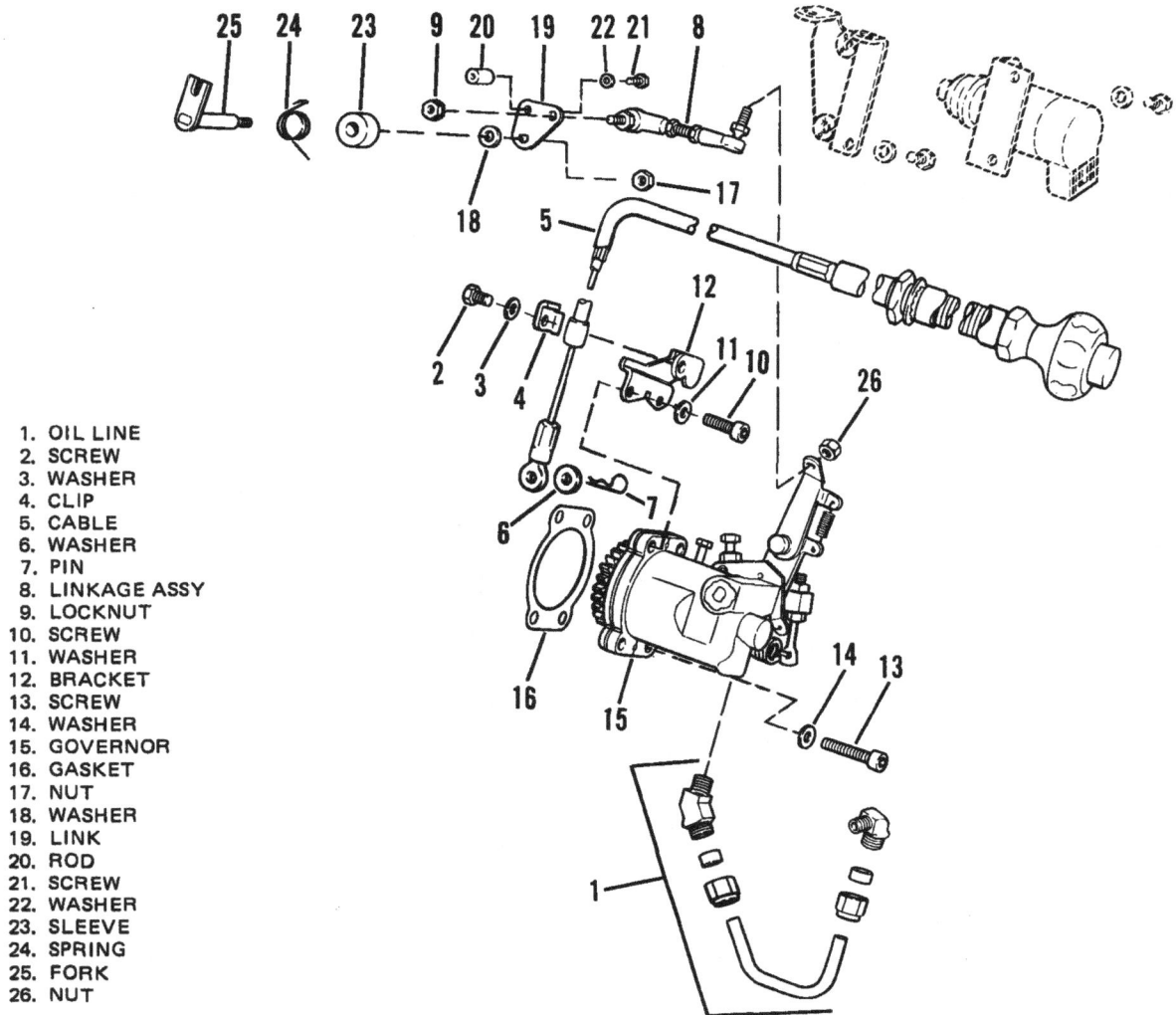

1. OIL LINE
2. SCREW
3. WASHER
4. CLIP
5. CABLE
6. WASHER
7. PIN
8. LINKAGE ASSY
9. LOCKNUT
10. SCREW
11. WASHER
12. BRACKET
13. SCREW
14. WASHER
15. GOVERNOR
16. GASKET
17. NUT
18. WASHER
19. LINK
20. ROD
21. SCREW
22. WASHER
23. SLEEVE
24. SPRING
25. FORK
26. NUT

Figure 4-34. Governor and Linkage, Exploded View.

(7) Remove two screws (10), washers (11), and bracket (12).

(8) Remove two screws (13) and washers (14).

(9) Remove governor (15) and gasket (16).

(10) Remove screw (21), washer (22) and rod (20).

NOTE

Spring (24) and fork (25) are inside the engine and be removed. They are shown cannot here for reference only.

(11) Remove nut (17), washer (18), link (19) and sleeve (23) from engine fork (25).

c.   Installation.

(1)   Place sleeve (23) and link (19) on engine fork (25 and secure with washer (18) and nut (17).

(2)   Secure rod (20) to link (19) with washer (22) and screw (21).

(3)   Place gasket (16) on governor (15).

(4)   Place governor (15) into position. Make sure governor drive gear meshes with camshaft gear.

NOTE

Be sure that two short screws (10) are installed closest to the crankcase. Apply sealing compound (Table 3-1, item 13) to screws (12) and sealing compound (Table 3-1, item 10) to screws (10) before installation.

(5) Install screws (10 and 13), washers (11 and 14) and bracket (12).

NOTE

Check governor linkage adjustment only if a new governor is being installed. If original governor is being installed do not check governor linkage adjustment.

(6) Check governor linkage adjustment as follows:
Check fuel cutoff solenoid adjustment (refer to paragraph 4-37).

(a)

(b)   Rotate external arm counterclockwise toward the governor as shown in Figure 4-35.

(c)   Governor linkage assembly (8, Figure external link 4-34) should fit easily into (19) hole and governor (15) link hole.

(d)   Adjust length of governor linkage (8) as required by loosening lock nuts, turning ends in or out, and tighten locknuts.

(7)   Install governor linkage (8) and secure with locknuts (9).

(8)   Connect throttle cable (5) with washer (6) and clip (7).

(9)   Secure throttle cable (5) to bracket (12) with clip (4), washer (3) and screw (2).

(10)   Connect governor oil line (1).

(11)   Rotate engine support up.
d.   Adjustment.   If a new governor is being installed refer to higher echelon of maintenance for correct maximum fuel stop adjustment.

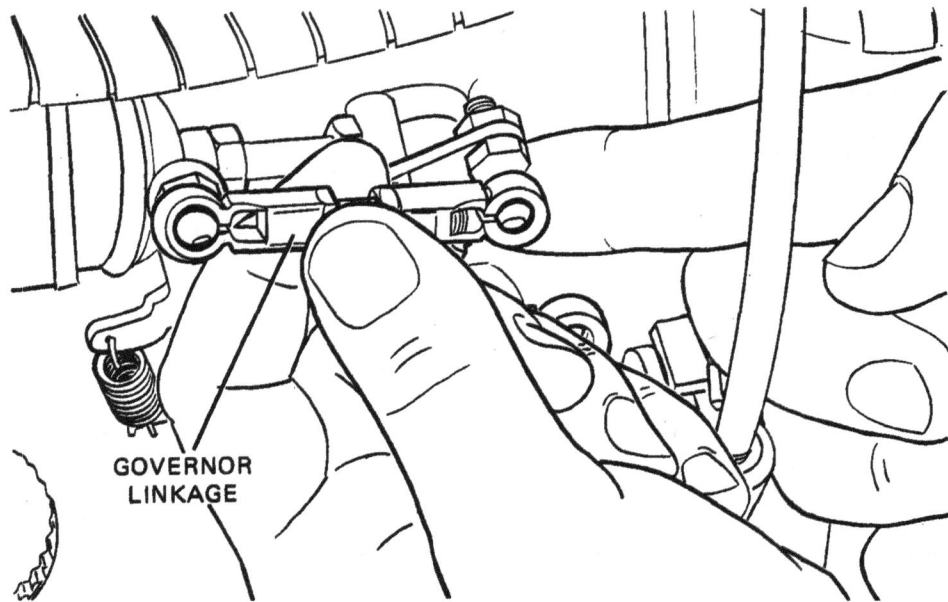

Figure 4-35.Governor Linkage Adjustment.

4-45.GOVERNOR DROOP ADJUSTMENT.   (See Figure 4-36).

NOTE

To properly set the governor droop requires the use of a three kilowatt (3 kw) load bank; however, approximate field adjustments can be made when necessary. For proper droop specification, the change in rated speed or frequency between no load and full load (3 kw) should be 1.5-3.0 percent.

(1)   Operate generator set at rated speed and load until it reaches operating temperature.

Adjust set to rated speed at full load using manual speed control.
(2)

Remove load and record new stabilized speed (Hz).
(3)

Calculate droop.Use the following formula to calculate droop:
(4)

$$\text{Droop} = \frac{\text{no load speed (Hz)-full load speed (Hz)}}{\text{full load speed (Hz)}} \times 100 \text{ percent}$$

(5) Adjust droop screw (1) and recheck droop.

NOTE

By adjusting droop screw, droop can be changed by moving governor spring closer to (less droop) or away from (more droop) the governor cross shaft.

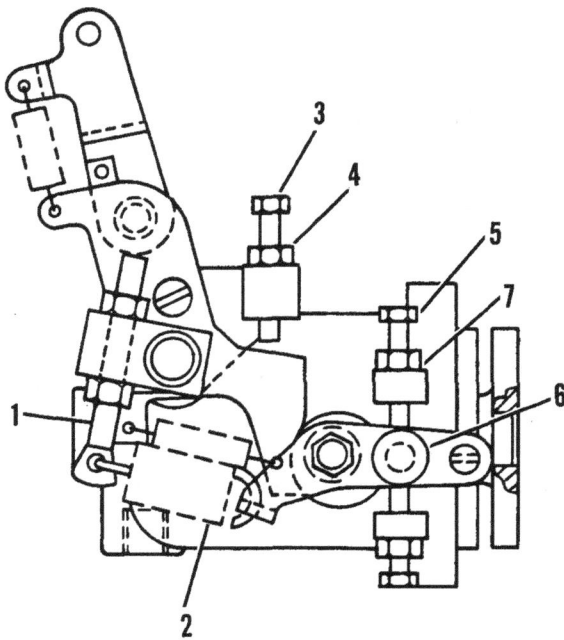

Figure 4-36.Governor Droop Adjustment.

NOTE

Check that flattened end of droop screw remains in line with spring (2) when adjusting nuts.   If a further adjustment of the maximum fuel screw or governor linkage length is required for any reason following droop adjustment, the droop adjustment must be rechecked.

If the required droop adjustment cannot be obtained with droop screw, shut-down generator set and go to step 6.

Disconnect governor linkage by removing ball stud nut at either governor or injection pump end (see Figure 4-35). Shorten linkage length by one complete turn and tighten adjusting nuts.

(6)   Reconnect governor linkage (see Figure 4-35). Use two wrench method when tightening nut on ball stud to prevent movement of external arm or governor
arm.

(7)   Loosen maximum fuel screw (3, Figure 4-36) using two wrench method and back out maximum fuel screw counterclockwise one quarter turn.Tighten locknut
(4) using two wrench method.

(8)   Adjust droop screw (1) to the center of its travel.

Check droop.Repeat steps (1) through (5).

(9)

(10)

MARINE CORPS TM 05926B/06509B-12
ARMY   TM 5-6115-615-12
NAVY   NAVFAC P-8-646-12

NOTE

The above adjustment of the maximum fuel screw is only approximate.For a more exact adjustment, notify higher echelon of maintenance.

(11)   Recheck no load high idle speed.The no load high idle speed should fall between 3850 and 3900 RPM (or equivalent frequency).   If adjustment is required, proceed to step (12).

(12)   With no load connected, adjust speed to 3850-3900 speed control. If speed cannot be obtained, back   RPM using the manual out high idle screw (5).

(13)   idle screw (5) until it contacts throttle lever stop pin (6) and tighten adjusting screw nut (7).

4-46.   OIL FILTER - REPLACEMENT.   (See Figure 4-37)

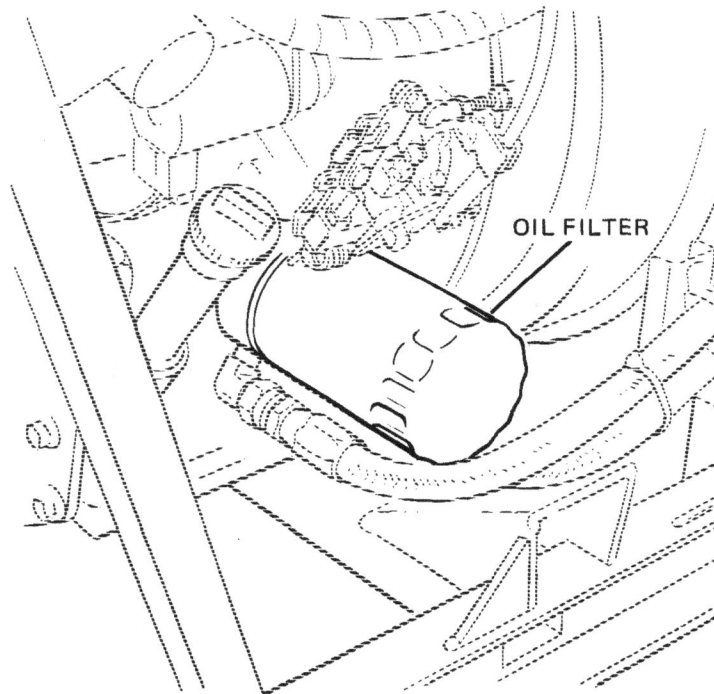

Figure 4-37.Oil Filter.

(1)   Unscrew filter from filter adapter.   Discard filter.

(2)   Apply a light coating of oil to the gasket surface on   the new filter.

(3)   Screw new filter onto filter adapter and hand tighten   (refer to
LI 05926/06509B-12, Note 4).

4-54

MARINE CORPS TM 05926B/06509B-12
ARMY   TM 5-6115-615-12
NAVY   NAVFAC P-8-646-12

4-47.   **OIL FILTER ADAPTER - INSPECTION.**   (See Figure 4-38)

Inspect oil filter adapter, lines and fittings for damage lubricant.   Notify higher level of   or leakage of engine connections, or to maintenance to tighten
damaged parts.

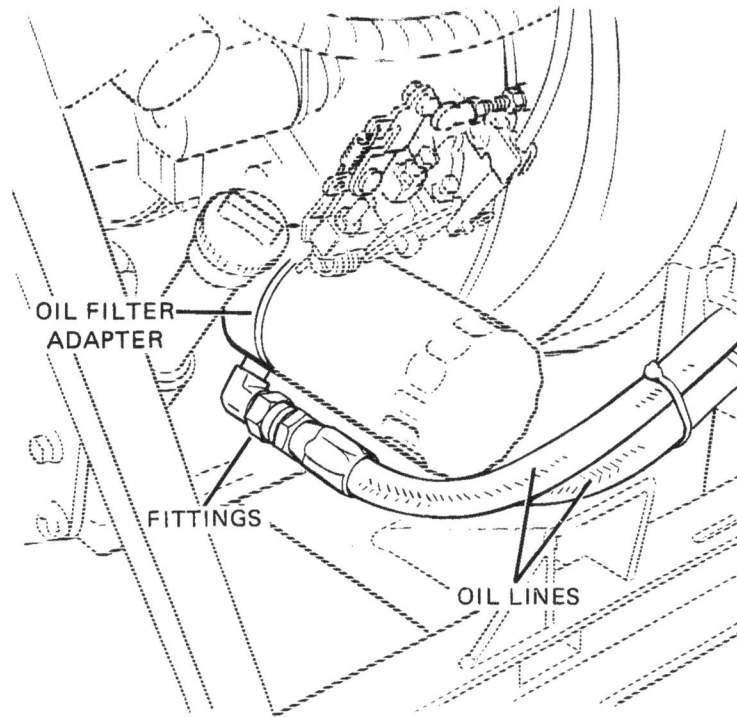

Figure 4-38. Oil Filter Adapter.

4-48.   **STARTER ASSEMBLY.**   (See Figures 4-39 and 4-40)

    a.   <u>Inspection.</u>

        (1)   Inspect starter assembly for secure mounting.

        (2)   Inspect wiring connected to starter assembly for damaged insulation and secure mounting. Inspect to see that all electrical connections are clean and tight.

        (3)   Check that starter engages to the START and turns flywheel when master switch is moved starter does not turn position. If Troubleshooting (Table 4-2).

        (4)   Repair or replace wiring as necessary. Replace starter if defective.

    b.   <u>Test.</u>

        (1) Make sure the batteries are fully charged and that all battery and starter cables are serviceable and properly installed.

(2)   Connect a voltmeter as shown in Figure 4-39 - Test A.   If voltage is indicated, solenoid is defective.

Momentarily connect a jumper as shown in Figure 4-39 - Test B.Voltmeter should indicate battery voltage and starter should crank the engine. If voltmeter does not read battery voltage, the solenoid is defective. If the voltmeter indicates battery voltage, but starter does not operate, the starter is defective.

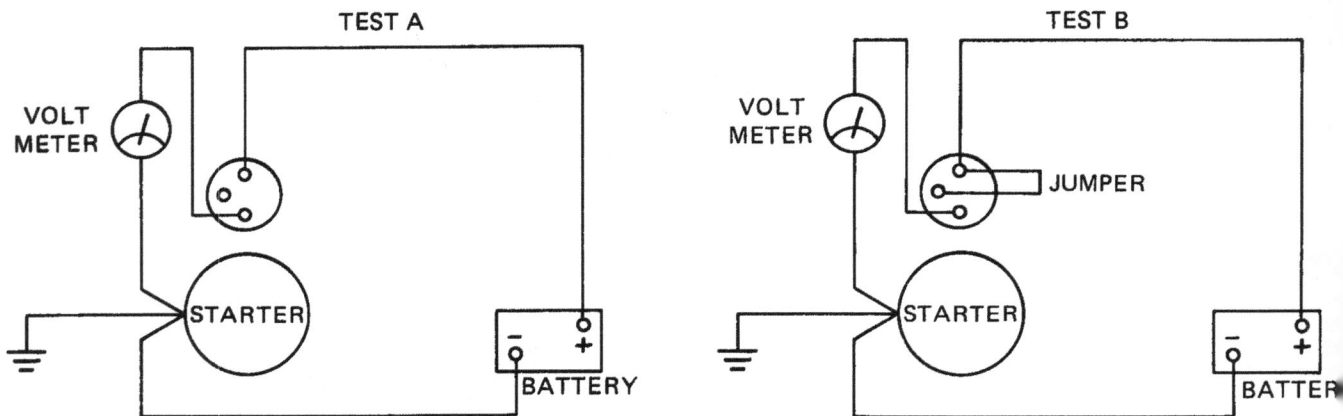

Figure 4-39. Starter Solenoid Test Circuit.

c.   Removal.

NOTE

Always disconnect the negative (-) battery first and   battery cable cable   at the the positive (+)

(1)   Disconnect negative (-) battery cable and positive (+) battery cable from battery (see paragraph 3-17).

(2)   Tag battery cables (1 and 2, Figure 4-40) and remove nuts (8), washers (7), bolt (17), washer (18), battery cables (1 and 2) and wires (3 and 19) connected to starter assembly.

Remove screw (6), washer (5) and wire (4).

(3)   Remove two bolts (12), washers (11), lockwashers (10) and nuts (9).

(4)   Remove nut (14) and washer (15).

(5)   Remove rear starter bracket (13).

(6)   Remove screw (20), washer (21), and starter (16).

(7)

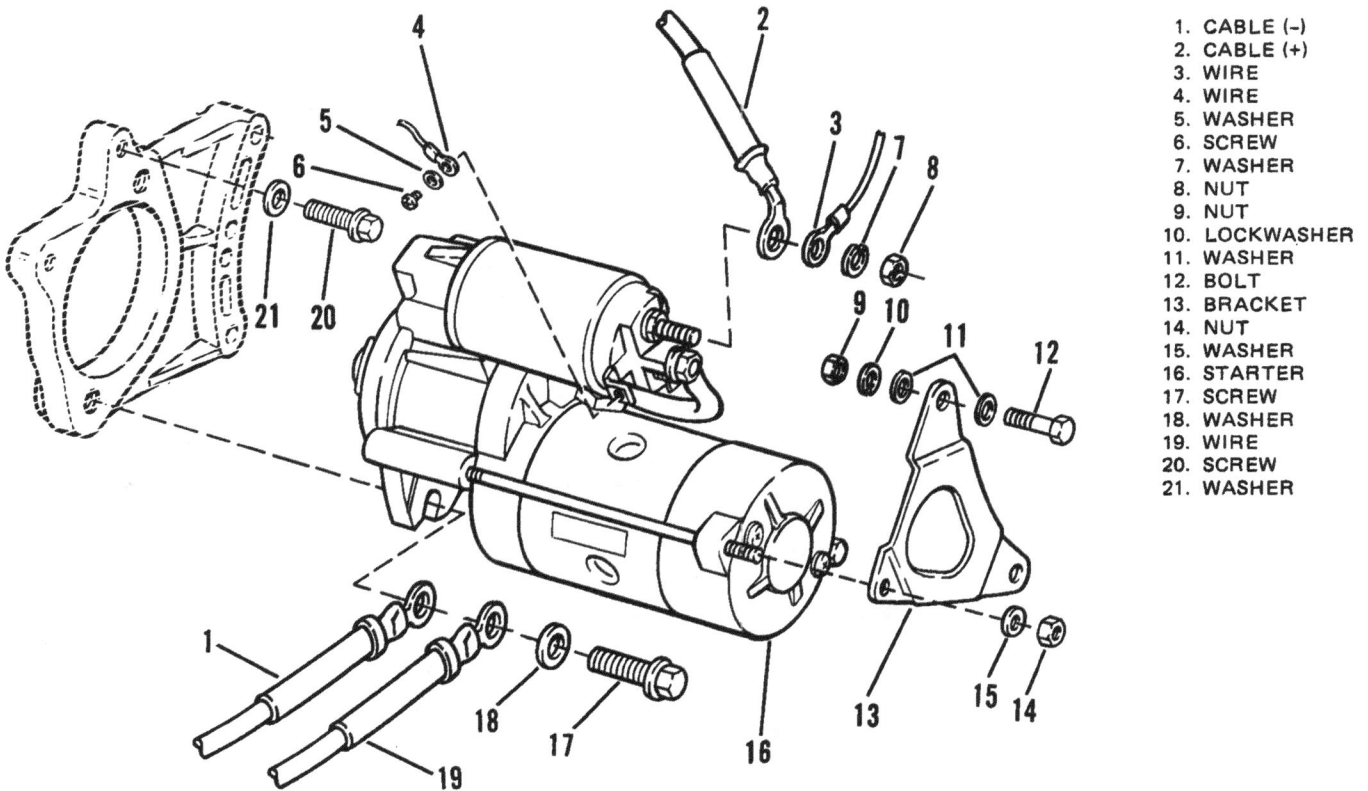

1. CABLE (−)
2. CABLE (+)
3. WIRE
4. WIRE
5. WASHER
6. SCREW
7. WASHER
8. NUT
9. NUT
10. LOCKWASHER
11. WASHER
12. BOLT
13. BRACKET
14. NUT
15. WASHER
16. STARTER
17. SCREW
18. WASHER
19. WIRE
20. SCREW
21. WASHER

Figure 4-40.Starter.

d.   Installation.

NOTE

Apply sealing compound (Table 3-1, item 10) to all screws prior to installation.

(1) Put starter (16) in position and install screw (20) and washer (21) in hole closest to crankcase.

(2) Put rear starter bracket (13) in position and install nut (14) and washer (15).

(3) Install two bolts (12), washers (11), lockwashers (10) and nuts (9).

(4) Install nut (8) wires (3 and 19), battery cables (1 and 2) washers (18 and 7), and screw (17).

(5) Connect wire (4) to starter solenoid with screw (6) and washer (5).

(6) Connect battery negative (−) battery cable and positive (+) battery cable to (see paragraph 3-17).

4-49.   INTAKE MANIFOLD.   (See Figure 4-41)

(1)   Remove air cleaner assembly (refer to paragraph 4-42), and remove clamp (1) and air intake hose (2) from intake

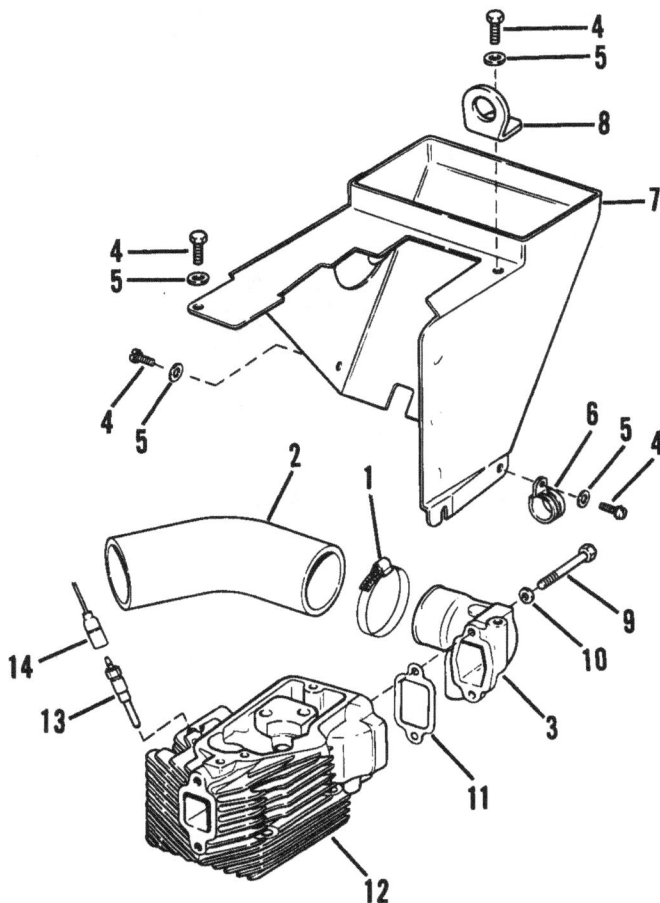

1. CLAMP
2. HOSE
3. MANIFOLD, INTAKE
4. SCREW
5. WASHER
6. CLAMP
7. EXIT DUCT
8. LIFTING EYE
9. SCREW
10. WASHER
11. GASKET
12. CYLINDER HEAD
13. GLOW PLUG
14. LEAD, ELECTRICAL

manifold (3).

Figure 4-41.   Intake Manifold and Glow Plug.

(2)
Remove five screws (4) washers (5), clamp (6) and remove exit duct (7) and lifting eye (8).

(3)   Remove two screws (9) and washers (10).

(4)   Remove intake manifold (3) and gasket (11) from cylinder head (12).

b. Inspection._____Inspect intake manifold for cracks and damage.

c.   Installation.

(1) Place intake manifold (3) and gasket (11) in position on cylinder head (12).

4-58

(2) Apply sealing compound (Table 3-1, item 10) to screws (9).   Install screws (9) with washers (10).

(3)

Place exit duct (7) and lifting eye (8) in position install screws (4), washers (5) and and clamp (6).

(4) Connect air intake hose (2) and clamp (1) to intake install air cleaner assembly manifold (3) and (see paragraph 4-42).

4-50.   GLOW PLUG. (See Figure 4-41)

a. Test.

(1) With electrical lead connected, use a multimeter to glow plug connection   check voltage between master and clean engine ground. With PREHEAT position, voltage should be 10 to 14 switch in If voltage is low, volts. check glow plug circuit.   If voltage is high, go to   Step 2.

Disconnect negative (-) battery cable at battery.
(2)
Remove electrical lead (14) from glow plug (13).
(3)

(4) Use a multimeter to test for continuity from terminal of glow plug to a clean engine ground.

An open circuit (infinite ohms) indicates a defective glow plug.
(5) c. Removal.

(1) Disconnect electrical lead (14) from glow plug (13).

(2) Use a 3/8 in. 6 point, deep socket (1/4 in. drive) to remove glow plug (13) from cylinder head (12).

d.   Installation.
Screw glow plug (13) into cylinder head (12) and tighten to 12-13 ft. lbs. (16-18 N.m).

(1) Connect electrical lead (14) to glow plug (13).

(2)

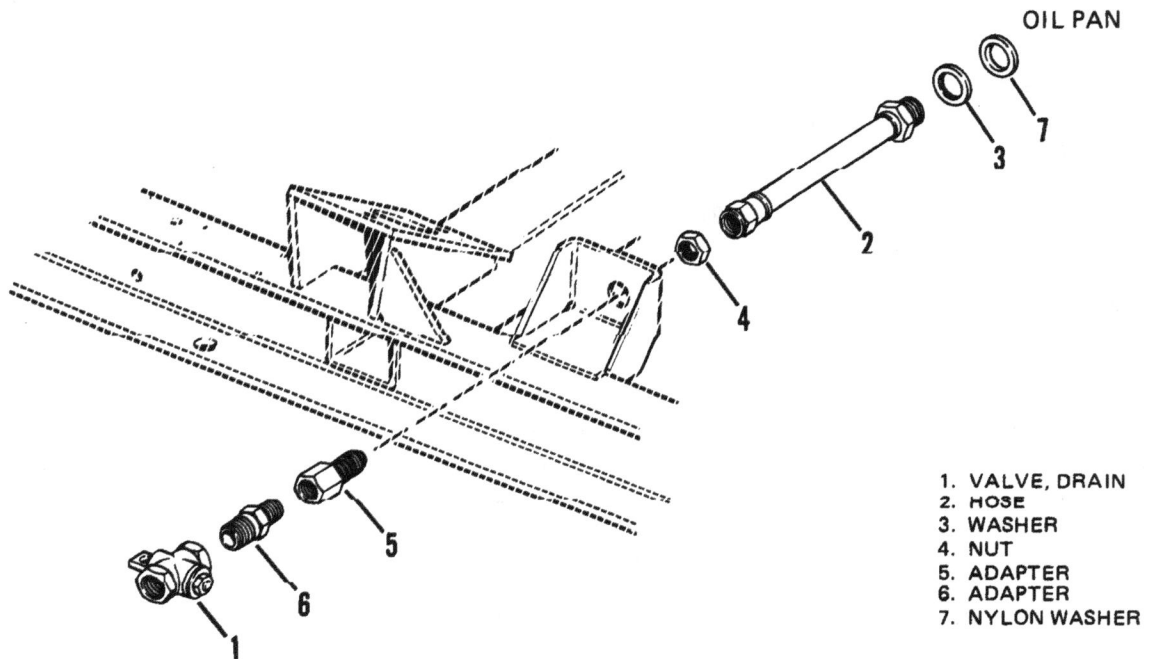

1. VALVE, DRAIN
2. HOSE
3. WASHER
4. NUT
5. ADAPTER
6. ADAPTER
7. NYLON WASHER

Figure 4-42.Engine Oil Drain Valve, Exploded View.

_____ **CAUTION** _____

Do not apply air pressure to the crankcase to speed the oil drain process.Air pressure can force the oil seals out of the crankcase.

(1) Drain engine oil into a suitable container (refer to paragraph 4-7).

(2) Disconnect hose (2) from fitting (5).

(3) Remove nut (4) and remove valve (1) with adapters (5 and 6).

(4) If necessary, disassemble adapters (5 and 6) from valve (1).

(5) Remove hose (2) and washers (3 and 7) from oil pan.

b. Installation.

(1) Install hose (2) and washers (3 and 7) on oil pan.

(2) If necessary assemble adapters (5 and 6) to valve (1).

(3) Put valve assembly in position on skid base and install nut (4).

(4) Connect hose (2) to fitting   (5).

4-52.   OIL COOLER.   (See Figure 4-43)

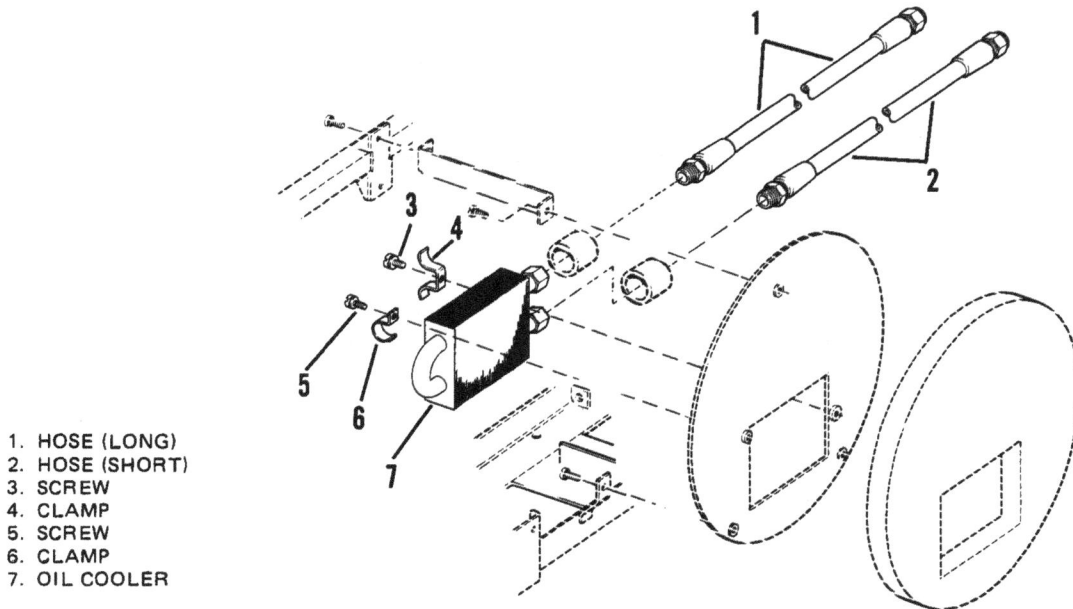

1. HOSE (LONG)
2. HOSE (SHORT)
3. SCREW
4. CLAMP
5. SCREW
6. CLAMP
7. OIL COOLER

Figure 4-43.Engine Oil   Cooler.

a. Inspection.

(1) Remove battery (see para 4-22).

(2) Inspect oil cooler for signs of leakage, bent cooling fins, or other
visible damage.

(3) Inspect cooling lines for signs of leakage or visible damage. Tighten or replace hoses and fittings as necessary.For
replacement of oil cooler, notify higher echelon of maintenance.

b.   Cleaning.
Avoid prolonged contact and inhalation of fumes of dry cleaning solvent.Use dry cleaning solvent

---

WARNING

---

only in a well ventilated area.   Do not direct pressurized air toward skin.Personal
injury could result.

Clean engine oil cooler with dry cleaning solvent (Table 3-1, item 9) or pressurized air (do not exceed 15
psi).

Section XIII. ENGINE EXHAUST

WARNING

Unless manual directs otherwise, do not attempt any of the following maintenance procedures when generator set is operating.   Do not touch exposed electrical connection
when a source of power such as utility
power or another generator set is connected to the load terminals.   Severe electrical
shock or death by electrocution may result.

4-53. DESCRIPTION AND FUNCTION. The engine exhaust system consists of the muffler, exhaust pipe and clamps. The muffler quiets the engine exhaust noises and the exhaust pipe routes the exhaust gases away from the operator.

4-54. MUFFLER. (See Figure 4-44)

WARNING

Make sure muffler is completely cooled off before performing any maintenance procedures.

(1)   Remove screws (1), nuts (2) and shield (4).   Remove heat shield (4) from frame.

(2)   Remove screws (5) and nuts (6).   holes or corrosion.

(3)   Inspect muffler (14) for dents,
       clamps (3) at three locations from heat

(4)   Check to see that clamps (8), (9) and (17) are securely attached and that they are not damaged.

(5)   Inspect exhaust pipe (20) for damage, obstructions and secure mounting.

(6)   Check mounting brackets (15) and (18)   (1), nuts (2), clamps (3), screws (5)

(7)   Install heat shield (4) with screws and nuts (6).
       for damage, and secure mounting.

b.   Removal.

(1) Remove screws (1), nuts (2) and clamps (3) at three locations
       on heat shield (4).

(2) Remove screws (5) and nuts (6). Remove heat shield (4) from frame.

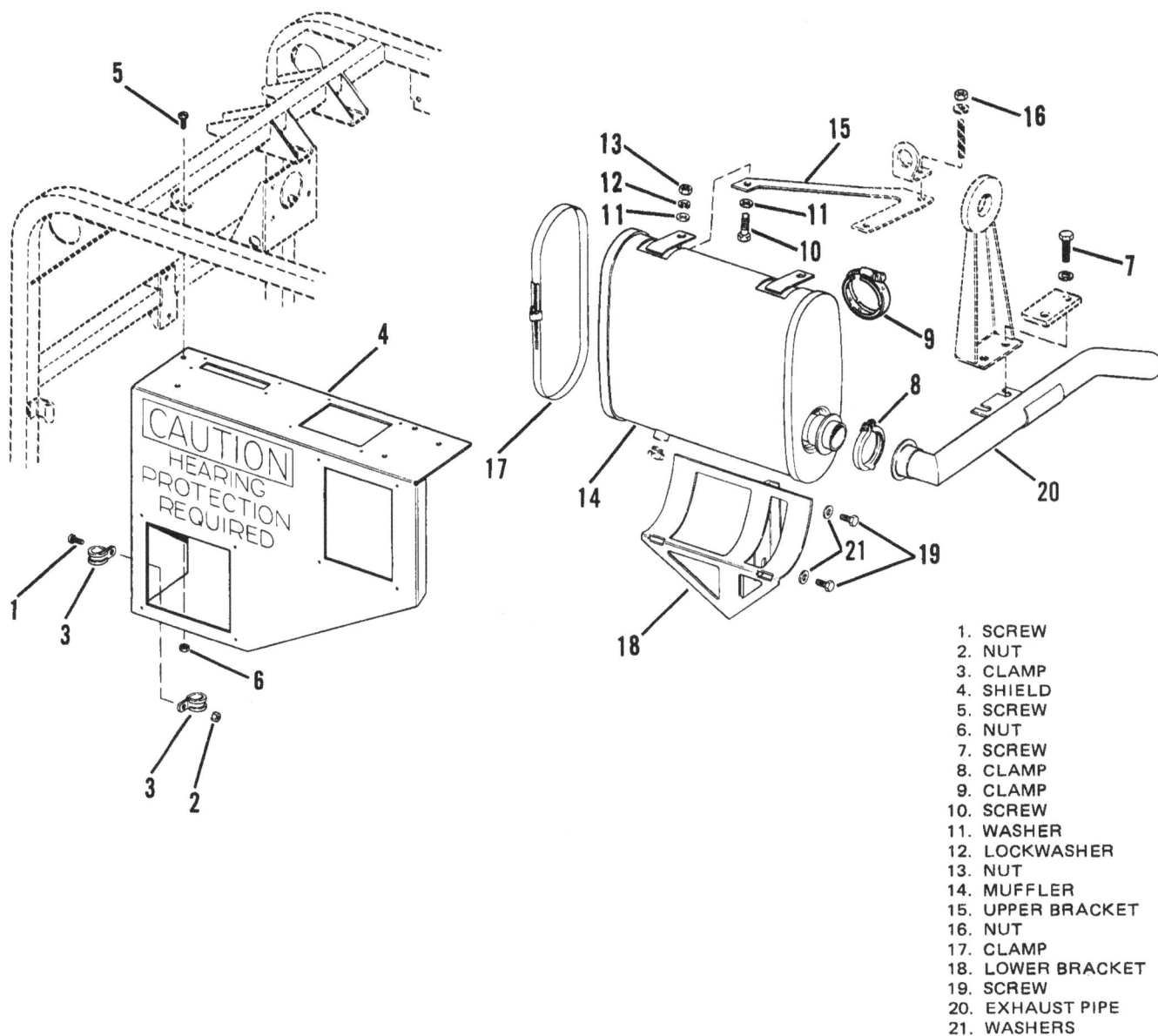

1. SCREW
2. NUT
3. CLAMP
4. SHIELD
5. SCREW
6. NUT
7. SCREW
8. CLAMP
9. CLAMP
10. SCREW
11. WASHER
12. LOCKWASHER
13. NUT
14. MUFFLER
15. UPPER BRACKET
16. NUT
17. CLAMP
18. LOWER BRACKET
19. SCREW
20. EXHAUST PIPE
21. WASHERS

Figure 4-44. Engine Exhaust.

(3) Remove clamp (8) and exhaust pipe (20).

(4) Loosen clamp (9) from muffler inlet.

(5)   Remove screws (10), washers (11), lockwashers (12), and nuts (13) that secure muffler (14) to upper bracket (15). Loosen nut (16).

(6) Remove two bands (17) that secure muffler (14) to lower bracket (18).
Remove muffler (14).

(7) If necessary, remove lower bracket (18) by removing screws (19) and washers
(21).

MARINE CORPS TM 05926B/06509B-12
ARMY   TM 5-6115-615-12
NAVY   NAVFAC P-8-646-12
AIR FORCE TO 35C2-3-386-31

CAUTION

Failure to install muffler in correct sequence result in premature failure will
of muffler.

(1)   If removed, loosely install lower bracket (18) with and washers (21).   mounting screws (19)

Position muffler (14) on exhaust stub of engine and

(2)   secure with clamp (9).

Tighten lower bracket mounting screws (19). Secure bracket (18) with bands

(3)   (17).   muffler (14) to lower

(4)   Install upper bracket (15) with screws (10), washers (11), lockwashers
(12), and nuts (13). Tighten nut (16).

(5)   Install exhaust pipe (20) and secure with clamp (8).

(6)   Install heat shield (4) on frame with screws (5) and nuts (6).

(7)   Put wire harness and fuel line in position on heat shield (4).   Install
                          (1) and nuts (2).

4-55.   **EXHAUST PIPE AND CLAMP.**   (See Figure 4-45)

a. Removal.

WARNING

Make sure exhaust before pipe is completely cooled off any maintenance
performing

(1) Remove clamp (1) that connects exhaust pipe (2) to muffler (not shown).

(2) Remove screws (3), washers (4), spacer (5) and lifting eye (6).

(3) Remove exhaust pipe (2).

1. CLAMP
2. EXHAUST PIPE
3. SCREW
4. WASHER
5. SPACER
6. LIFTING EYE

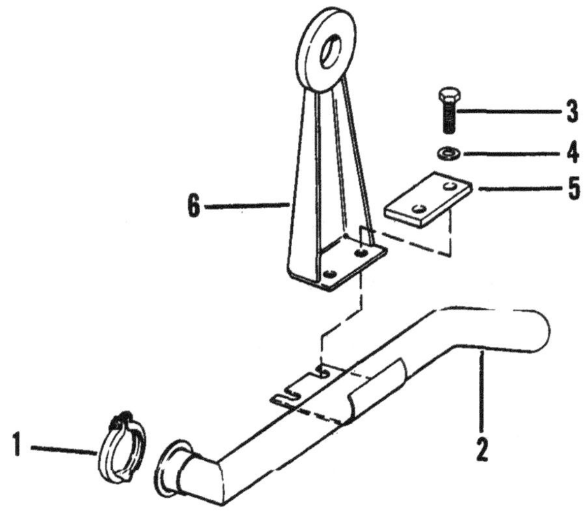

Figure 4-45. Exhaust Pipe and Clamp.

MARINE CORPS TM 05926B/06509B-12
ARMY   TM 5-6115-615-12
NAVY   NAVFAC P-8-646-12

(1)   Place exhaustpipe (2) in position.

(2)   Align fittings and install clamp (1) that connects exhaust pipe (2) to muffler (not shown).

        Secure exhaust pipe (2) to lifting eye (6), with spacer (5), washer (4) and screws (3).

(3)

Section XIV. ENGINE CONTROLS AND INSTRUMENTS.

WARNING

Unless manual directs otherwise, do the following   not attempt any of when
maintenance proedures is operating. Do not touch   generator set

connections when a source of power such as utility
power or another generator set is connected to the load terminals.   Severe
electrical shock or death by electrocution may result.

4-56.   **THROTTLE AND BRACKET.**   (See Figure 4-46)

a.   Removal.

1. SCREW
2. WASHER
3. CLIP
4. NUT
5. LOCKWASHER
6. CABLE
7. BRACKET
8. SCREW
9. NUT

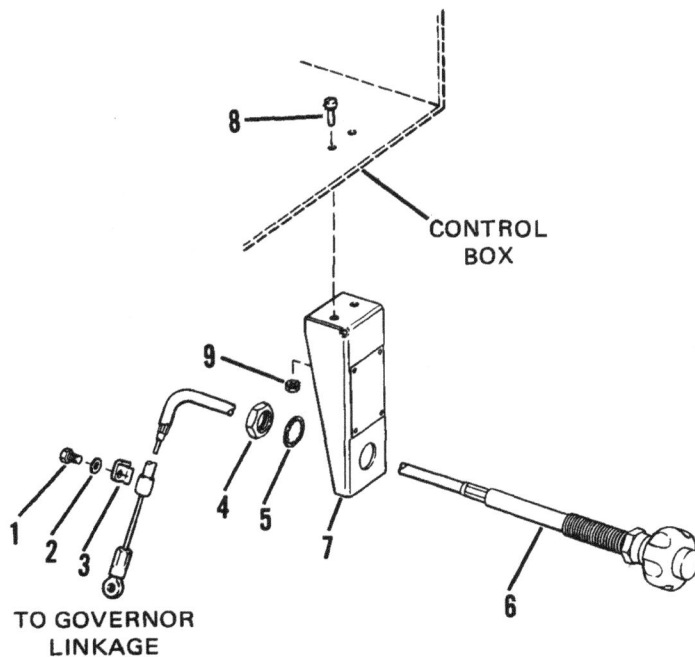

CONTROL
BOX

TO GOVERNOR
LINKAGE

Figure 4-46. Throttle Assembly.

(1) Remove screw (1), washer (2) and cable clip (3).

(2) Disconnect throttle cable from governor (refer to paragraph 4-44).

(3) Remove nut (4), lockwasher (5), and throttle cable (6).

(4) Remove screw   (8), nut (9) and bracket (7) from control box.

b.   Installation.

(1) Place bracket (7) in position on control nut (9).   throttle bracket (7) with

(2) Install throttle cable (6) and secure to lockwasher (5) and nut (4).
box and secure with screw (8) and

(3) Connect throttle cable to governor (refer to paragraph 4-44).

(4) Install cable clip (3), washer (2) and screw (1).

4-57. HOURMETER. (See Figure 4-47)

a.   Test.

(1) Start and run generator set.

(2) With an accurate watch or clock, check the elapsed time reading on the hourmeter.

b.   **Removal.**

1. HOURMETER
2. NUT
3. SCREW

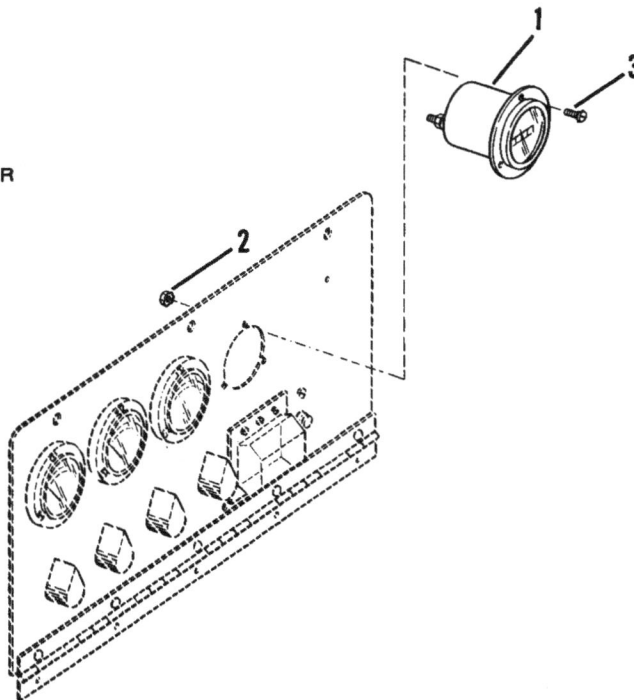

Figure 4-47. Hourmeter (Model MEP-021B Shown).

NOTE

Procedures for all three sets, MEP-016B, MEP-021B and MEP-026B are the same.Figure 4-47 illustrates the Model MEP-021B (400 Hz) set.

(1)   Disconnect the negative (-) battery cable from the battery.

(2)   Tag and disconnect wiring from the hourmeter (1, Figure 4-44

(3)   Remove nuts (2), screws (3) and hourmeter (1).

c.   Installation.

(1)   Secure hourmeter (1) to front panel with screws (3) and nuts   (2).

(2)   Using tags for identification, connect wiring to hourmeter.

(3)   Connect negative (-) battery cable to the battery.

4-58.   CONTROL BOX ASSEMBLY.   (See Figure 4-48)

a.   Test.

(1)   Perform operationaltest on control box, checking for proper operation.

(2)   Check that generator set does come to an emergency stop when DC control circuit breaker is pulled out.

(3)   Test individual components of control box as indicated in paragraphs 4-59 through 4-73.

(1)   Shut down set and disconnect load cables from load terminal board on back of control box.

(2)   Disconnect negative (-) battery cable from battery.

(3)   Refer to Table 4-3, tag and disconnect the wires that run from the generator to the control box.These wires must be fed through the holes in the bottom of the control box as the box is removed from the frame.

(4)   Unscrew electrical connector (1) from back of control box.

(5)   Remove two screws (2) and nuts (3), and remove throttle control bracket (4) from bottom of control box.

(6)   Remove screw (5),   nut (6), and ground strap (7) from control box.

(7)   Remove bolts (8),   lockwashers (9), and washers (10) that hold control box assembly to frame(four locations).

Table 4-3. Generator Wires and Connection Points.

Models MEP-016B and MEP-021B			Model MEP-026B		
Wire #	Terminal Board	Terminal	Wire #	Terminal Board	Terminal
F1	A1-TB2	1	F1	A1-TB1	1
F2	A1-TB2	2	F2	A1-TB1	2
T1	A1-TB1	1	T1	A3-TB1	4
T2	A1-TB1	2	T2	A3-TB1	2
T3	A1-TB1	3	T3	A3-TB1	3
T4	A1-TB1	4			
T5	A1-TB1	5			
T6	A1-TBI	6			

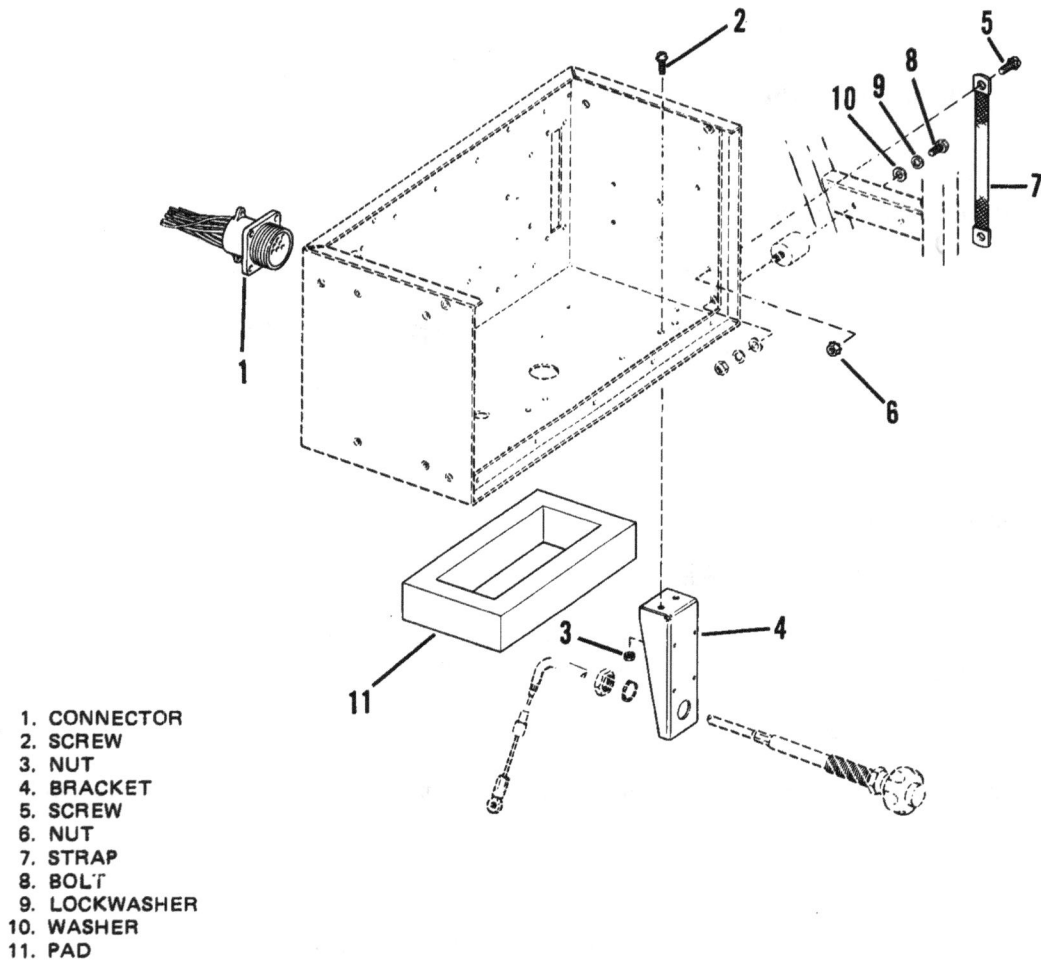

1. CONNECTOR
2. SCREW
3. NUT
4. BRACKET
5. SCREW
6. NUT
7. STRAP
8. BOLT
9. LOCKWASHER
10. WASHER
11. PAD

Figure 4-48. Control Box Removal.

CAUTION

Remove control box slowly to avoid damaging the wires that run from control box. generator through the bottom of the

(8)   Remove control box

assembly.

(9)   Remove pad (11) if

c.   Installation.

(1) Install pad (11) if removed.

(2) Feed wires from generator through grommets in bottom of control box as the box is positioned on the generator.

(3) Refer to Table 4-3 for connection of wires.

(4) Install washers (10), lockwashers (9), and bolts (8) that hold control box assembly to frame (four locations).

(5) Connect ground strap (7) to control box with screw (5) and nut (6).

(6) Connect throttle control bracket (4) to control box with two screws (2) and nuts (3).

(7) Connect electrical connector (1) to back of control box.

(8) Connect the negative (-) battery cable to the battery.

4-59.MASTER SWITCH.   (See Figure 4-49)

a.   Test.

(1)   Tag and disconnect wiring from master switch.

(2)   Refer to Table 4-4 and use a multimeter to check the continuity of the switch contacts for each position on the switch.Continuity will be indicated when the terminals are connected.

Replace switch if continuity requirements are not met.

(3)

Table 4-4.Blaster Switch Connections.

Models MEP-016B and MEP-021B			Position	Contacts Closed	
Position	Contacts Closed		Preheat	11-15 off	
			None Run/Aux Fuel 11-13-14-17		
Pre-Heat	11-15 off None		Run	11-13-14	
Run/Aux Fuel	11-13-14-17		Start	11-13-14-15 25-27-28	
Run	11-13-14				
Start	11-13-14-15	25-27-28			
Model MEP-026B					

b. **Removal.**

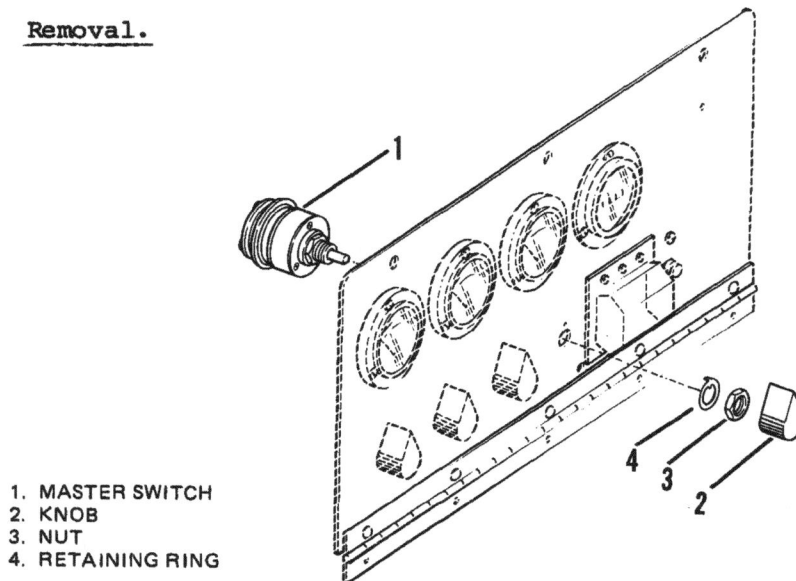

1. MASTER SWITCH
2. KNOB
3. NUT
4. RETAINING RING

Figure 4-49. Master Switch (Model MEP-021B Shown).

(1) Disconnect the negative (-) battery cable.

(2) Tag and disconnect   wiring from master switch   (1).

(3) Loosen setscrew in   knob (2) and remove knob.

(4) Remove retaining nut (3) and retaining ring (4).

(5) Remove master switch.

c.   Installation.

(1)   Secure master switch (1) to front panel   with retaining ring (4) and nut
(3).

(2)   Install knob (2) and tighten setscrew.

(3)   Using tags for identification, connect wiring to master switch.

(4)   Reconnect the negative (-) battery cable.

4-60.   **CURRENT SELECTOR SWITCH.**   (Models MEP-016B and MBP-021B only)
(See Figure 4-50)

a.   Test.

(1) Disconnect the negative (1) battery cable.

(2) Tag and disconnect wiring from master switch.

(3) Refer to Table 4-5 and use a multimeter to check the continuity of the switch contacts for each position on the switch.   Continuity will be indicated when terminals are connected.

(4) Replace   switch if continuity requirements are not met.

(5) Connect   wiring to master switch.

(6) connect   the negative (-) battery cable.

Table 4-5.Current Selector Switch Connections.

Position	Closed	Contacts
1-1		11-12 I-2
11-13		
I-3	11-14	

b.   <u>Removal.</u>

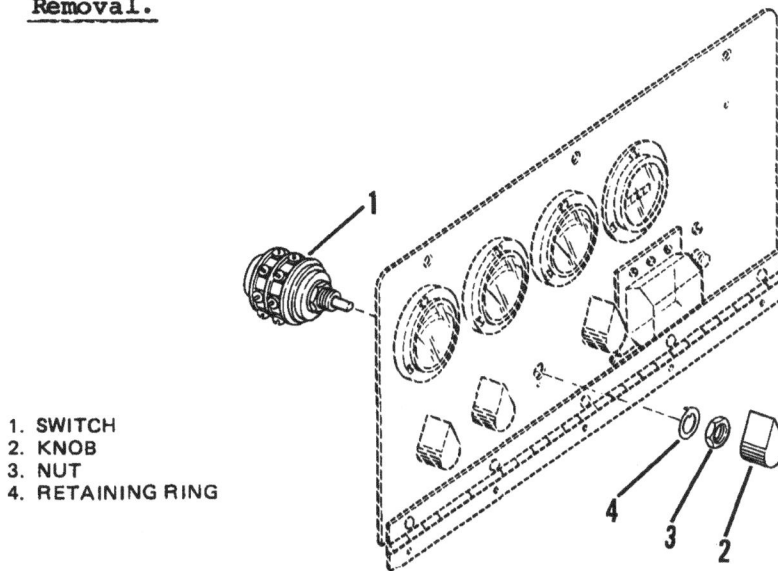

1. SWITCH
2. KNOB
3. NUT
4. RETAINING RING

Figure 4-50.Current Selector Switch (Model MEP-021B Shown).

(1)   Disconnect the negative (-) battery cable.

(2)   Tag and disconnect wiring from current   selector switch (l).

(3)   Loosen   setscrew in knob (2) and remove   knob .

(4)   Remove   retaining nut (3) and retaining   ring (4).

(5)   Remove   current selector switch (l).

c.Installation.

(1) Secure current selector switch (1) to front panel with retaining ring (4)
and nut (3).

(3) Using tags for identification, connect   wiring to current selector switch.

(4) Connect the negative (-)   battery cable.

4-61.   VOLTAGE SELECTOR SWITCH. (Models MEP-016B and MEP-021B only)
(See Figure 4-51)

a.Test.

(1)   Disconnect the negative (-) battery cable.

(2)   Tag and disconnect wiring from voltage selection switch.

(3)   Refer to Table 4-6 and use a multimeter to check the continuity of the switch contacts for each position on the
switch.Continuity will be indicated when terminals are connected.

(4)   Replace   switch if continuity requirements are not met.

(5)   Connect   wiring to the voltage selection switch.

(6)   Connect   the negative (-) battery cable.

Table 4-6.Voltage Selector Switch Connections.

Position	Contacts Closed	
V1-0	11-12-15	21-22
V2-0	11-13	21
V3-0	11-14	21
V1-2	11-12-15	21-25
V2-3	11-16-17	21-26
V3-1	11-17	21-27

(1)   Disconnect the negative (-) battery cable.

(2)   Tag and disconnect wiring from voltage selector switch (1).

(3)   Loosen setscrew in knob (2) and remove knob.

(4)   Remove retaining nut (3) and washer (4).

1. SWITCH
2. KNOB
3. NUT
4. WASHER

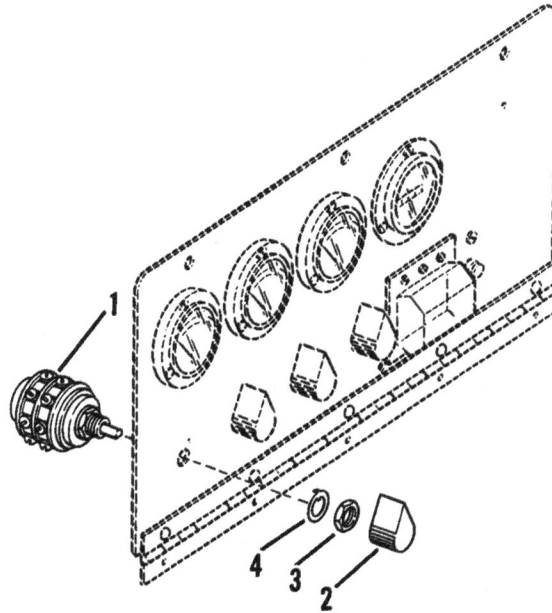

Figure 4-51. Voltage Selector Switch (Model MEP-021B Shown).

(5) Remove voltage selector switch (1).

c. Installation.

(1) Secure voltage selector switch (1) to front panel with washer (4) and nut (3).

(2) Install knob (2) and tighten setscrew.

(3) Using tags for identification, connect wiring to voltage selector switch.

(4) Connect the negative (-)   battery cable.

4-62.   OUTPUT SELECTION SWITCH   (Models MEP-016B and MBP-021B Only)
         (See Figure 4-52)

a. Test.

(1) Disconnect the negative   (-) battery cable.

NOTE

Removal of switch will make testing easier but is not required.

(2) Tag and disconnect wiring from output selection switch.

(3)   Refer to Table 4-7 and use a multimeter to check the continuity of the switch contacts for each position on the switch.Continuity will be indicated when terminals are connected.

      Replace switch if continuity requirements are not met.

(4)

Table 4-7.Voltage Reconnection Switch Connections.

Position		Contacts Closed				
120/208V	3 Phase	1-2 5-6 7-8		9-10	17-18	
120V	3 Phase	1-2 3-4		11-12 13-14 19-20		
240V	1 Phase	3-4	9-10 21-22 None		None	
120V	1 Phase	3-4	11-12 13-14 15-16 23-24			

(5) Connect wiring to the output selection switch.

(6) Connect the negative (-) battery cable.

b.   Removal.

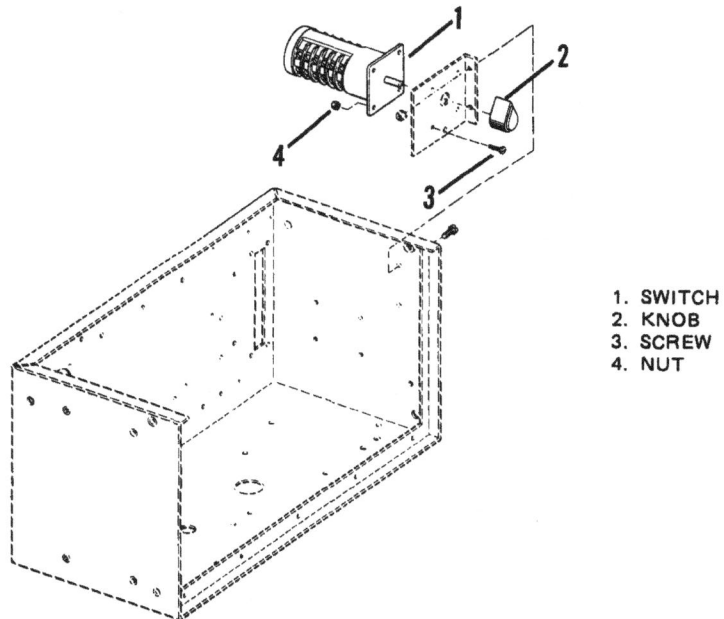

1. SWITCH
2. KNOB
3. SCREW
4. NUT

Figure 4-52.Output Selection Switch.

(1)   Disconnect the negative (-) battery cable.

(2)   Tag and disconnect wiring from output selection switch (1).

(3)   Loosen setscrew in knob (2) and remove knob.

(4) Remove four screws (3), nuts (4) and output selection switch (1).

(1) c. <u>Installation.</u>
Secure output selection switch (1) to control box with screws (3) and nuts
(4).

(2)

Install knob (2) and tighten setscrew.

(3)

Using tags for identification, connect wiring to output selection switch.

(4)

4-63. VOLTMETERS.   (Figure 4-53)

a. <u>Test.</u>

NOTE

Models MEP-016B and MEP-021B use a 0-250 VAC meter.
Model MEP-026B has a 0-50 VDC meter.

(1) Adjust screw on bottom of voltmeter until needle reads zero.

(2) Connect an accurate multimeter across the generator set voltmeter terminals.

(3) Start and run generator set.

(4) While varying the output, observe both the multimeter voltmeter.   and the generator set

(5) Readings on generator set voltmeter multimeter.   should agree with   readings on

NOTE

Procedures for all three sets,   MEP-016B, MEP-021B and 4-53
MEP-026B are the same. Figure   illustrates the
MEP-021B (400 Hz) set.

(1) Disconnect the negative (-) battery   cable.

(2) Tag and disconnect wiring from voltmeter (1).

(3) Remove nuts (2), screws (3) and voltmeter (1).

c.   <u>Installation.</u>

(1) Secure voltmeter (1) to front panel with screws   3) and nuts (2).

(2) Using tags for identification, connect wiring to   voltmeter.

MARINE CORPS TM 05926B/06509B-12
ARMY   TM 5-6115-615-12
NAVY   NAVFAC P-8-646-12

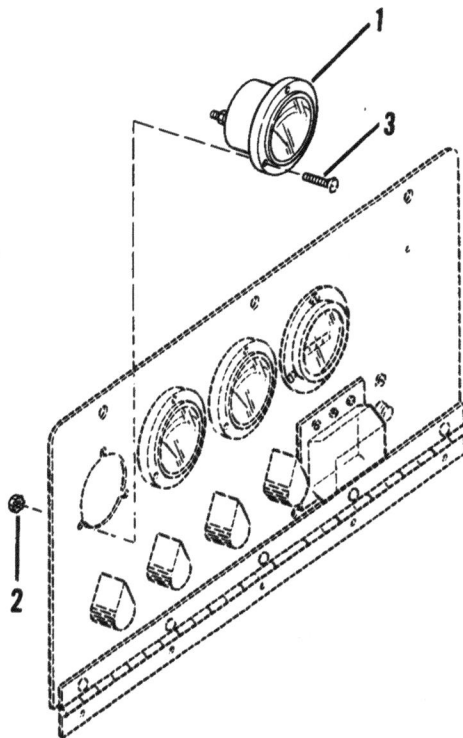

1. VOLTMETER
2. NUT
3. SCREW

Figure 4-53. Voltmeter(Model MEP-021B Shown).

(3) Connect the negative (-) battery cable.

4-64.   **CURRENT INDICATING METER.**   (See Figure 4-54)

a. Test.

(1)   Adjust screw on bottom of current indicating meter until needle reads zero.

(2)   If a load meter is suspected of being faulty, replace meter with a meter that is known to be serviceable.

(3)   If problem persists, original meter was performing properly and should be re-installed.   Further troubleshooting should be done to locate the problem.

NOTE

Procedures for all three sets, MEP-016B, MEP-021B, and MBP-026B are the same. Figure 4-54 illustrates the Model MEP-021B (400 Hz) set.

1. METER
2. NUT
3. SCREW

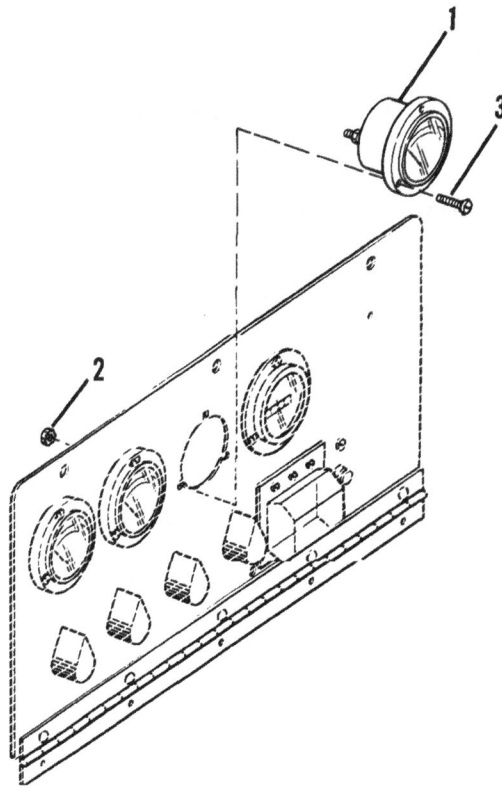

Figure 4-54.   Current   Indicating Meter (Model MEP-021B Shown).

(1)   Disconnect the negative   (-) battery cable.

(2)   Tag and disconnect wiring from load meter (1).

(3)   Remove nuts (2), screws (3) and load meter (1).

c.   Installation.

(1) Secure load meter to front panel with screws (3) and nuts (2).

(2) Using tags for identification, connect wiring to load meter.

(3) Connect the negative (-) battery cable.

4-65. RHEOSTAT.   (See Figure 4-55)

a.   Test.

(1)   Disconnect the negative (-) battery cable.

(2)   Tag and disconnect wiring from rheostat.

(3)   Connect a multimeter across the terminals that the wires were removed from.

(4)   Viewing rheostat from shaft end, turn rheostat fully counterclockwise.
Multimeter should indicate low (approximately O) ohms.

(5)   Slowly turn rheostat fully clockwise while observing multimeter.
Multimeter should indicate a gradual steady   increase in resistance.

(6)   If multimeter indicates an erratic increase be replaced.   in resistance, rheostat should

(7)   Connect   wiring to rheostat.

(8)   Connect   the negative (-) battery cable.

NOTE

Procedures for all three sets, MEP-016B, MEP-021B, and
MEP-026B are the same.Figure 4-55 illustrates the
Model MEP-021B (400 Hz) set.

1. RHEOSTAT
2. KNOB
3. NUT
4. RING, RETAINING

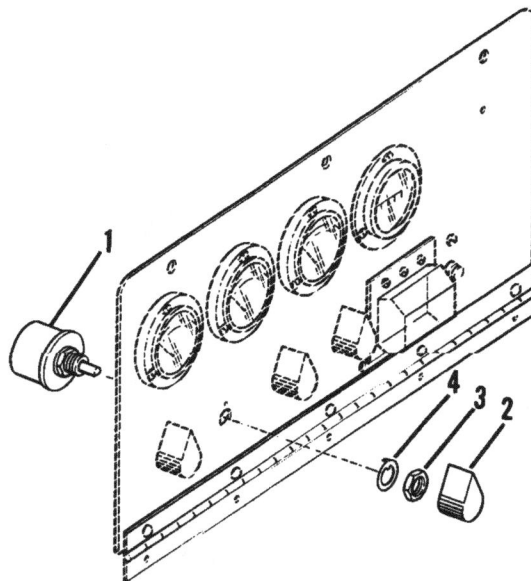

Figure 4-55. VoltageAdjust Rheostat (Model MEP-021B Shown).

(1)   Disconnect the negative   (-) battery cable.

(2)   Tag and disconnect wiring from voltage adjust rheostat (l).

(3)   Loosen knob setscrew and remove knob (2).

(4)   Remove nut (3), retaining ring (4) and rheostat (1).

c.   Installation.

(1)   Secure rheostat (1) with retaining ring (4) and nut (3).

(2)   Secure knob (2) on rheostat shaft with knob setscrew.

(3)   Using tags for identification, connect wiring to rheostat.

(4)   Connect the negative (-) battery cable.

4-66.   FREQUENCY METER.   (Models   MEP-016B and MEP-021B Only) (See Figure 4-56)

a. Removal.

(1) Disconnect the negative   (-) battery cable.

NOTE

Procedures for the   MEP-016B and MEP-021B sets are
the same. Figure 4-56 illustrates the MEP-021B
set.

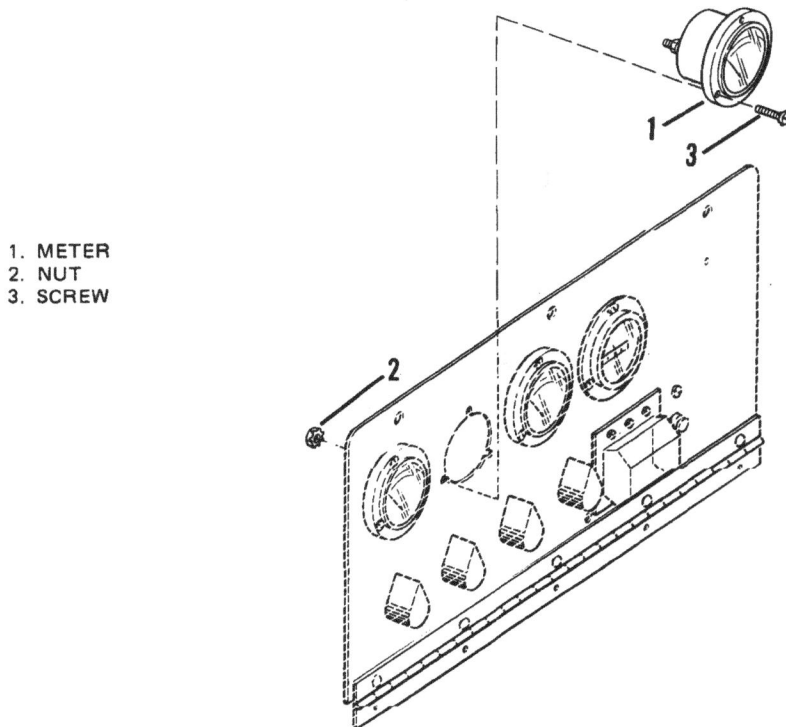

1. METER
2. NUT
3. SCREW

Figure 4-56.   Frequency Meter (Model MEP-021J3 Shown).

(2) Tag and disconnect   wiring from frequency meter (1).

(3) Remove nuts (2), screws (3), and meter (1).

MARINE CORPS TM 05926B/06509B-12
ARMY   TM 5-6115-615-12
NAVY   NAVFAC P-8-646-12
AIR FORCE   TO 35C2-3-386-31

(1) Secure frequency meter to front panel with screws (3) and nuts (2).

(2) Using tags for identification, connect wiring to frequency meter.

(3) Connect the negative (-) battery cable.

4-67.   **TACHOMETER (RPM INDICATOR).**   (Model MEP-026B   Only) (See Figure 4-57)

a. Test.

(1)   Adjust screw on   bottom of tachometer until   needle reads zero.

(2)   Connect a known ter- accurate tachometer across   the generator set tachometer
minals.

(3)   Start and run   generator set.

(4)   While varying gener- engine speed observe both the test tachometer and the tachometer.

(5)   Readings on generator set tachometer should agree with readings on test tachometer.

1. TACHOMETER
2. NUT
3. SCREW

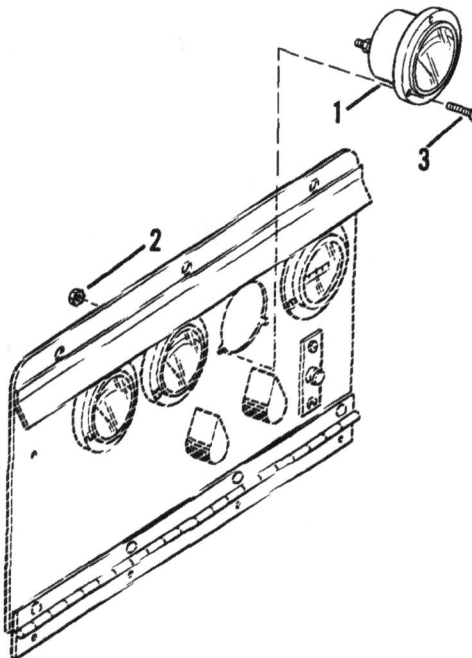

Figure 4-57. Tachometer (Model MEP-026B Only).

(1)   Disconnect the negative (-) battery cable.

(2)   Tag and disconnect wiring from tachometer (1).

(3)   Remove nuts (2), screws (3) and tachometer (1).

c.   Installation.

(1) Secure tachometer (1) to front panel with screws (3) and nuts (2).

(2) Using tags for identification, connect wiring to tachometer.

(3) Connect the negative (-) battery cable.

4-68.   FREQUENCY TRANSDUCER. (See Figure 4-58)

a.   Inspection.

(1) Inspect that   frequency transducer is securely mounted.

(2) Inspect that   frequency transducer connections are tight.

(3) Inspect frequency transducer for cracks or other visible damage.

b.   Removal.

NOTE

Procedures for all three sets, MEP-016B, MEP-021B, and MEP-026B are the same.Figure 4-58 illustrates the MEP-016B and MBP-021B set.

(1) Disconnect the negative (-) battery cable.

(2) Tag and disconnect wiring from frequency transducer (1).

(3) Remove screws (2) and   remove transducer   (1).

c.   Installation.

(1) Secure transducer (1)   with screws (2).

(2) Using tags for identification, connect   wiring to frequency transducer.

(3) Connect the negative (-) battery cable.

4-69.RECTIFIER BRIDGE.   (Model MEP-026B Only)   (See Figure 4-59)

a. Inspection.

(1) Inspect that   bridge is securely mounted   and that all wiring is tight.

(2) Inspect heat   sink fins for cracks, bent   fins or other visible damage.

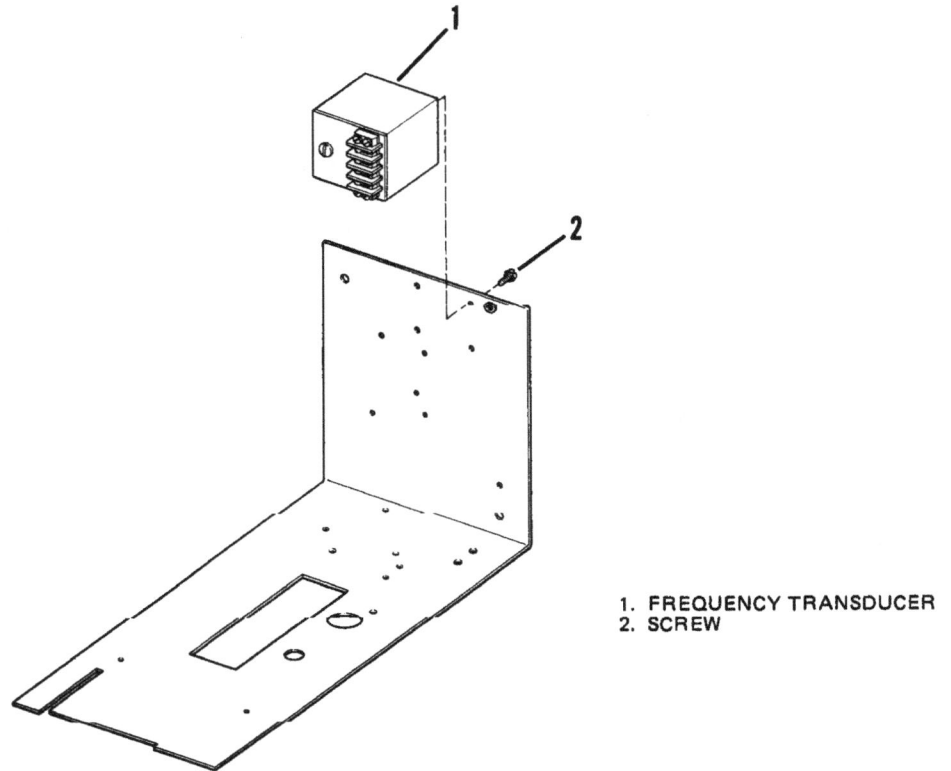

1. FREQUENCY TRANSDUCER
2. SCREW

Figure 4-58.Frequency Transducer.

b. Removal

(1)   Disconnect negative (-) battery cable from battery.

(2)   Tag and disconnect eight wires from front of rectifier bridge.Note locations of mounting hardware as wires are removed.

(3)   Remove screws (1) and nuts (2).Carefully remove rectifier bridge assembly (3 through 31) from control box.

(4)   Remove screws (3) and nuts (4).Remove rear support (5).

(5)   Remove screws (6) and nuts (7).

(6)   Remove nuts (8)and washers (9).

NOTE

Note location of straps before removal.

(7)   Remove screws (10), nuts (11), lockwashers (12), and washers (13). Remove two ground straps (14).

MARINE CORPS TM 05926B/06509B-12
ARMY   TM 5-6115-615-12
NAVY   NAVFAC P-8-646-12
AIR FORCE   TO 35C2-3-386-31

1. SCREW
2. NUT
3. SCREW
4. NUT
5. SUPPORT, REAR
6. SCREW
7. NUT
8. NUT
9. WASHER
10. SCREW
11. NUT
12. LOCKWASHER
13. WASHER
14. STRAP
15. DIODE
16. DIODE
17. NUT
18. LOCKWASHER
19. WASHER
20. NUT
21. LOCKWASHER
22. WASHER
23. WASHER
24. NUT
25. LOCKWASHER
26. WASHER
27. TERMINAL
28. SUPPORT, FRONT
29. VIBRATION DAMPER
30. HEAT SINK
31. HEAT SINK

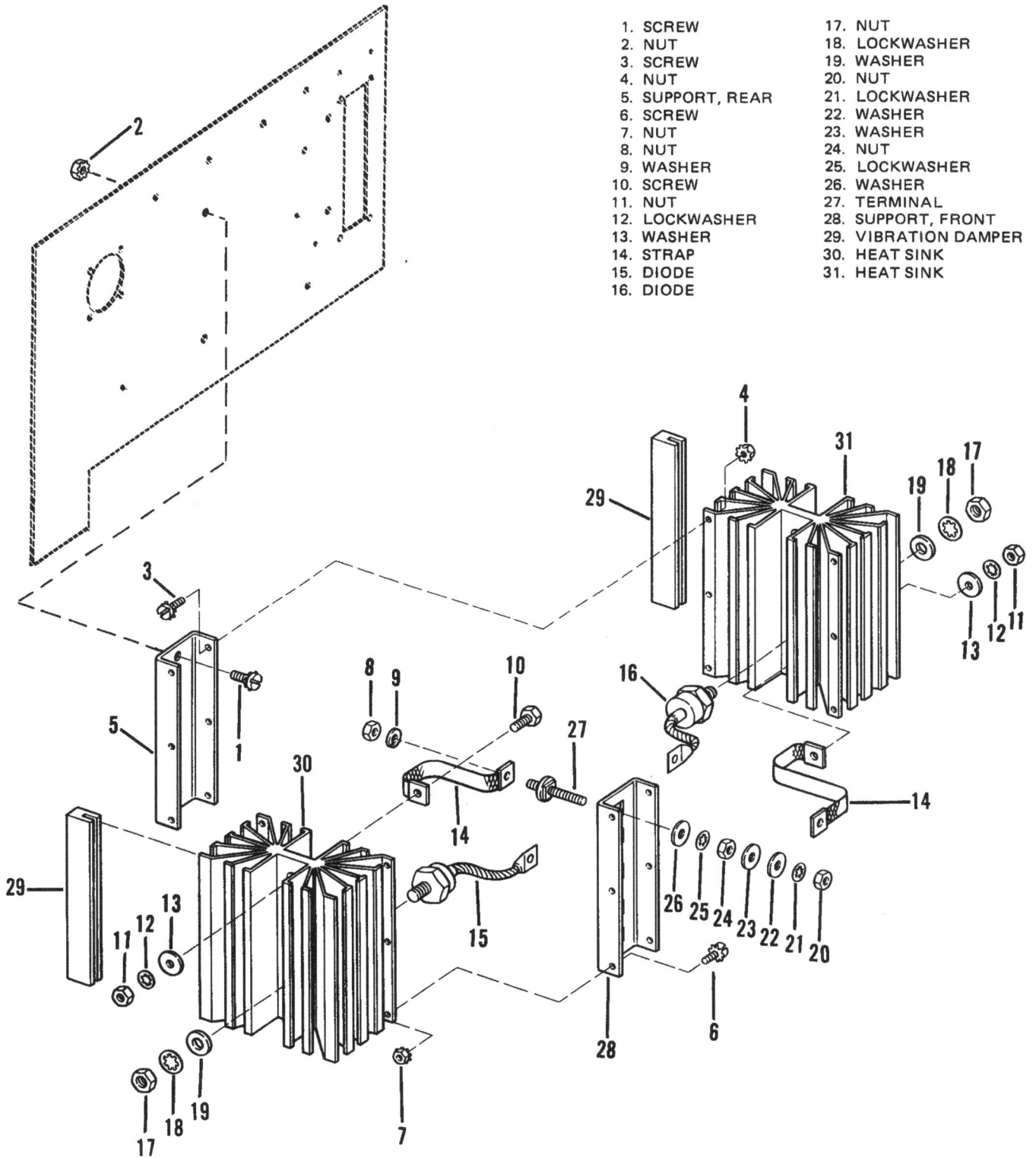

Figure 4-59. Rectifier Bridge (Model MEP-026B Only).

NOTE

Diodes (15 and 16) are not the same.Be sure they are installed in their origi-

(8)   Remove nuts (17), lockwashers (18), and washers (19) to remove diodes (15 and 16).

(9)   If necessary, remove nuts (20), lockwashers (21), washers (22 and 23), nuts (24), lockwashers (25) and washers (26) to remove terminals (27) and front support (28).

(10)   Remove vibration dampers (29) from neat sinks (30 and 31).

c.   Test.

(1)   Check diodes (15 and 16) with an ohmmeter.

(2)   Measure the resistance between the terminal end and the threaded base (heat sink) end of the diode.

(3)   Reverse ohmmeter leads and repeat the resistance measurement.

(4)   A diode in good condition will have a very high resistance for one measurement and a resistance near zero when ohmmeter probes are reversed.

(5)   Failure to obtain these two extremes in resistance measurement indicates a defective diode that should be removed and replaced.

d.   Installation.

(1)   Install vibration dampers (29) on heat sinks (30 and 31).

(2)   Put terminals (27) in position in front support (28) and install washers (26), lockwashers (25), nuts (24), washers (23 and 22), lockwashers (21), and nuts (20).

(3)   Put diodes (15 and 16) in their original locations and install washers (19), lockwashers (18) and nuts (17).

(4)   Put ground straps (14) in position and install screws (10), washers (13), lockwashers (12) and nuts (11).

(5)   Connect ground straps (14) to terminals (27) and install washers (9) and nuts (8).

(5)   Put support (28) in position on heat sink (30) or (31) and install screws (6) and nuts (7).

(6)   put rear support (5) in position and install screws (3) and nuts (4).

(7)   Position rectifier bridge assembly in control box and install screws (1) and nuts (2).

(8)

(9) Using tags for identification, connect eight wires to terminals on front of rectifier bridge.

(l0) Connect negative (-) battery cable to battery.

4-70. MAIN CIRCUIT BREAKER.   (Models MEP-016B and MEP-021B) (See Figure 4-60)

a. Test

(1) Tag and disconnect wiring from circuit breaker.

(2) With circuit breaker OFF, use a multimeter to check for resistance terminals A1 and B1, A2 and B2, and between A3 and B3.

(3) Multimeter should have indicated infinity (high ohms) between each terminals.   pair of

(4) Flip circuit breaker to the ON position. Recheck resistance between terminals A1 and B1, A2 and B2, and A3 and B3.

(5) Multimeter should now indicate continuity (0 ohms) between each pair terminals.
Replace the circuit breaker if the resistance requirements are not met.   of

(6)

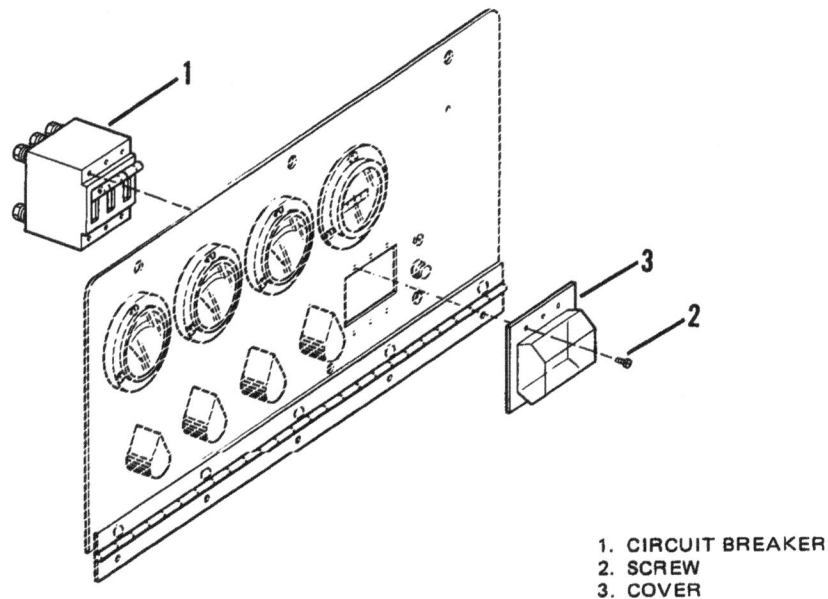

1. CIRCUIT BREAKER
2. SCREW
3. COVER

Figure 4-60. AC Circuit Breaker (Model MEP-021B Shown).

NOTE

Procedures for the MEP-016B and MEP-021B, (60 and 400 Hz) sets are the same. Figure 4-60 illustrates the MEP-021B (400 Hz) set.

(1) Disconnect the negative (-) battery cable.

(2) Tag and disconnect wiring from circuit breaker (1).

(3) Remove six screws (2), cover (3) and circuit breaker (1).

c. Installation.

(1) Install circuit breaker (1) and cover (3) with six screws (2).

(2) Using tags for identification, connect wiring to circuit breaker.

(3) Connect the negative (-) battery cable.

4-71. MAIN CIRCUIT BREAKER. (Model MEP-026B) (See Figure 4-61)

a. Test.

(1) Disconnect the negative (-) battery cable.

(2) Disconnect one cable, two wires, and three bus bars from circuit breaker.

(3) With circuit breaker off, use an ohmmeter to check for resistance between terminals A- and B- and between terminals A+ and B+.

(4) Ohmmeter should have indicated no continuity (high ohms) between both pairs of terminals.

(5) Switch circuit breaker to the on position. Recheck resistance between terminals A- and B- and between terminals A+ and B+.

(6) Ohmmeter should now indicate continuity terminals. (low ohms) between each pair of

(7) Connect one cable, two wires, and three bus bars to the circuit breaker.

(8) Connect the negative (-) battery cable.

(1) Disconnect the negative (-) battery cable.

(2) Remove nuts (1) and lockwashers (2) and washer (3). Disconnect three bus bars (4), cable (5) and two wires from circuit breaker (6).

(3) Remove four screws (7), cover (8), and circuit breaker (6).

1. NUT
2. LOCKWASHER
3. WASHER
4. BUS BAR
5. CABLE
6. CIRCUIT BREAKER
7. SCREW
8. COVER

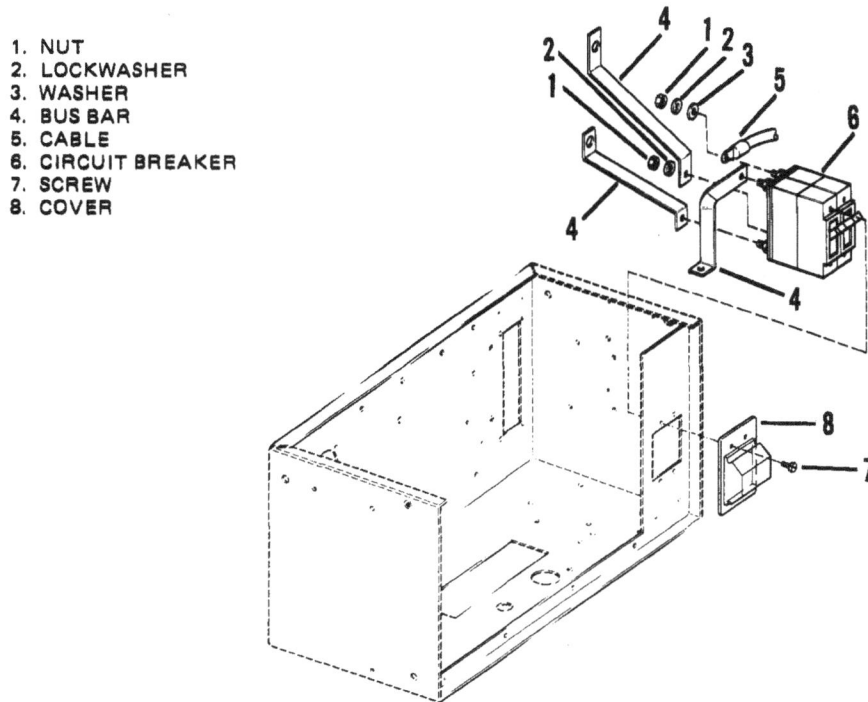

Figure 4-61.DC Circuit Breaker (Model MEP-026B Only).

(1)   Secure circuit breaker (6) and cover (8) with four screws (7).

(2)   Connect three bus bars (4), cable (5), and two wires to circuit breaker (6) with nuts (1), lockwashers (2) and washer (3).

(3)   Connect the negative (-) battery cable.

4-72. WIRINGHARNESS.

a. Inspection.

(1) Inspect that all connections are tight.

(2) Inspect harness for damaged wires,loose or missing ties, missing insulation or other visible damage.

b.   Test.   Test wiring harness by measuring continuity of wires from point of origin to point of termination. Continuity (low ohms) indicates a good wire. No continuity (high ohms) indicates a broken wire that must be replaced.

c.   Removal.

(1) Disconnect the negative (-) battery cable.

(2)   Tag and disconnect wiring from all terminals.

Remove ties and clamps securing wiring harness.

(4)   Carefully remove harness.

d. Repair.   Repair consists of wrapping deteriorated insulation with electrical tape, or replacing damaged wires or terminals as necessary.Be sure to use the same size, color and length of wire when making repairs.   Solder connections where applicable.Tag each wire and corresponding terminal at time of removal to ensure correct reassembly. The following tables and illustrations give the wire number, gauge, color, and terminal lugs for each wire in the harnesses.

(1) Control Harnesses.Models MEP-016B and MEP-021B are listed in Table 4-8 and illustrated in Figures 4-62, 4-63, and 4-64. Model MEP-026B is listed in Table 4-9 and illustrated in Figures 4-65, 4-66, and 4-67.

(2) Engine Harness (All Models). Listed in Table   4-10 and illustrated in Figure 4-68.

e.   Installation.

(1) Using tags for   identification, attach leads to components.

(2) Secure harness with ties and clamps.

Table 4-8.Control Wiring Harness (Mbdels MEP-016B and MEP-021B).

NOTE

Notes concerning repair and installation of this harness, and the key to the "terminal find numbers" are found at the end of the table.

Wire Ref No.	From	Term Find No.	TO	Term Find No.	Wire Find No.	Wire Find Stamp No.	Wire stamp Color
1	A1-J1-A	.25 strip	A3-TB2-6	4	1	P1A16	BLACK
2	A1-J1-J	.25 strip	A3-TB2-5	4	11	P2B16	BLACK
3	A1-J1-N	.25 strip	A1-TB2-9	4	1	P3B16N	BLACK
4	A1-J1-L	.25 strip	A3-TB2-2	4	11	P4A16	BLACK
5	A1-J1-Z	.25 strip	A1-TB2-8	44	1	P5B16N	BLACK
6	A1-J1-X	.25 strip	A3-TB2-1	4	2	P6A16	BLACK
7	A1-J1-W	.25 strip	A2-S1-17	11	11	P7A16	BLACK
8 9	A1-J1-T	.25 strip	A1-R2-1	4	2	H8A12	BLACK
10	A1-J1-E	.25 strip	A1-TB2-4	4	2	x9B16V	BLACK
11	A1-J1-G	.25 strip	A1-TB2-3	4		X10B16V	BLACK
12	A1-J1-V	.25 strip	A1-TB2-5	4		P11B12	BLACK
	A1-J1-P	.25 strip	A1-TB2-8			P5E12N	BLACK

Table 4-8. Control Wiring Harness (Models MEP-016B and MEP-021B), Continued.

Wire Ref	Term Find		Term Find		Wire Find	Wire Find Stamp	Wire stamp
13	A1-K1-X2	9	A2-S1-14	4	1	P12A16	BLACK
14	A1-K1-X1	4	A2-M2-(-)	4	1	P13A16	BLACK
15	A1-K1-A1	.25 strip	A2-S3-21	9	2	P11E12	RED
16	A1-K1-A2	4	A2-S3-11	4	1	K14A16	BLACK
17	A1-J1-S	10 4	A2-M3-(+)	11	2	K14C12	BLACK
18	A1-TB2-9	10	A2-M3-(-)	4 4	11	P3E16N	BLACK
19	A2-M4-(+)	10 4	A1-TB2-6	10 4	1	P15A16	BLACK
20	A3-TB1-7	4	A1-TB2-11	4	1	P38A16	BLACK
21	A2-M1-(+)	4	A2-S1-25	10	1	D17A16	BLACK
22	A2-M1-(-)	4	A2-S2-11	10 4	1	D16A16	BLACK
23	A1-FC-(+)	6	A4-CT1-C2	4	1	D18A16	BLACK
24	A1-FC-(-)	4	A1-TB2-6	4	11	D19A16	BLACK
25	A1-FC-TL	10	A1-TB2-1O	4 6	1	X20A16A	BLACK
26	A1-FC-BL	6	A2-R1-c	4	1	X49D16A	BLACK
27	A2-CB2-2	4	A2-R1-B	4	1	P22A16	RED
28	A1-S1-18	4	A2-S1-11	.375s	1	P23A16	BLACK
29	A2-M2-(+)	4	A1-TB2-2	.375s 4	1	P37A16	BLACK
30	A1-TB1-4	6	A1-TB2-1	4	1	Z20816A	BLACK
31	A1-VR1-27	4	AI-TB2-10	4	1	X25D16A	BLACK
32	A1-VR1-28	4	A1-TB3-6	4	1	P24A16	BLACK
33	A1-VR1-26	4	A1-R2-1		1	P26A16	BLACK
34	A2-CB2-2	4		.25S	1	P22B16	BLACK
35	A1-VR1-21	10	A2-S1-15		1	V27A16	BLACK
36	A1-VR1-20		A3-TB1-1	4 4	1	V28A16	BLACK
37	A2-S3-12	4			1	X25E16A	BLACK
38	A2-S3-24	4			1	X35C16N	BLACK
39	A1-R1-1		A3-TB1-2	4	1	H8C16	BLACK
40	NOT USED		A3-TB1-9	4	1		
41	A3-TB1-11	4	A1-TB2-7	4		K31A16	BLACK
42	A1-TB2-4	4	A2-M4-(-)	10 6		X9C16V	BLACK
43	NOT USED	4	A2-CB2-1	4	1		
44	NOT USED	4	A3-TB1-18	4	1		
45	A1-TB2-3	4	A3-TB1-13	11	1	X10C16V	BLACK
46	A1-TB2-1	4	A1-R2-2	4	1	V28B16	BLACK
47	A1-K1-X2	4	A3-TB1-19	4	11	P21A16N	BLACK
48	A1-TB2-9	9	A2-S1-27	4	1	P3c16N	BLACK
49	A3-TB1-13	.25 strip	A2-S1-13	10	2	P11F16	RED
50	A1-TB2-2	4	A1-LO	10	1	V27B16	BLACK
51	A1-TB2-5	4	A1-L2		1	P11D12	RED
52	A1-K1-A2	4 4			1	K14B16	BLACK
53	A1-R2-2	A3-TB1-14			1	H32A16	BLACK
54	A3-TB1-17	A3-TB1-1O			1	P33A16	BLACK
55	A3-TB1-15	A1-TB2-5			1	P34A16	BLACK
56	A1-TB3-6	A3-TBI-8				X35B16N	BLACK
57	A1-TB3-2 4 4	A1-K1-A2 A1-TB2-GRD				X36c 16B	BLACK
58	A1-TB3-5	4	A1-L1	10	1	x39C16A	BLACK

4-89

Table 4-8. Control Wiring Harness (Models MEP-016B and MEP-021B), Continued.

Wire Ref	Term Find		Term Find	Wire Find	Wire Find Stamp	Wire Stamp
59	A1-TB3-1	.25 strip	A3-TB1-12	10 4	1	X40B16C	BLACK
60	A1-J1-K	.25 strip	A1-TB1-3	4	1	P4C16	BLACK
61	A1-J1-H	7	A1-TB1-6	5	1	P5D16N	BLACK
62	A1-TB1-5		A1-TB1-4	4	3	X29A10C	BLACK
63	A1-S1-B	5		5	1	P63A16	BLACK
64	A1-S1-2	5		5	3	X41A10B	BLACK
65	A1-S1-1O	5	A3-TB1-5	5	3	x42A10B	BLACK
66	A1-S1-11		A3-TB1-4		3	X20D10A	BLACK
67	NOT USED		A3-TB1-3				
68	NOT USED	4	A3-TB1-6	4			
69	A1-S1-23	4	A2-F1-1	4	11	P43A16	BLACK
70	A1-S1-19	4	A1-TB1-1	4 4	1	P11A16	BLACK
71	A1-S1-17	4	A2-CB1-A2	.375s	1	P45A16	BLACK
72	A1-S1-21	6	A1-TB2-12	7	2	P46A16	BLACK
73*	A1-TB1-4	5	A2-CB1-A3	11 5	3	X20C12A	BLACK
74	A1-S1-13	5	A1-LO	11	3	x25A10A	BLACK
75	A1-TB1-2	5	A1-TB2-10	11 5	3	X47B10C	BLACK
76	A1-TB1-2	5	A1-TB2-11	5	3	X47C10C	BLACK
77	A1-S1-1	5	A2-F2-1	.375s	3	x48A10C	BLACK
78	A1-S1-5	7	A2-J1-L	25s	3	X35A10N	BLACK
79	A1-TB1-1	7	A1-ground	7	3	x25B10A	BLACK
80	A1-TB1-1	5	A2-J1-L	9	2	x25C10A	BLACK
81*	A1-TB2-11	.375 strip	A1-L1	11	2	X49A12A	BLACK
82*	A2-F2-2	9	A1-L1	11	2	X50A12A	BLACK
83*	A2-J1-gnd	.375 strip	A2-CB1-A1	11	2	x51A12N	BLACK
84*	A2-F1-2	11	A2-CB1-A1	11	3	X52A12A	BLACK
85	A2-CB1-B1	11 5			3	X39A10A	BLACK
86	A2-CB1-B1	5			3	X39B10A	BLACK
87	A1-TB2-1O		A2-S2-12	4	3	X49B10A	BLACK
88	A1-TB2-11		A2-S2-13	4		X49C10A	BLACK
89	NOT USED	6	A2-CB1-A2	11			
90	NOT USED	6	A1-L2	11	1		
91	A4-CT1-A1	5	A1-L2	11	1	P53A16	BLACK
92	A4-CT2-B1	11	A1-L3	11	3	P54A16	BLACK
93	A1-TB2-12	11	A2-S3-17	4	3	X47D10C	BLACK
94	A2-CB1-B2	11	A2-S3-26	4	3	X36A10B	BLACK
95	A2-CB1-B2	10 4	A1-TB1-2	7	3	X36B10B	BLACK
96	A2-CB1-B3	5	A2-S2-14	4	1	X4OA10C	BLACK
97	A2-CB1-A3	6	A1-VR1-24	4	1	X48B16C	BLACK
98	A1-TB2-12	4			3	X47E16C	BLACK
99	A1-S1-12	Al-L3			1	X47A1OC	BLACK
100	A4-CT1-C1	A3-TB2-2			1	P55A16	BLACK
101	A1-TB2-6	A1-TB2-8				X20F16A	BLACK
	4	Al-S1-3					

* Used on Model MEP-016B (60 Hertz) harness only.

Table 4-8. Control Wiring Harness, Models MEP-016B and MEP-021B, Continued.

Key For Table 4-8

Find	Part Or	Qty	Nomenclature		
1	MS5086/2-16-9	A/R	Wire, elect, 600V aircraft white, #16 AWG		MIL-W-5086/2C
2	MS5086/2-12-9	A/R	Wire, elect, 600V aircraft white, #12 AWG		MIL-W-50862C
3	MS5086/2-10-9	A/R	Wire, elect, 600V aircraft white, #10 AWG		MIL-W-50862C
4	MS25036-107	101	Terminal lug, AwG-#6 stud	crimp, #16	MIL-T-7928/5B
5	MS17143-15	26	Terminal lug, AWG-#6 stud	crimp, #10/12	MIL-T-7928/5B
6	MS-25036-153	13	Terminal lug, - #8 stud	crimp, #16 AWG	MIL-T-7928/5B
7	MS-25036	6	Terminal lug, AWG - #8 stud	crimp, #10/12	MIL-T-7928/5B
8	MS-25036-108	1	Terminal lug, - #10 stud	crimp, #16 AWG	MIL-T-7928/5B
9	MS-25036-112	5	Terminal lug,	crimp, #10-12 AWG	MIL-T-7928/5B
10	MS-25036-154	15	- #10 stud		MIL-T-7928/5B
11	MS-25036-157	19	Terminal lug, crimp, #16 AWG .25 stud		MIL-T-7928/5B
12	MS3100R32-7P	1	Terminal lug, crimp, #10/12 AWG - .25 stud		
13	MS-3367-4	120	Connector		MIL-5-23190
14	SN60WPR2	A/R	Tie wrap		QQ-S-571
15	MS3368-1-9A	1	Solder, tin aly		MIL-S-23190
16	13213E4128	1	Tag, identification		

Table 4-8.Control Wiring Harness (Models MEP-016B and MEP-021B), Continued.
NOTES FOR TABLE 4-8

1.   and

   Position tie-down straps (Find No.13) at approximately 3 inch intervals at branch-offs.

2. For electrical wiring diagram (Models MEP-016B and MEP-021B) see Figure 4-63.

   For electrical schematic (Models MEP-016B and MEP-021B) see Figure 4-64.
3. Installed crimp style terminals shall meet the performance requirements of MIL-T-7928.

4. Leads not terminating in a crimp style terminal shall be stripped 0.50 inch and tinned in accordance with MIL-STD-454, require-merit No. 5.

5. Mark each wire with appropriate wire number and color.Permanency, legibility, and type of lettering shall be in accordance with MIL-STD-130. Marking shall be repeated at 15 inch maximum intervals.The final 15 inches at each end shall be marked at 3 inch maximum intervals.Lengths of less than 3 inches need not be marked.

6.
   Solder in accordance with MIL-STD-454, requirement No. 5, using solder, Find No. 14.

   Dimensions shown are for general routing and may be altered to facilitate installation.

   For interpretation of drawing, DOD-STD-100 applies.

7.
   Mark item 16 "30554-84-13154" in accordance with MIL-STD-130.Locate approximately as shown.

8. Connect battery charging voltage regulator leads only after the harness is installed in the control box.

   Wire reference No's 73, 81, 82, 83, and 84 are used on Model MEP-016B (60
9. Hertz) harness only.

10.

11.

*

4-92

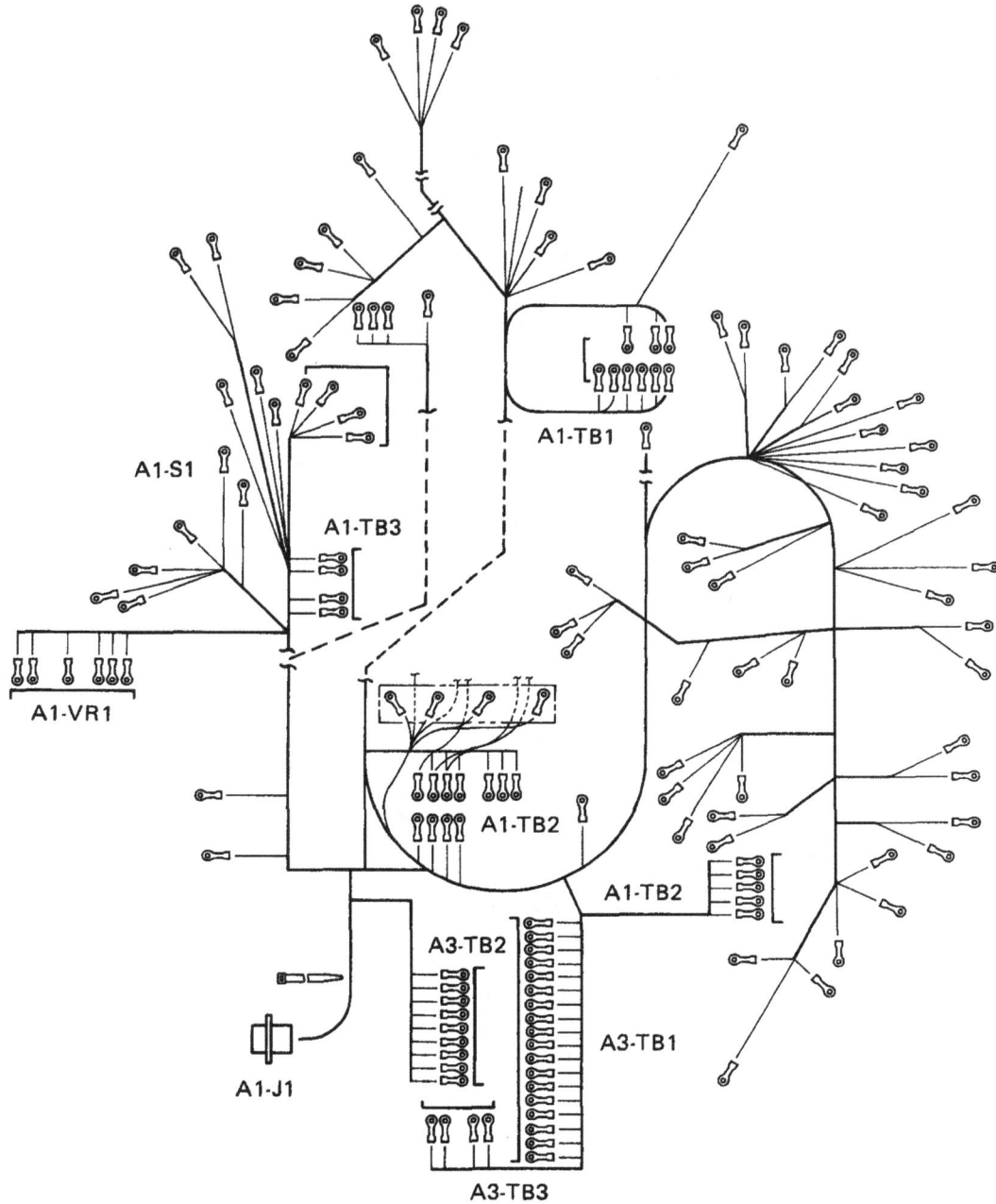

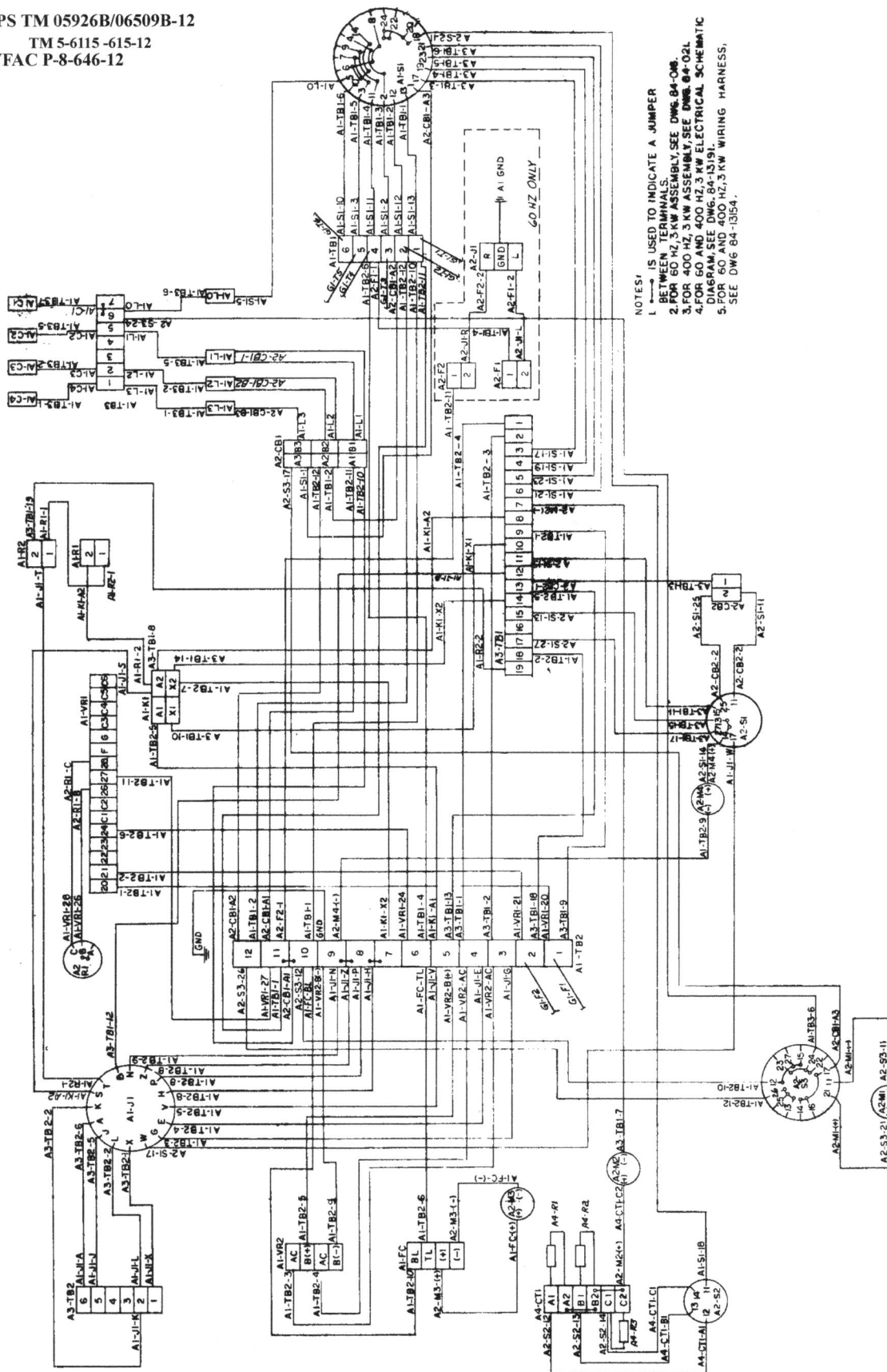

Figure 4-63.   **Wiring Diagram (Models   MEP-016B and MEP-021B).**

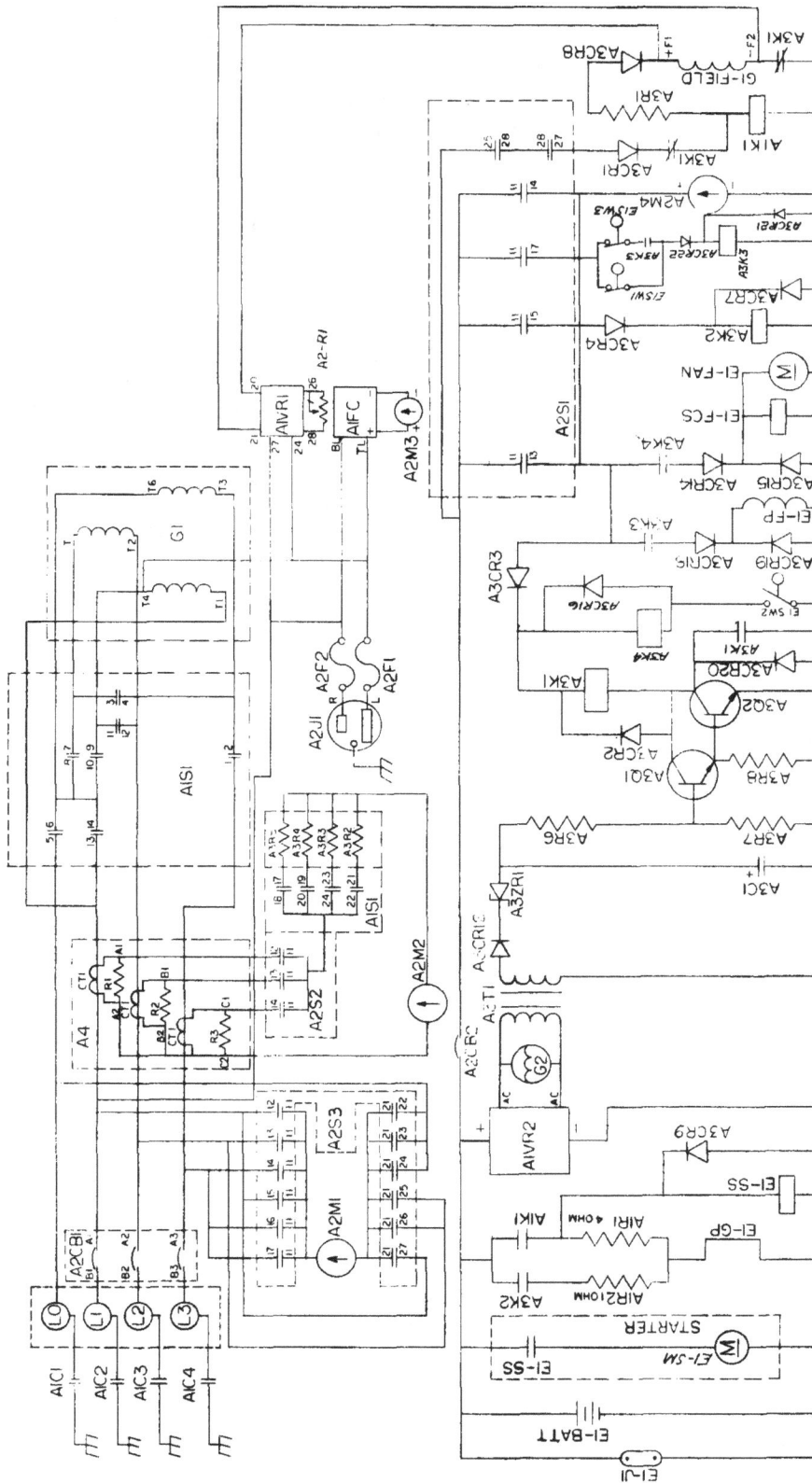

Figure 4-64.Schematic Diagram (Models MEP-016B and MEP-021B) (Sheet 1 of 2).

A1-CONTROL BOX ASSEMBLY

C1, C2, C3, C4 . . . . EMI CAPACITORS
K1 . . . . AUXILIARY START RELAY
L0, L1, L2, L3 . . . . OUTPUT TERMINALS
FC . . . . FREQUENCY CONVERTER
S1 . . . . VOLTAGE RECONNECTION SW
TB1, TB2, TB3 . . . . TERMINAL BOARDS
VR1 . . . . . . GEN SET VOLTAGE REGULATOR
VR2 . BATT. CHARGING VOLTAGE REGULATOR
R1 . . . . PRE-HEAT RESISTOR
R2 . . . . AFTER GLOW RESISTOR

A2-SUB-ASSEMBLY-FRONT PANEL

CB1 . . . . LOAD CIRCUIT BREAKER
CB2 . . . . DC CONTROL BREAKER
U1 . . . . AC RECEPTACLE
M1 . . . . AC VOLTMETER
M2 . . . . LOAD CURRENT METER
M3 . . . . FREQUENCY METER
M4 . . . . RUNNING TIME METER
R1 . . . . VOLTAGE ADJUST POT
S2 . . . . CURRENT SELECTOR SWITCH
S3 . . . . VOLTAGE SELECTOR SWITCH
S1 . . . . MASTER SWITCH
F1, F2 . . . . AC RECEPTACLE FUSE

A3-SUB-ASEMBLY-CIRCUIT BOARD

R2, R3, R4, R5 . . . . . VOLTMETER RESISTORS
K2 . . . . PRE-HEAT RELAY
T1 . . . . . START DISCONNECT TRANSFORMER
R6, R7, R8 . . . START DISCONNECT RESISTORS
ZR1 . . . . START DISCONNECT ZENER
C1 . . . . START DISCONNECT CAPACITOR
Q1, Q2 . . . . START DISCONNECT TRANSISTORS
K1 . . . . START DISCONNECT RELAY
K4 . . . . LOW FUEL SHUTDOWN RELAY
K3 . . . . FUEL LEVEL RELAY
CR1-23 . . . . DIODES
R1 . . . . FIELD FLASHING RESISTOR

A4-SUBASSEMBLY-CURRENT TRANSFORMER

CT1, CT2, CT3 . . . . CURRENT TRANSFORMERS
R1, R2, R3 . . . . BURDEN RESISTORS

E1-ENGINE ASSEMBLY

E1-BATT – BATTERY
E1-GP – GLOW PLUGS
E1-SS – STARTER SOLENOID
E1-FCS – FUEL CUTOFF SOLENOID
E1-FAN – NOISE KIT FAN
E1-FP – FUEL PUMP
E1-SW1 – FUEL LEVEL SWITCH
E1-SW2 – LOW FUEL SWITCH
E1-J1 – AUXILIARY BATTERY JACK

-GENERATOR ASSEMBLIES

G1-T1, T2, T3, T4, – STATOR FIELD WINDINGS
T5, T6
G1-F1, F2 – GENERATOR FIELD
G2 – BATTERY CHARGING ALTERNATOR

A2S3	VOLTAGE SELECTOR SWITCH
POSITION	TERMINALS CONNECTED
V1-∅	11 AND 12 AND 15, 21 AND 22
V2-∅	11 AND 13, 21 AND 23
V3-∅	11 AND 14, 21 AND 24
V1-2	11 AND 12 AND 15, 21 AND 25
V2-3	11 AND 16, 21 AND 26
V3-1	11 AND 17, 21 AND 27

A2S2	CURRENT SELECTOR SWITCH
POSITION	TERMINALS CONNECTED
I-1	11 AND 12
I-2	11 AND 13
I-3	11 AND 14

A1S1	VOLTAGE RECONNECTION SWITCH
POSITION	TERMINALS CONNECTED
120/208 3∅	1 AND 2, 5 AND 6, 7 AND 8, 9 AND 10, 17 AND 18
120 3∅	1 AND 2, 3 AND 4, 11 AND 12, 13 AND 14, 19 AND 20
240 3∅	3 AND 4, 9 AND 10, 21 AND 22
120 1∅	3 AND 4, 11 AND 12, 13 AND 14, 15 AND 16, 23 AND 24

A2S1	MASTER SWITCH
POSITION	TERMINALS CONNECTED
PRE-HEAT	11 AND 15
OFF	NONE
RUN/AUX. FUEL	11 AND 13, 14 AND 15 AND 16 AND 17
RUN	11 AND 13 AND 14
START	11 AND 13 AND 14 AND 15, 25 AND 27 AND 28

THE 400 Hz MODEL DOES NOT CONTAIN THE AC RECEPTACLE A2J1 OR THE FUSES A2F1 OR A2F2.

FRONT PANEL SWITCH A2S3 HAS JUMPERS INSTALLED:

– BETWEEN A2S3-16 AND A2S3-17
– BETWEEN A2S3-22 AND A2S3-23 AND A2S3-24
– BETWEEN A2S3-25 AND A2S3-26

FRONT PANEL SWITCH S2S1 HAS JUMPER INSTALLED BETWEEN A2S1-13 AND A2S1-13.

Figure 4-64. Schematic Diagram (Models MEP-016B and MEP-021B) (Sheet 2 of 2).

Table 4-9. Control Wiring Harness (Model MEP-026B) .

NOTE

Notes concerning repair and installation of this harness, and the key to the "terminal find numbers" in the table are found at the end of the table.

Wire Ref No.	From	Term Find No.	To	Term Find No. No.	Wire Find No.	Wire Stamp No.	Wire stamp Color
1	A1-J1-A	.25 strip	A1-R3-1	4 4	1	P1A16	BLACK
2	A1-J1-J	.25 strip	A1-TB1-4	4	1 1	P2B16	BLACK
3	A1-J1-N	.25 strip	A1-TB1-3	4 4	1	P3B16N	BLACK
4	A1-J1-L	.25 strip	A1-TB1-5	4 4	1	P4A16	BLACK
5	A1-J1-Z	.25 strip	A1-TB1-1O	10 4	1	P5B16N	BLACK
6	A1-J1-X	.25 strip	A4-TB1-14	4	1	P6A16	BLACK
7	A1-J1-W	.25 strip	A1-TB1-10	5	2	P7A16	BLACK
8	A1-J1-T	.25 strip	A1-TB1-5	5	1	H8A12	BLACK
9	A1-J1-E	.25 strip	A4-TB1-8	4	1	X9B16V	BLACK
10	A1-J1-G	.25 strip	A1-R3-2	4	2 2	X10B16V	BLACK
11	A1-J1-V	.25 strip	A1-TB1-mtg	4	1	P11B12	RED
12	A1-J1-P	4	A2-S1-14	4	1 1	P29B12N	BLACK
13	A1-K1-X2	4 7	A2-M2-(+)	10 4	1	P12A16	BLACK
14	A1-K1-X1	4	A2-M2-(-)	4	2	P13A16	BLACK
15	A1-K1-A1	.25 strip	A2-M2-(+)	9 9	1	P11E16	RED
16	A1-K1-A2	4	A2-M3-(+	9	1	K14A16	BLACK
17	A1-J1-S	4 6	A2-M3-(-)	9	1	K14C12	BLACK
18	A1-TB1-8	6	A3-TB1-3	9	1	P29A16N	BLACK
19	A2-M4-(+)	9	A3-TB1-2	9	1	P15A16	BLACK
20	A1-R1-S+	4	A2-S1-25	9	1	D16A16	BLACK
21	A1-R1-S-	4 4		4	1	D17A16	BLACK
22	A2-M1-(-)	4			1	D16B16	BLACK
23	A1-FC-(+)	6			1 1	D18A16	BLACK
24	A1-FC-(-)		A1-FC-BL			D19A16	BLACK
25	A1-FC-TL		A3-TB1-4	4		X20A16C	BLACK
26	A1-FC-BL	4	A1-CB1-A-	9		X21B16B	BLACK
27	A2-CB1-2	4 4	A2-R1-B	9	1 1	P22A16	BLACK
28	NOT USED	4	A2-S1-11	.375s 4	1		
29	NOT USED	6 4	A1-TB1-2	4 4	1		
30	A1-VR1-22	4	A1-TB1-1	.375s 9	1	X21C16B	BLACK
31	A1-VR1-27	9	A2-R1 -A	9	1	X25B16A	BLACK
32	A1-VR1-F	9	A2-M1-(+)		1 1	P24A16	BLACK
33	A1-VR1-26	.25 strip	A1-R3-1		1	P26A16	BLACK
34	A2-CB1-2	A4-TB2-6			1	P22B16	BLACK
35	A1-VR1-21	A4-TB2-5				V27A16	BLACK
36	A1-VR1-20	A1-TB1-11				V28B16	BLACK
37	A2-M1-(+)	A4-TB2-2				P23B16	BLACK
38	A1-CB1-A+	A1-TB1-10				P23A16	BLACK
39	A1-R2-2	A4-TB2-1				H8C16	BLACK
	.25 strip	A2-S1-17					

Table 4-9.Control Wiring Harness (Model MEP-026B), Continued.

Wire Ref No.	From	Term Find No.	To	Term Find No.	Wire Find No.	Wire Stamp No.	Wire Stamp Color
40	A1-TB1-11	4	A2-TB1-3	4	1	P3C16N	BLACK
41	A4-TB1-11	4	A2-S1-15	4	1	P30A16	BLACK
42	A1-TB1-4	4	A4-TB1-1	4	1	X9E16V	BLACK
43	A1-TB1-4	4	A2-TB1-1	4	1	X9C16V	BLACK
44	A1-TB1-3	4	A2-TB1-2	4	1	X10C16V	BLACK
45	A1-TB1-3	4	A4-TB1-2	4	1	X10E16V	BLACK
46	A1-TB1-1	4	A4-TB1-9	4	1	V28C16	BLACK
47	A1-K1-X2	4	A1-TB1-9	4	1	P12B16N	BLACK
48	A1-TB1-9	4	A2-M4-(-)	4	1	P12C16N	BLACK
49	A1-K1-A1	7	A2-CB1-1	6	1	P11D16	RED
50	A1-TB1-2	4	A4-TB1-18	4	1	V27C16	BLACK
51	A1-TB1-5	4	A4-TB1-13	4	1	P11F16	RED
52	A1-K1-A2	8	A1-R3-2	10	2	K14B12	BLACK
53	A1-R2-1	.25 strip	A4-TB1-19	4	1	H31A16	BLACK
54	A4-TB1-17	4	A2-S1-27	4	1	P32A16	BLACK
55	A4-TB1-15	4	A2-S1-13	4	1	P33A16	BLACK
56	NOT USED						
57	NOT USED						
58	NOT USED						
59	NOT USED						
60	A1-J1-K	.25 strip	A4-TB2-2	4	1	P4C16	BLACK
61	A1-J1-H	.25 strip	A1-TB1-10	4	1	P5D16N	BLACK
62	A1-J1-C	.25 strip	A1-R3-1	9	1	H8D16	BLACK

Key For Table 4-9

Find No.	Part Or Identifying No.	Qty Reqd	Nomenclature Or Description	Specification
1	MS5086/2-16-9	A/R	Wire, elect, 600V aircraft white, #16 AWG	MIL-W-5086/2C
2	MS5086/2-12-9	A/R	Wire, elect, 600V aircraft white, #12 AWG	MIL-W-5086/2C
3	MS25036-107	4	Terminal lug, crimp, #16 AWG-#6 stud	MIL-T-7928/5B
4	MS25036-111	2	Terminal lug, crimp #12 AWG-#6 stud	MIL-T-7928/5B
5	MS25036-153	5	Terminal lug, crimp #16 AWG-#8 stud	MIL-T-7928/5B

Key For Table 4-9, Continued

Find No.	Part Or Identifying No.	Qty Reqd	Nomenclature Or Description	Specification
6	MS25036-108	2	Terminal lug, crimp #16 AWG-#10 stud	MIL-T-7928/5B
7	MS25036-112	1	Terminal lug, crimp #12 AWG-#10 stud	MIL-T-7928/5B
8	MS25036-154	17	Terminal lug, crimp #16 AWG-#1/4" stud	MIL-T-7928/5B
9	MS25036-157	3	Terminal lug, crimp #12 AWG-1/4" stud	MIL-T-7928/5B
10	MS3100R32-7P	1	Connector, recp, elect	
11	MS3367-4	125	Tie wrap	MIL-S-23190
12	SN60WRP2	A/R	Solder, tin aly, wire, resin core 2.2 percent flux	QQ-S-571
13	MS3368-7-9A	1	Tag, identification	MIL-S-23190

NOTES FOR TABLE 4-9

1. Position tie-down straps (Find No. 14) at approximately 3 at branch-offs.   inch intervals and

2. For electrical wiring diagram (Model MEP-026B) see Figure   4-65.

   For electrical schematic (Model MEP-026B) see Figure 4-66.

3.

4. Installed crimp style terminals shall meet the performance requirements of MIL-T-7928.

5. Leads not terminating in a crimp style terminal shall be stripped .50 inch and tinned in accordance with MIL-STD-454, requirement No. 5.
   Mark each wire with appropriate wire number and color.   Permanency, legibility, and

6. type of lettering shall be in accordance with MIL-STD-130. Marking shall be repeated at 15 inch maximum intervals. The final 15 inches at each end shall be marked at 3 inch maximum intervals. than 3 inches need not be marked.

   Lengths of less

7. Solder in accordance with MIL-STD-454, requirement No. 5,   using solder, Find No. 15.

NOTES FOR TABLE 4-9, Continued.

8.   Dimensions shown are for   general routing and may be altered to facilitate installation.

9.   For interpretation of drawing, DOD-STD-100 applies.

10.   Mark item 16 "30554-84-13156" in accordance with MIL-STD-130.   Locate approximately as shown.

11.   Connect battery charging voltage regulator leads only after the harness is installed in the control box.

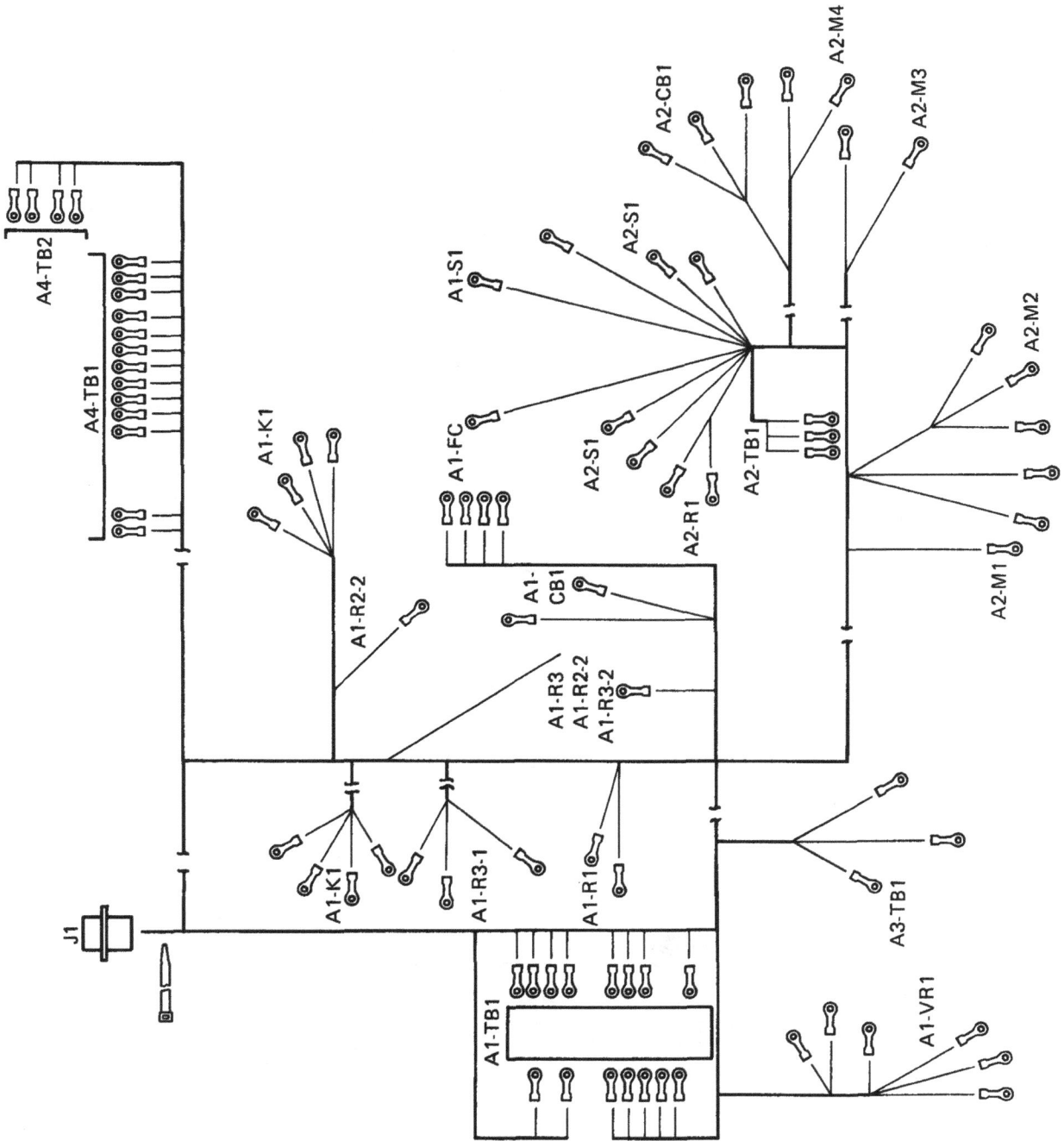

Figure 4-65.Control Wiring Harness (Model MEP-026B).

MARINE CORPS TM 05926B/06509B-12
ARMY   **TM 5-6115-615-12**
NAVY   **NAVFAC P-8-646-12**

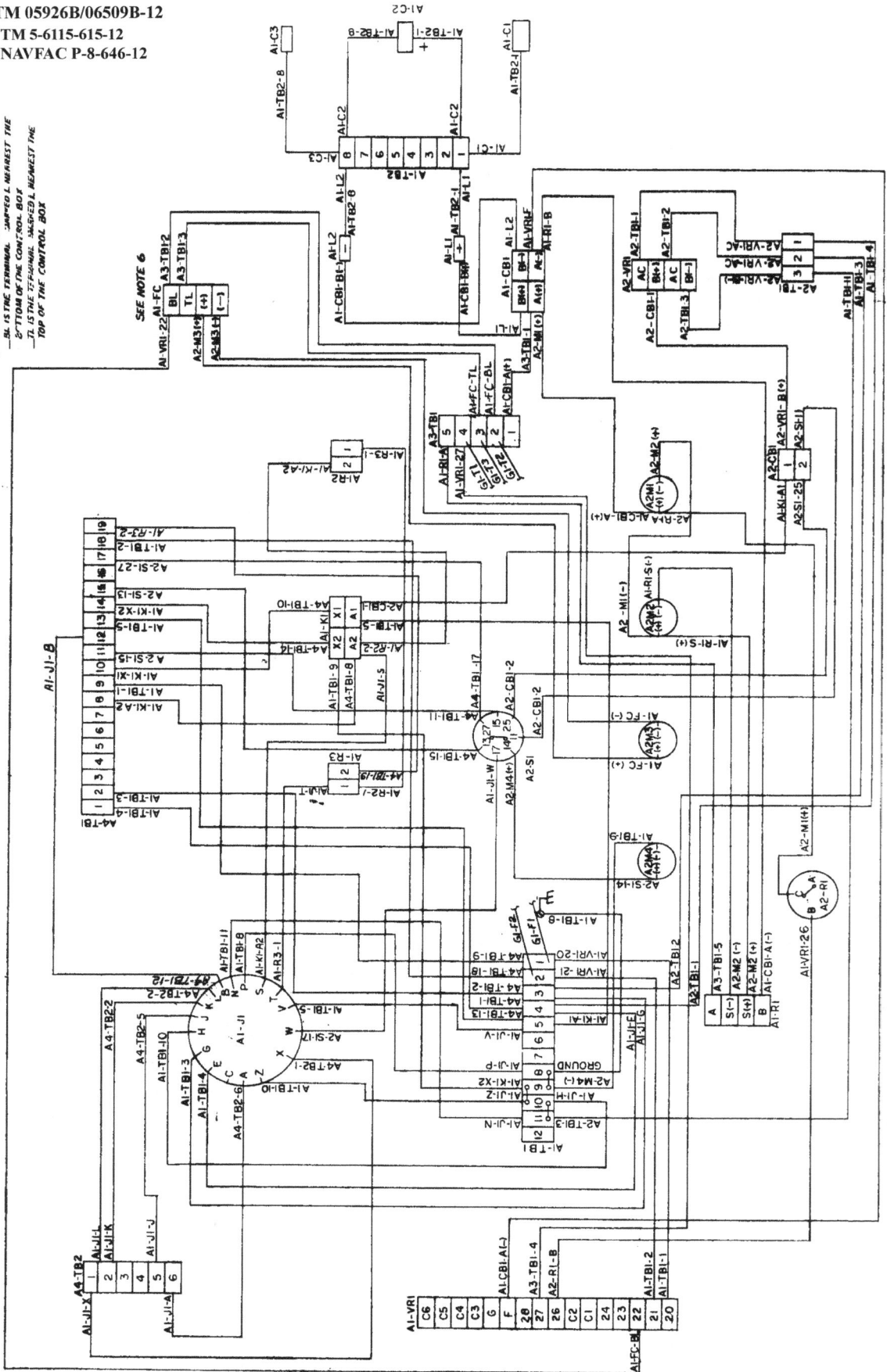

Figure 4-66. Wiring Diagram (Model MEP-026B).

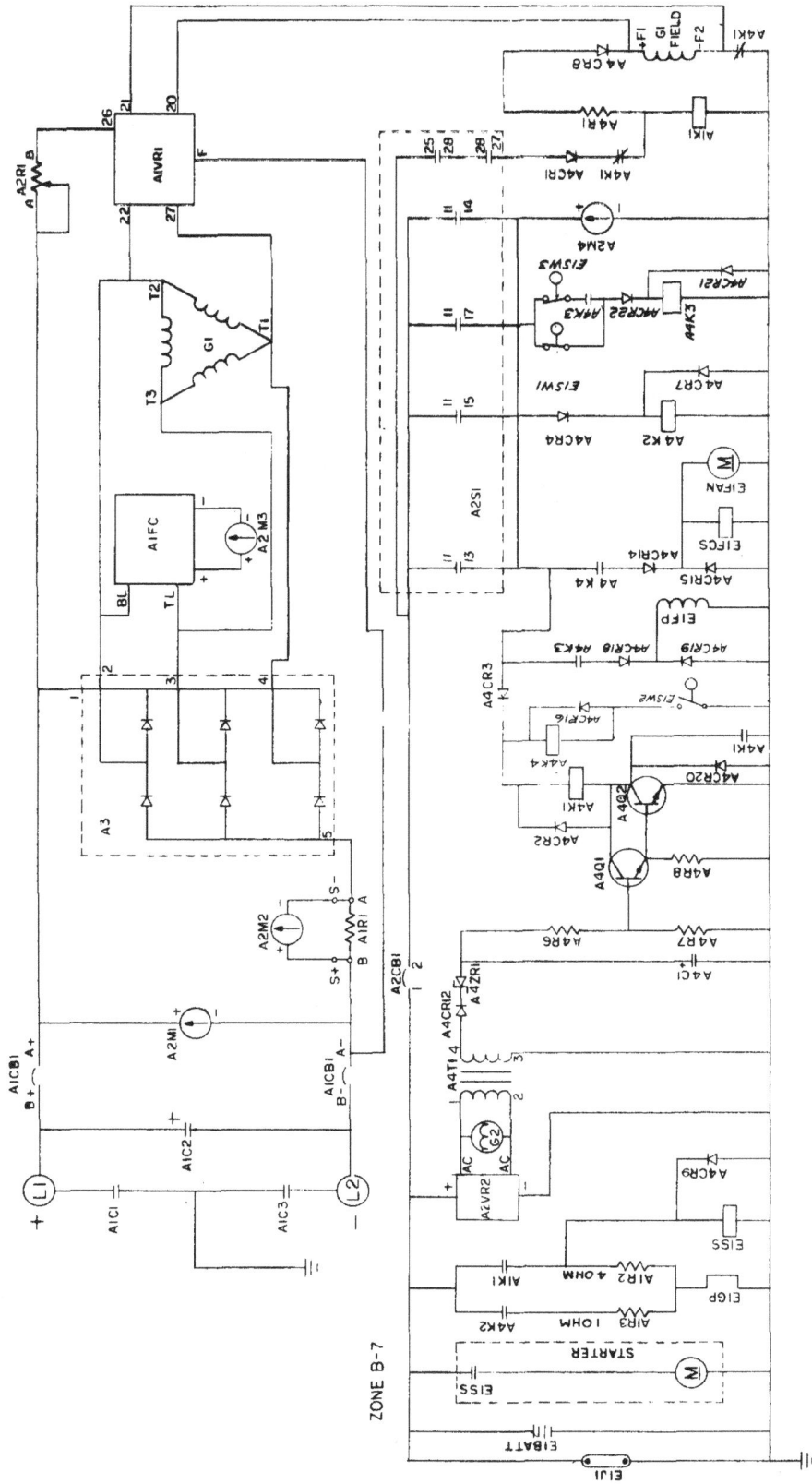

Figure 4-67.   Schematic Diagram(Model MEP-026B) (Sheet 1 of 2).

A1 . . . . CONTROL BOX ASSY
A1, C1, C2, C3 . . . . EMI CAPACITOR
K1 . . . . AUXILIARY START RELAY
L1, L2 . . . . OUTPUT TERMINALS
FC . . . . FREQUENCY CONVERTER
VR1 . GENERATOR SET VOLTAGE REGULATOR
CB1 . . . . LOAD CIRCUIT BREAKER
R1 . . . . LOAD CURRENT SHUNT
R2 . . . . PRE-HEAT RESISTOR
R3 . . . . AFTER GLOW RESISTOR

A2-SUB-ASSEMBLY-FRONT PANEL

CB1 . . . . DC CIRCUIT BREAKER
M1 . . . . DC VOLTMETER
M2 . . . . DC LOAD CURRENT METER
M3 . . . . RPM METER
R1 . . . . VOLTAGE ADJUST POT
S1 . . . . MASTER SWITCH
VR2 . . . BATTERY CHARGING VOLTAGE REG.
TB3 . . . . TERMINAL BOARD
M4 . . . . RUNNING TIME METER

A3-SUB-ASSEMBLY-RECTIFIER STACK

A4-SUB-ASSEMBLY-CIRCUIT BOARD
R2, R3, R4, R5 . . . . . . VOLTMETER RESISTORS
K2 . . . . PRE-HEAT RELAY
T1 . . . . . START DISCONNECT TRANSFORMER
R6, R7, R8 . . . START DISCONNECT RESISTORS
ZR1 . . . . START DISCONNECT ZENER
C1 . . . . START DISCONNECT CAPACITOR
Q1, Q2 . . . . START DISCONNECT TRANSISTORS
K1 . . . . START DISCONNECT RELAY
K4 . . . . LOW FUEL SHUTDOWN RELAY
K3 . . . . FUEL LEVEL RELAY
CR1-23 . . . . DIODES
R1 . . . . FIELD FLASHING RESISTORS

E1 . . . . ENGINE ASSEMBLY
E1BATT — BATTERY
E1GP — GLOW PLUG
E1SS — STARTER SOLENOID
E1FCS — FUEL CUTOFF SOLENOID
E1FAN — NOISE KIT FAN
E1FP — AUXILIARY FUEL PUMP
E1SW1 — FUEL LEVEL SWITCH
E1SW2 — LOW FUEL SWITCH
E1J1 — AUXILIARY BATTERY JACK

GENERATOR ASSEMBLIES
G1-T1, T2, T3 . . . . . STATOR FIELD WINDINGS
G1-F1, F2 . . . . FIELD WINDING
G2 — BATTERY CHARGING ALTERNATOR

A2S1	MASTER SWITCH
POSITION	TERMINAL CONNECTIONS
PRE-HEAT	11 AND 15
OFF	NONE
RUN/AUX. FUEL	11 AND 13 AND 14 AND 15 AND 17
RUN	11 AND 13 AND 14
START	11 AND 13 AND 15, 25 AND 27 AND 28

Figure 4-67.Schematic Diagram (Model MEP-026B) (Sheet 2 of 2).

Table 4-10. Engine Wiring Harness (All Models).

NOTE

Notes concerning repair and installation of this harness, and the key to the "terminal find numbers" are located at the end of the table.

Wire Ref	Term Find		Term Find	Wire Find	Wire Stamp	Wire stamp
1	A1-P1-A	.25 strip	E1-P6	3	1	P1B16
2	A1-P1-Z	.25 strip	E1-P4-8	12,13 1		P5F16N
3	A1-P1-L	.25 strip	E1-P4-A	12,13 1		P4B16
4	A1-P1-N	.25 strip	E1-P5-A	.25s 1		P3A16N
5	A1-P1-X	.25 strip	E1-P5-B	.25s 1		P6B16
6	A1-P1-J	.25 strip	E1-P5-C	.25s 1		P2A16
7	A1-P1-W	.25 strip	E1-P5-D	.25s 1		P7B16
8	A1-P1-H	.25 strip	E1-P8	4	1	P5C16N
9	A1-P1-K	.25 strip	E1-P7	3	1	P4D16
10	A1-P1-T	.25 strip	E1-P3	4	1	H8B16
11	A1-P1-G	.25 strip	E1-P1	3	1	X10A16V
12	A1-P1-E	.25 strip	E1-P2	3	1	X9A16V
13	A1-P1-P	.25 strip	E1-GND	7	2	P29C12N
14	A1-P1-S	.25 strip	E1-SS	9	2	K14D12
15	A1-P1-V	.25 strip	E1-BATT	8	2	P11A12

Key For Table 4-10

Find No.	Part Or Identifying No.	Qty Reqd	Nomenclature Or Description	Specification
1	MS5086/2-16-9	A/R	Wire, elect, 600V aircraft white, #16 AWG	
2	MS5086/2-12-9	A/R	Wire, elect, 600V aircraft white, #12 AWG	
3	MS27144-2	4	Connector, electrical	
4	MS27142-3	2	Connector, electrical	
5	MS3106R145-25	1	Connector, electrical	
6	MS3106R32-75	1	Connector, electrical	
7	MS25036-158	1	Terminal, lug, crimp style cop, ins, cli, 12-10 AWG No.1/2 stud size, blue	

Table 4-10.Engine Wiring Harness (All Models), Continued.
Key For Table 4-10, Continued

Find	Part Or	Qty	Nomenclature

No. Specification	Identifying No.	Reqd	Or Description
8	MS25036-1 13	1	Terminal, lug, crimp style cop, ins, cli, 12-10 AWG No. 5/16 stud size, yellow
9	MS25036-112	1	Terminal, lug, crimp style cop, ins, cli, 12-10 AWG No. 10 stud size, yellow
10	MS3367-4-9	50	Strap, tie-down, adj, type I class I, 5/8 max. dia.
11	SN60WRP2	A/R	Solder, tin sly, wire, resin core,2.2 percent flux
12	84-13120	1	Body, connector
13	84-13121	2	Terminal, lock receptacle
14	MS3368-1-9A	1	Tag, identification
15	8724494	1	Connector, shell, electrical

NOTES FOR TABLE 4-10

1. Position tie-down at branch-offs. straps (Find No. 10) at approximately 3 inch intervals and

2. For control box wiring diagrams, see Figures 4-62 and 4-65.

3. For control box schematic diagrams, see Figures 4-63 and 4-66.

4. Installed crimp MIL-T-7928. style terminals shall meet the performance requirements of

5. P6 = Fuel Pump
P4 = Fuel Cut-Off Solenoid
P5 = Fuel Tank Jack
P8 = Extra
P7 = Extra
P3 = Glow Plug
P1 = Battery Charge
P2 = Battery Charge

Table 4-10. Control Wiring Harness (All Models), Continued.
NOTES FOR TABLE 4-10, Continued

5.
Leads not terminating in a crimp style terminal shall be stripped .50 inch and tinned in accordance with MIL-

6.
STD-454, requirement No. 5.
Mark each wire with appropriate wire number and color.Permanency, legibility, and   with MIL-STD-130. The final
type of lettering shall be in accordance Marking shall be repeated at 15 inch maximum   15 inches Lengths of less
intervals. at each end shall be marked at 3 inch maximum intervals. than 3 inches need

7.
not be marked.
Solder in accordance with MIL-STD-454, requirement No. 5, using solder, Find No. 11.

8.
Dimensions shown are for general routing and may be altered to facilitate installation.

For interpretation of drawing, DOD-STD-1OO applies.

9.
Mark item 16 "30554-84-13118" in accordance with MIL-STD-130. Locate approximately as shown.

10.
Upon completion of assembly, install Find No. 16 into Find No. 3.

11.
Upon completion of assembly, install Find No. 18 into Find No. 4.

12.

E1-BATT

E1-SS

E1-GND

E1-P2

E1-P1

E1-P6

E1-P5

E1-P4

E1-P3

E1-P8 NOTE 12

E1-P7 NOTE 13

A1-P1

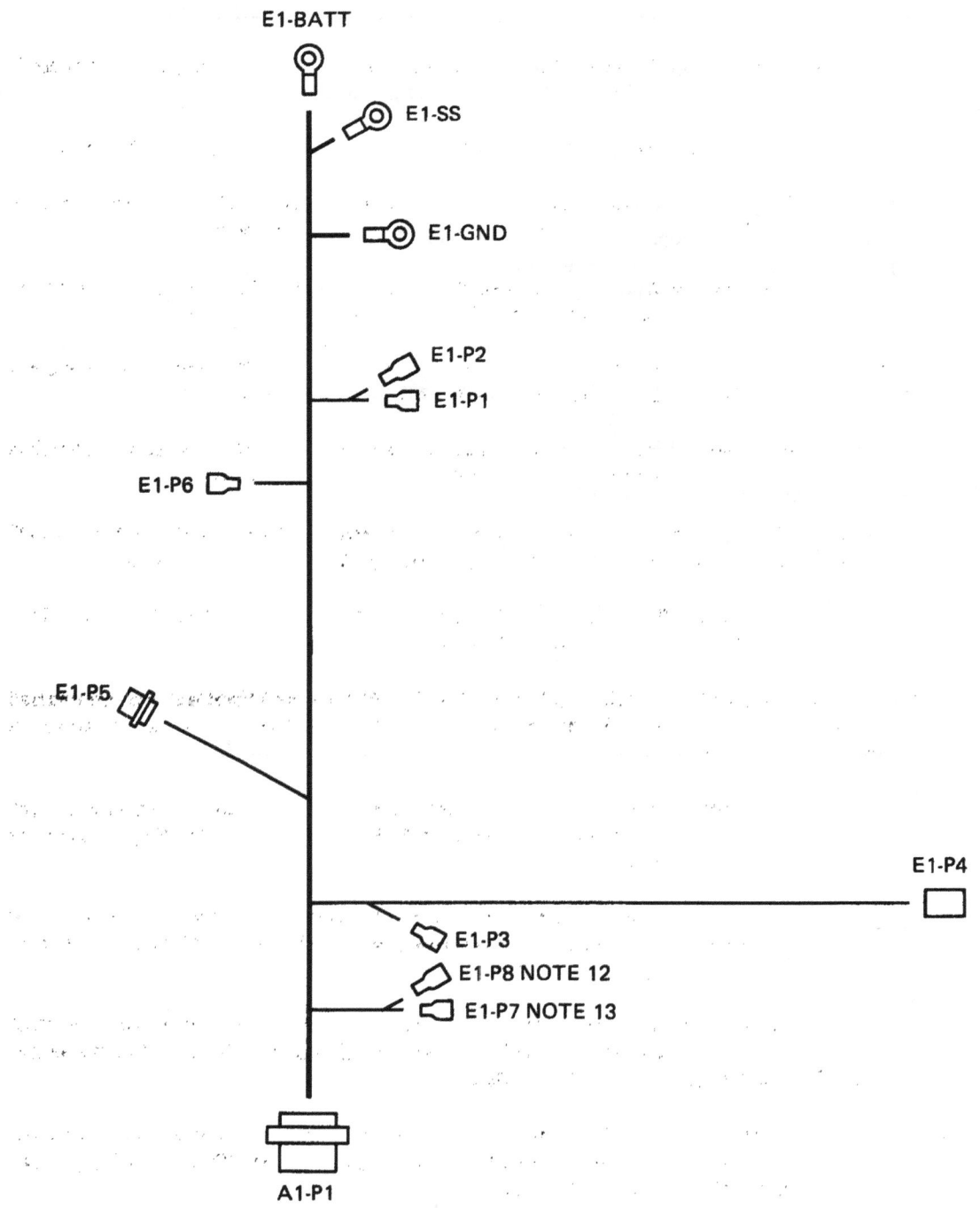

Figure 4-68. Engine Wiring Harness (All Models).

4-73.   ENGINE CONTROL CIRCUIT BOARD.

a.   Testing.

(1) Locate terminal board TB1 and TB2 on the engine control circuit board
    assembly.

(2) Check the ammeter scaling resistors R2, R3, R4, and R5 as follows:

   (a) Connect an ohmmeter across terminals #6 and #7 of TB1 to test resistor
       R2.   R2 should have a resistance of 7,604 - 7,756 ohms.

   (b) Connect an ohmmeter across terminals #5 and #7 of TB1 to test resistor
       R3.   R3 should have a resistance of 17,346 - 17,594 ohms.

   (c) connect an ohmmeter across terminals #4 and #7 of TB1 to test resistor R4.
       R4.   R4 should have a resistance of 9,000 = 9,180 ohms.

   (d) Connect an ohmmeter across terminals #3 and #7 of TB1 to test resistor
       R5.   R5 should have a resistance of 4,277 - 4,363 ohms.

(3)   Check resistor R1 the   by connecting an ohmmeter across terminal #10 of TB1 and Resistance of R1 should
      anode of CR8.   be 24.4 - 25.4 ohms.

(4)   Check diodes CR1,   CR2, CR3, CR4, CR7, CR8, CR9, CR12, CR14, CR15, CR16,

(a)   Check CR9 by connecting the red (+) lead of the of TB1 and the black (-)voltmeter to terminal #14 TB1.
      lead to terminal #8 of CR9 should be 0.5 - 1.1 VDC.

      Check CR8 by connecting the red (+) lead of the of TB1 and the black (-)
(b)   lead to terminal #9 of CR8 should be 0.5 - 1.1 VDC.   Voltage drop across

      Check CR1 by connecting the red (+) lead of the of TB1 and the black (-)
      lead to terminal #10 of across CR1 should be 0.5 - 1.1 VDC.
(c)   voltmeter to terminal #10 TB1.
      Check CR15 by connecting the red (+) lead of the #14 of TB1 and the
      black (-) lead to terminal #2 across CR15 should be 0.5 - 1.1 VDC.

(d)   Check CR19 by connecting the red (+) lead of the #14 of TB1 and the   Voltage drop across
      black (-) lead to terminal #6 across CR19 should be 0.5 - 1.1 VDC.   voltmeter to terminal #17 TB1.
                                                                              Voltage drop
      Check CR18 by connecting the red (+) lead of the   voltmeter to terminal of TB2.
(e)   Voltage drop

                                                                            voltmeter to terminal of TB2.
(f)   Voltage drop
      CR18 and the black (-) lead to terminal #6 of TB2. Voltage drop across CR18 should be 0.5 - 1.1 VDC.

      Check CR17 by connecting the red (+) lead of the voltmeter to terminal #3 of TB2 and the black (-) lead to
(g)   terminal #4 of TB2. Voltage drop across CR17 should be 0.5 - 1.1 VDC.

Check CR12, CR20, CR3, CR4, CR7, CR2, CR14, CR21, CR22 and CR16 by connecting the red (+) lead of the voltmeter to the anode and the black

(-) lead to the cathode of the diode. The voltage drop across the diodes should be 0.5 - 1.1 VDC.

(5) Check continuity of normally open contacts A1 and A2 of relay K2.

Continuity of contacts is checked with an ohmmeter across terminal #13 of TB1 and terminal #19 of TB1. With relay K2 de-energized, discontinuity (resistance greater than 100,000 ohms) should be indicated across contacts A1 and A2.

(a)   Apply 18 VDC to relay K2. Positive voltage lead is connected to terminal #11 of TB1 and negative (ground) lead is connected to terminal #14 of TB1.   Ohmmeter should now indicate continuity (resistance less than 5 ohms) across contacts A1 and A2.

(b)
Check continuity of normally open contacts A1 and A2 of relay K4.
Continuity of contacts is checked with an ohmmeter red (+) lead to terminal #15 of TB1 and the black (-) lead to terminal #2 of TB2.
With relay K4 de-energized, discontinuity (resistance greater than 100,000 ohms) should be indicated across
(6)   contacts A1 and A2.

Apply 18 VDC to relay K4.   Positive voltage lead is connected to terminal #15 of TB1 and negative

(a)   (ground) lead is connected to terminal #1 of TB2. Ohmmeter should now indicate continuity (resistance less than 5 ohms) across contacts A1 and A2.

(b)

(7) Check the start disconnect circuit.

(a)   Connect an ohmmeter across terminal #18 of TB1 and terminal #14 of TB1. Ohmmeter should indicate continuity (resistance less than 5 ohms) across the terminals.

With ohmmeter still connected, apply 18 VDC to the circuit.   Positive voltage
(b)   lead should be connected to terminal #15 of TB1 and negative
(ground) lead is connected to terminal #14 of TB1. Ohmmeter should still indicate continuity (resistance less than 5 ohms).

With the ohmmeter and 18 VDC supply still connected, apply an AC ramp voltage to terminal #1 of TB1 and
(c)   terminal #2 of TB1.   The ramp voltage applied
should be 10 - 20 Vrms at 540 Hz.

Slowly increase the ramp voltage from 10 Vrms to 20 Vrms while observing the ohmmeter.
(d)
The ohmmeter should change from a reading of continuity (resistance less than 5 ohms) to a reading of discontinuity (resistance greater than 100,000 ohms) when the ramp voltage is between 14.8 and 17 Vrms.

(e)
Remove the ramp voltage from the circuit. The ohmmeter should still indicate discontinuity (resistance greater than 100,000 ohms).

(f)

(g) Remove the 18 VDC supply.   The ohmmeter should now indicate continuity
    (resistance less than 5 ohms).

b.   Removal.

    (1) Tag and disconnect wiring from engine circuit board (1, Figure 4-69) for
        MEP-016B and MEP-021B (60/400 Hz) sets or (1, Figure 4-70) for MEP-026B (28

    (2) Remove screws   (2) and remove engine circuit   board.

c.   Installation.

    (1) Secure engine   circuit board (1) with screws   (2).

    (2) Using tags as   lead identification, connect wiring to circuit board
        terminals.

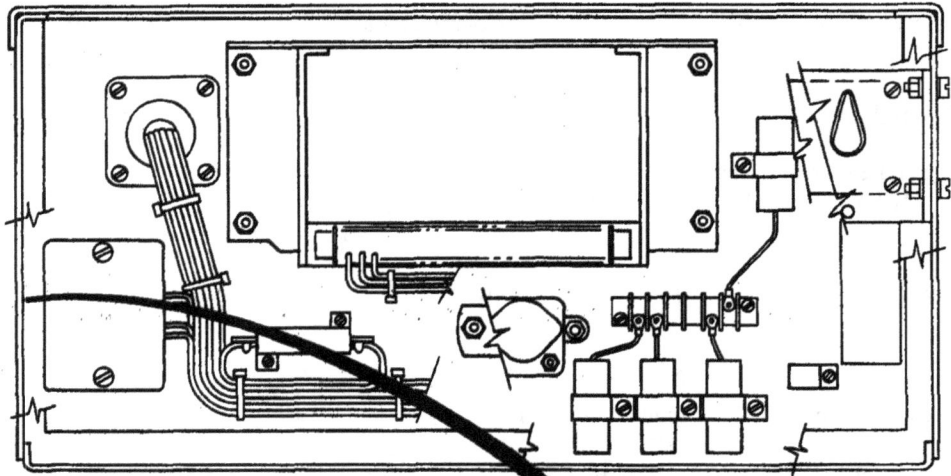

1. ENGINE CIRCUIT BOARD
2. SCREW

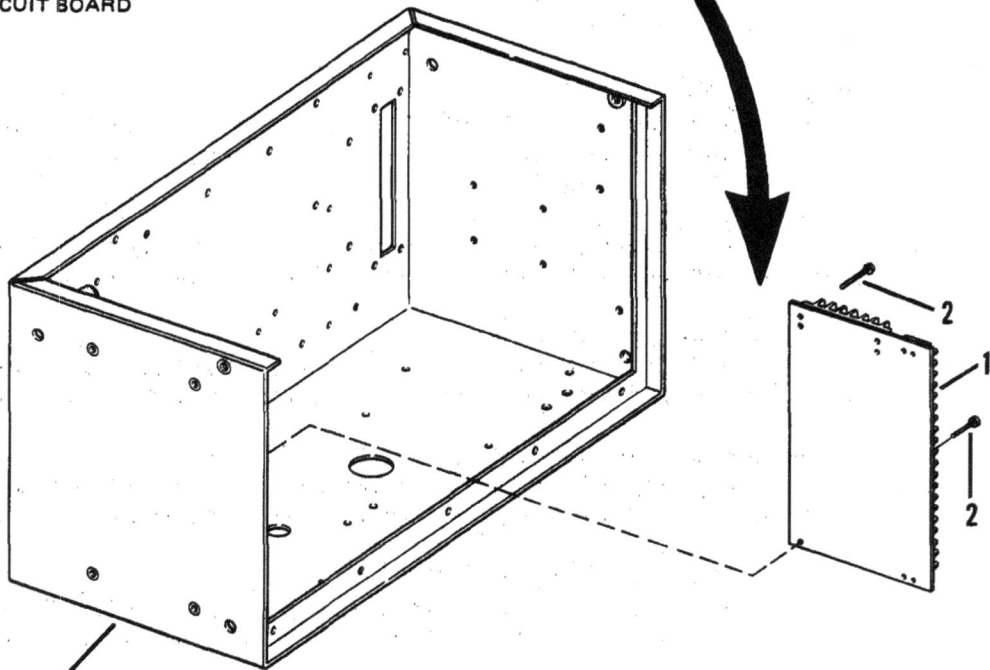

CONTROL BOX CHASSIS

Figure 4-69. Engine Circuit Board (MEP-O16B and MEP-021B).

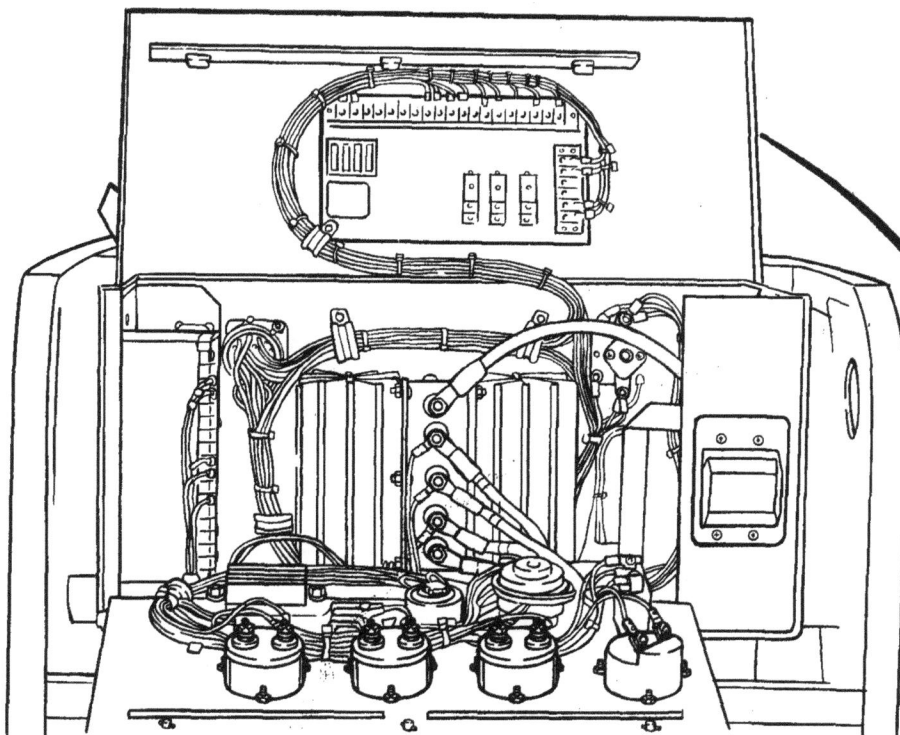

1. ENGINE CIRCUIT BOARD
2. SCREW

CONTROL BOX COVER

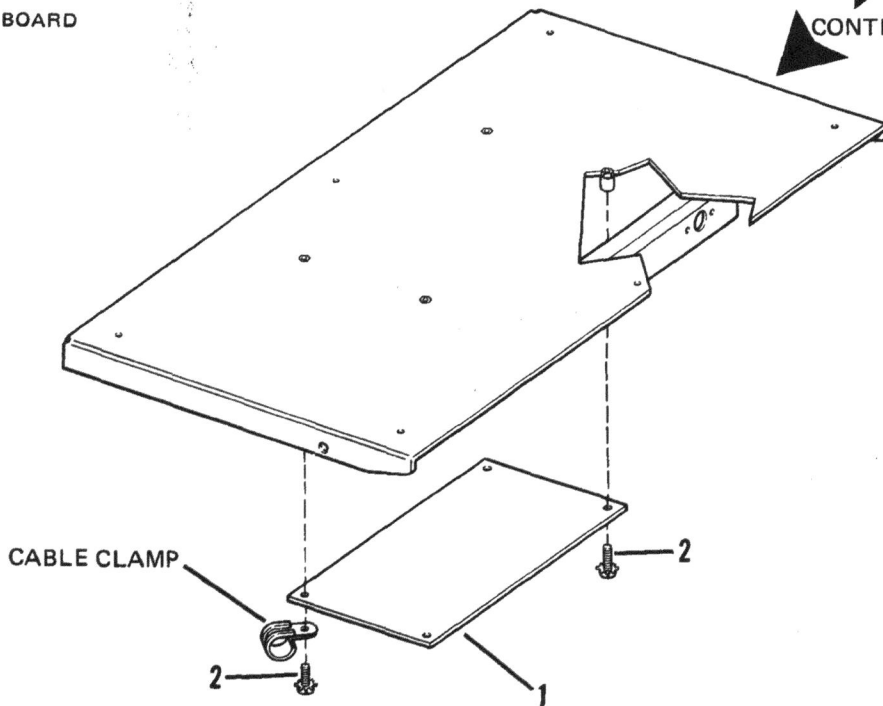

CABLE CLAMP

2

2

1

Figure 4-70.Engine Circuit Board (Model MEP-026B).

CHAPTER 5

(A) GENERATOR SET ACOUSTIC SUPPRESSION KIT

Section I. GENERAL.

5-1.   INFORMATION.   This chapter contains information on the Acoustic Suppression Kit (ASK) which is part of the modified 3 KW generator set Model MEP-016B only.   The ASK is intended to suppress the high noise level inherent in a diesel-driven generator. Any data contained in previous chapters 1 thru 4 of this manual which will undergo a change due to the addition of this kit will be shown in the tabulated data.

Section II. DESCRIPTION AND DATA.

5-2.   DESCRIPTION.   The ASK consists of six individual sound-insulated panels (Figure 5-1) labeled as panels 1 thru 6, two panel-mounted fans, and fan wiring harness.   The panels are made of lightweight aluminum bonded on one side with a fiberglass insulation which is coated to protect it against fuel spills and absorption. The insulating material is enclosed by an aluminum mesh to protect it against puncturing and to provide rigidity to the sound panels.   This ASK is only used on the MEP-016B generator set and is mounted to the frame of the generator.   The ASK and the modified generator, together, comprise Model MEP-701A.

5-3.   TABULATED DATA.

5-3.1.Identification and Instruction Plates.   All identification and instruc-tion plates are ex-plained in Table 5-1 with illustrations shown in Figures 5-2 and 5-3.

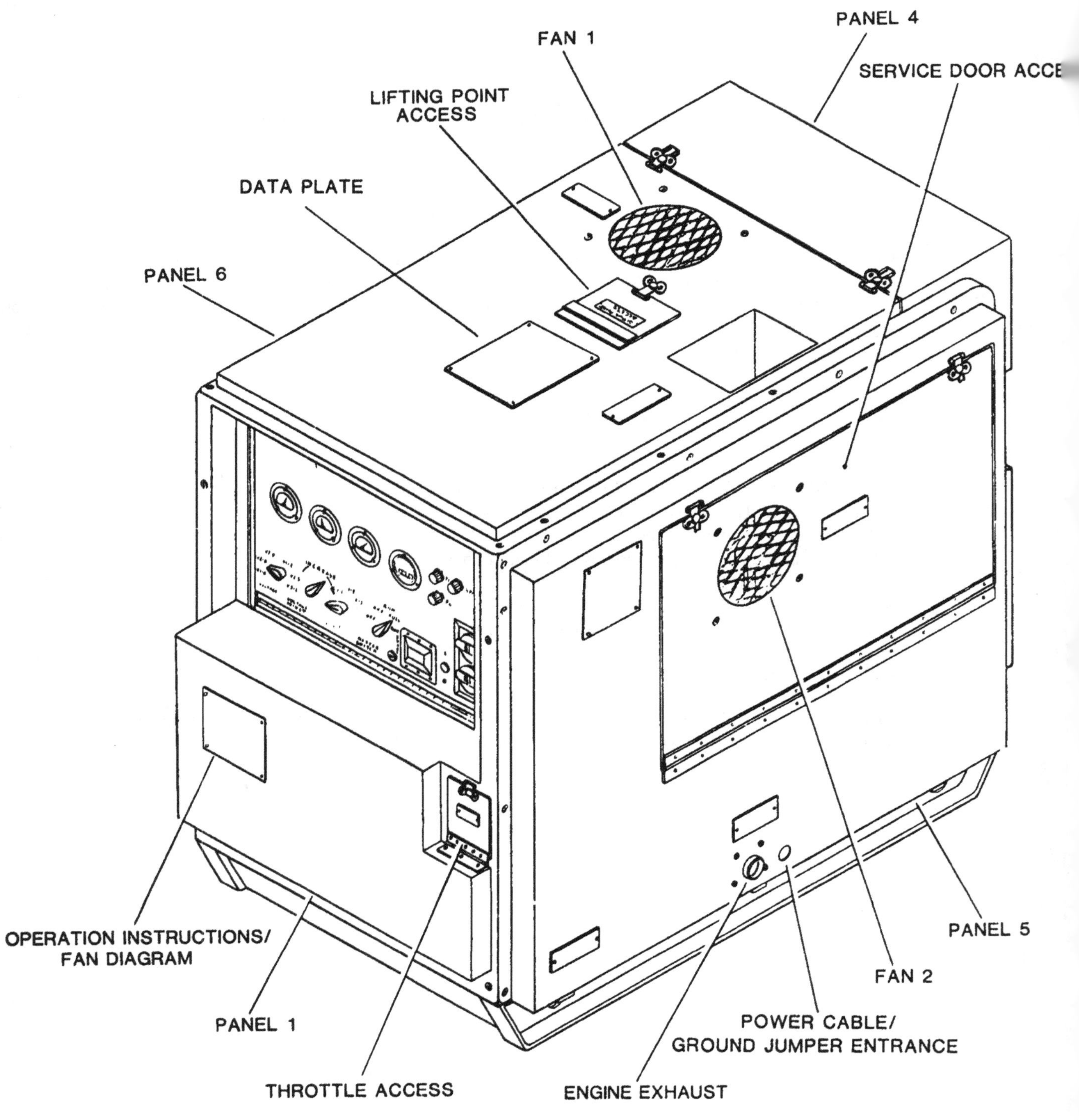

Figure 5-1.   Acoustic Suppression Kit (Sheet 1 of 2).

HOT AIR EXHAUST

AIR FILTER INTAKE

PANEL 2

PANEL 3

'AIR INLET/ OIL DRAIN

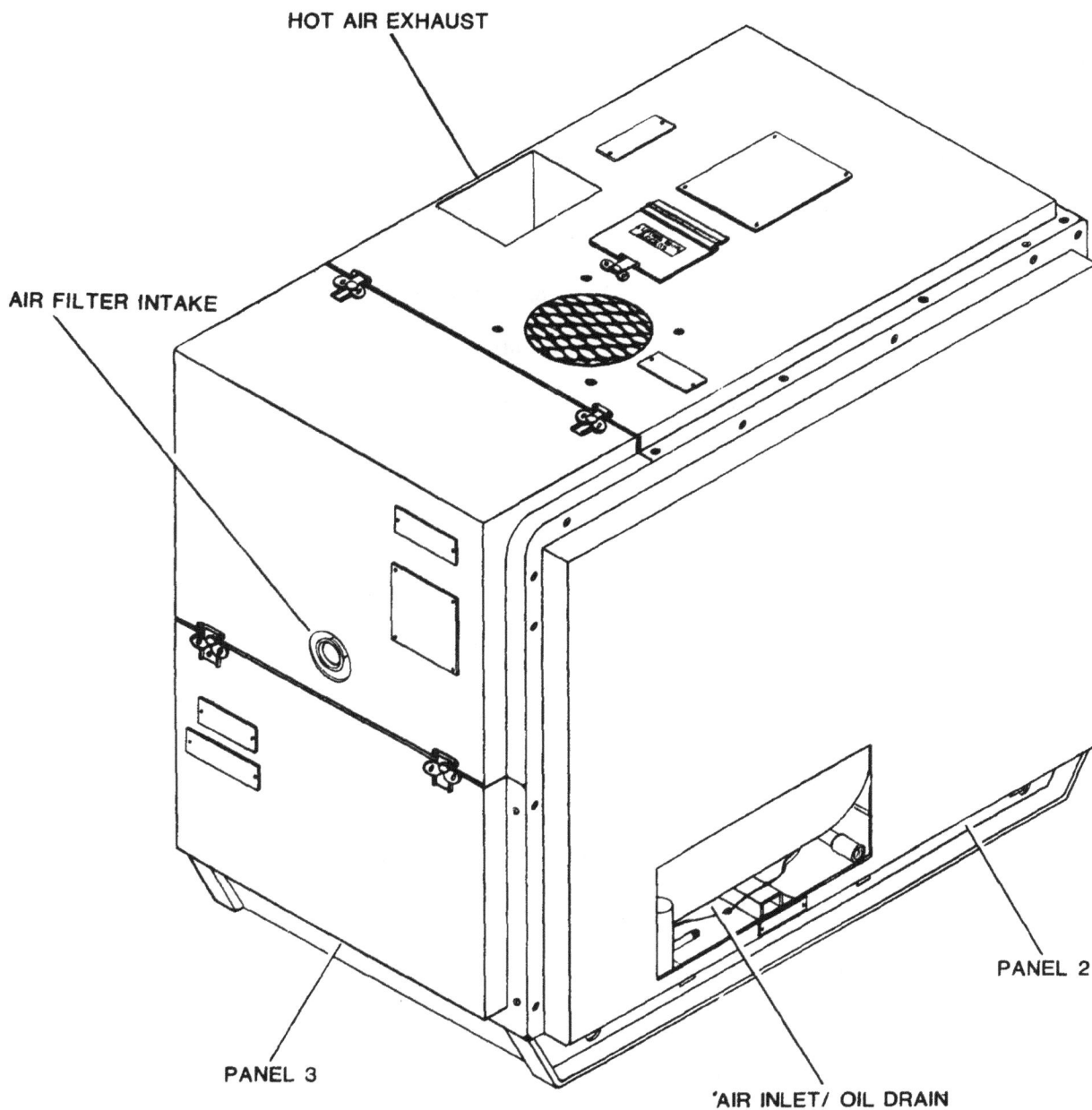

Figure 5-1.   Acoustic Suppression Kit (Sheet 2 of 2).

Table 5-1.   Identification and Instruction Plates.

Location	Type	Description
Panel 1	Schematic/ Instruction	Fan schematic/Operation instructions
Access door	Identification	Identifies throttle location
Panel 2	Identification	Identifies oil drain
Panel 3	Identification disconnecting and	Identifies battery location Instruction Instruction for connecting the battery
Panel 4	Diagram Identification	Block diagram of fuel system Identifies fuel tank location
Panel 5	Warning Danger -hot exhaust (engine) Identification ground terminal Instruction	Identifies Instructions for connecting load cables
Access door	Identification	Identifies area of oil level check and filter
Panel 6	Caution Caution Identification	Hot air exhaust Disconnect fan connector U.S. Department of Defense Data Plate. Contains model, serial number, and rating information for the set
Access Door	Identification	Identifies equipment lifting point

----

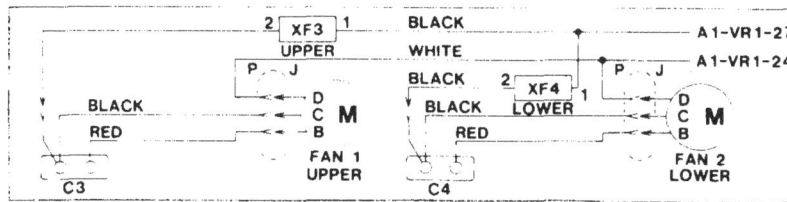

OPERATION INSTRUCTIONS

BEFORE STARTING SET

1. FILL CRANKCASE (DO NOT OVERFILL) AND FUEL TANK.
2. WARNING: GROUND SET TO AVOID SHOCK HAZARD.
3. CONNECT LOAD TERMINALS.
4. USE WINTERIZATION KIT FOR STARTING SET AT TEMPERATURES BELOW –25°F.

STARTING SET

1. MOVE "MASTER SWITCH" TO "PREHEAT" POSITION AND HOLD FOR 30 SECONDS.
   NOTE: PREHEAT IS NOT REQUIRED WHEN ENGINE IS HOT.
2. MOVE "MASTER SWITCH" TO "START" POSITION AND HOLD UNTIL ENGINE FIRES
   CONTINUOUSLY. (IF ENGINE DOES NOT START WITHIN 30 SECONDS, REPEAT
   STEPS 1 AND 2).
3. "MASTER SWITCH" WILL RETURN TO "RUN" POSITION WHEN RELEASED.
   NOTE: IF RUNNING FROM AUXILIARY FUEL SOURCE, MOVE "MASTER SWITCH" TO
   "RUN AUX" POSITION.
4. AFTER WARMING UP, ADJUST FREQUENCY AND VOLTAGE IF NECESSARY.
5. MOVE "CIRCUIT BREAKER" TO "ON" POSITION.

STOPPING SET

1. MOVE "CIRCUIT BREAKER" TO "OFF" POSITION.
2. MOVE "MASTER SWITCH" TO "OFF" POSITION.
3. FOR EMERGENCY STOP, PULL "DC CONTROL CIRCUIT BREAKER".

MARINE CORPS TM05926B/06509B-12
ARMY   **TM 5-6115-615-12**
NAVY   **NAVFAC P-8-646-12**
AIR FORCE   **TO 35C2-3-386-31**

5-3.2.   Tabulated Data.   The tabulated data for the ASK is contained in Table 5-2 with overall dimensions and weights of the ASK generator.

5-3.3.   Installation Plan.   The installation plan with the ASK is shown in Figure 5-4.

Table 5-2.   **ASK Tabulated Data.**

Dimensions and Weights

Panel 1

Overall Height	25" (63.5 cm)
Overall Width	23.5" (59.6 cm)
Overall Depth	2.09" (5.3 cm)
Overall Weight	8.5 lb (3.9 kg)

Panel 2

Height	25.5" (64.7 cm)
Width	39.0" (99 cm)
Depth	2.0" (5.1 cm)

Overall   Overall Weight   16.7 lb (7.6"
Overall   kg)
Overall

Panel 3

	Overall Heiqht	13.0" (33 cm)
Overall	1 Width	23.5" (59.6 cm)
Overall	1 Depth	2.0" (5.1 cm)
Overall	1 Weight	5.9 lb (2.7 kg)

Panel 4

	Height	13.0" (33 cm)
Overall	Width	23.5" (59.6 cm)
Overall	Depth	2.0" (5.1 cm)
Overall	Length	8.87" (22.5 cm)
Overall	Weight	7.2 lb (3.3 kg)
Overall		

Table 5-2.   ASK Tabulated Data, Continued.

Dimensions and Weights

    Panel 5 (including fan assembly)

        Overall Height.................................   25.5" (64.7 cm)
        Overall Width..................................   39.0" (99 cm)
        Overall Depth . . . . 2.0" (5.1cm)
        Overall Weight . . . .   22.3 lb (10.1 kg)

    Panel 6 (including fan assembly)

        Overall Height. ....   32.0" (81.2 cm)
        Overall Width...................................   23.5" (59.6 cm)
        Overall Depth . . . . 1.0" (2.5cm)
        Overall Weight..................................   13.1 lb (5.9 kg)

    <u>ASK Generator</u>

        Overall Height. ....   28.0" (71.1 cm)
        Overall Width...................................   28.0" (71.1 cm)
        Overall Depth...................................   43.0" (109.2 cm)
        Net Weight Empty . . . .   532 lb (241.3 kg)
        Net Weight Filled.............................   590 lb (267.6 kg)
        Shipping Weight . . . .   Not available at this time
        Cubage................................19.4 ft^3(552 dm^3)

MARINE
CORPS
ARMY

TM 05926B/06509B-12
TM 5-6115-615-12
NAVFAC P-8-646-12
TO 35C2-3-386-31

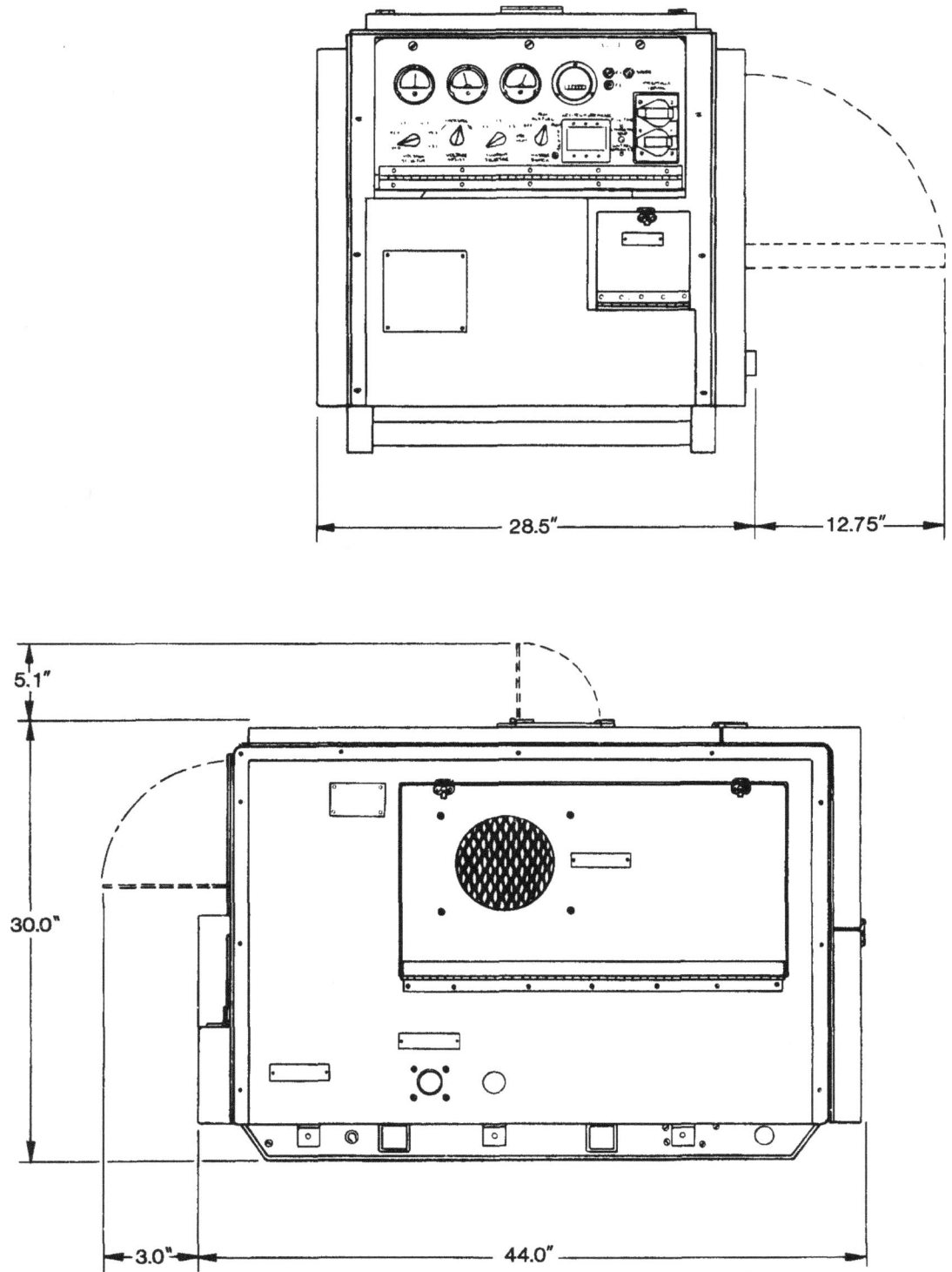

Figure 5-4.   Acoustic Suppression Kit   Installation plans

Section III.   PREVENTIVE MAINTENANCE CHECKS AND SERVICES (PMCS).

5-4.   GENERAL.   To ensure that the ASK does not hamper the overall operation of the generator set,the ASK should be inspected systematically so that defects may be discovered and corrected before they result in any serious damage or failure to the generator.   The necessary PMCS to be performed by the operator and organizational personnel are listed and described in paragraph 5-6.

5-5. CORRECTING AND REPORTING DEFICIENCIES.   Defects discovered during operation will be noted for future correction.   Stop operation immediately if a deficiency is noted which could damage the generator or present a safety hazard. All deficiencies will be recorded together with the corrective actions on
DA Form 2404.   Refer to current issue of DA Pam 738-750.

5-6.   OPERATOR AND ORGANIZATIONAL PREVENTIVE MAINTENANCE CHECKS AND SERVICES.

Table 5-3 contains a tabulating listing of PMCS to be performed by the operating personnel.   Table 5-4 contains a tabulating listing of the PMCS to be performed by the organizational personnel.   The item numbers are listed con-secutively and indicate the sequence of minimum requirements.

WARNING

The noise level of the generator set with the
Acoustic Suppression Kit has been found to exceed the allowable limits for unprotected personnel. Wear ear muffs or ear plugs.

WARNING

The generator should be shut down before performing any maintenance on the
Acoustic Suppression Kit.   _____

NOTE

Operator personnel are only allowed to remove panel 4 or open service/access doors for PMCS

Table 5-3.   Operator Preventive Maintenance Checks and Services.

Interval	B -Before Operation	A- After Operation	Total M/H 0.2
	D - During Operation Daily- 8 Hours		

Operator Daily BDA	Items to Be Inspected Inspection Procedures	Equipment Is Not Ready or Available If:

GENERAL

1			Make a visual inspection of the entire ASK for loose or missing hardware and for any bent, cracked, or broken parts. Tighten all loose hardware.	Damaged components, loose or missing hardware.
2	6		Check air intake on Panel 2 (See figure 5-1). Make sure that intake area is free from obstructions or accumulation of dirt, grease, etc. which would hinder air flow.	Air flow is re- stri cted.
			Check fans for proper operation. Make sure that fan area is free from dust, dirt, or grease that would hinder air flow.	
3	7		Check dual fan wiring harness inside service door access (Figure 5-1).	Fan is inoperative; or fan fan area is damaged.
4			Check flexible hose, clamps, modified exhaust pipe, and exhaust pipe assembly.	Wiring is chafed, frayed, broken, or burnt.
5	8			Any exhaust leak exists due to missing or damaged items.

Table 5-4.   Organizational Preventive Maintenance Checks and Services.

Interval W= Weekly (40 Hrs) M=Monthly (100

Organization WMS	Work Items to Be Inspected Inspection Procedures	Time M/H
	GENERAL	
1	Make a visual inspection of the entire ASK for loose or missing hardware and for any bent, cracked, or broken parts. Inspect all wires and terminals for damage and/or loose connectors.	0.3
2	Check air intake on Panel 2 (See figure 5-1) and remove any obstructions or accumulation of dirt, grease,etc. which would hinder air flow.	0.2
3	Check fans for proper operation and remove any dust, dirt,or grease from fan area that would hinder air flow.	0.2
	Check flexible hose, lamps, modified exhaust pipe and exhaust pipe for any exhaust leaks.	0.2
4	Check that panel gaskets are in place and not missing or deteriorated.	0.5
5		

Section IV.   TROUBLESHOOTING.

5-7.   GENERAL.   This section contains information for locating and correcting troubles which may develop in the ASK.   Table 5-5 is confined to troubleshooting involving the fans located in panels 5 and 6. The malfunction is followed by a list of tests or inspections which will help you to determine probable causes and corrective actions to take.   The first mention of the malfunction will be directed towards Operator/Crew (C); second mention towards Organizational (0). Perform the tests/inspections and corrective actions in the order listed.

Table 5-5.   Troubleshooting.

Malfunction
        Test or Inspection
                Corrective Action

----

1.   FAN(s) INOPERATIVE (C).

        Open Panel 5 service door access (paragraph 5-21) for Fan 2.

        Visually inspect Fan 1 without removing panel 6.

                Tighten loose connections on Fan 2;   if wiring is

                broken, notify higher echelon of mai   ntenance.   Notify

                higher echelon of maintenance if Fan.   1 is inoperative

                (blades not rotating).

2.   FAN(s) INOPERATIVE (0).

        Step 1.   Open the generator controls and instrument panel and check

                        the ASK wiring on voltage regulator terminal strap A1-VRI

                        (Figure 5-5).

                        Tighten loose connections.

Malfunction

    Test or Inspection

        Corrective Action

---

2. FAN(s) INOPERATIVE (O).

    Step 2. Check appropriate fan fuse in fuse block assembly. or remove ace. If replaced fuse continues to blow, proceed to Step 3.

    Step 3. Open Panel 5 service door access (paragraph 5-21) for Fan 2 and remove Panel 6 (paragraph 5-22) for Fan 1 and make voltage or continuity checks. See Figure 5-6. Replace fan,cabling, or capacitor.

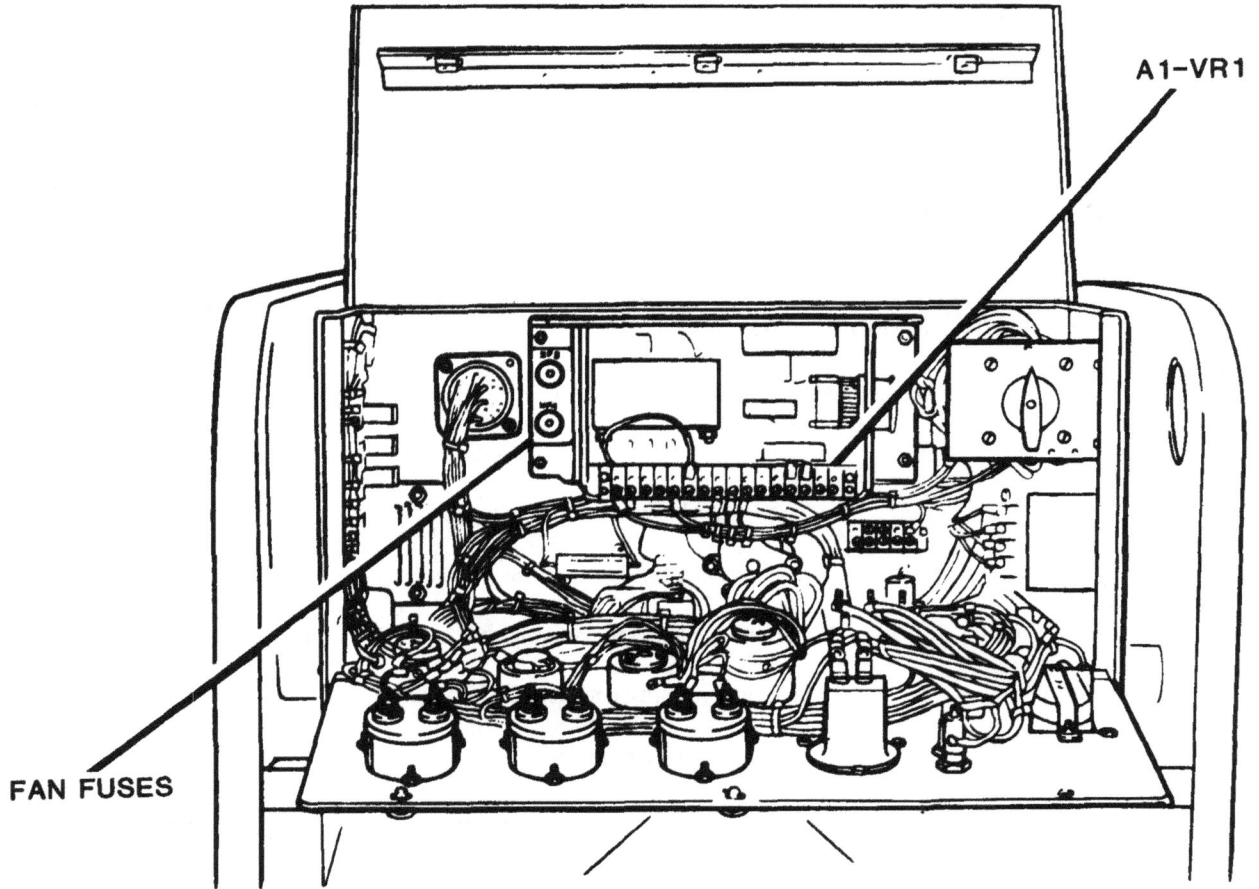

A1-VR1

FAN FUSES

Figure 5-5. Fuse Assembly, Fans.

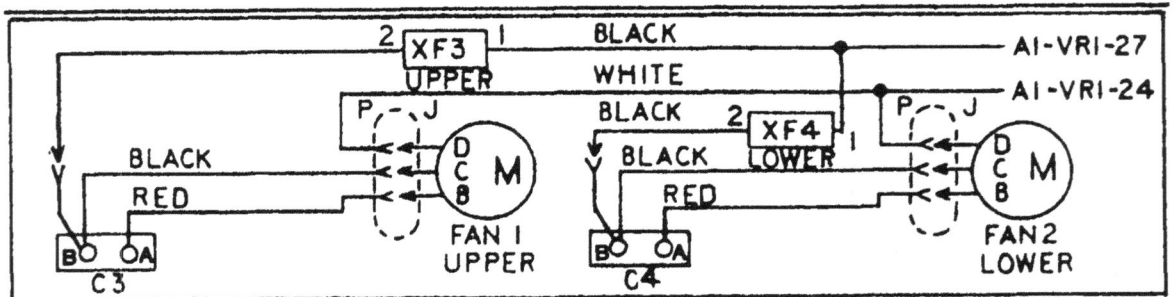

Section V.GROUND JUMPER AND LOAD CABLES INSTALLATION.

5-8. GENERAL. This section contains information on the installation of the ground jumper and load cables,as mentioned in Chapters 2 and 4, for the ASK.

NOTE

The ground jumper and load cables are not part of the ASK. Installation procedures are for information only.

5-9.   INSTALLATION.

5-9.1. Ground Jumper.

a.   Open the service door access (as described in paragraph 5-21a(1)) on Panel 5.

b.   Reaching in through the door, guide the ground jumper through the power cable entrance (Figure 5-1) on Panel 5.

c.   Loosen the two 3/4-inch clamps (Figure 5-7) and route the ground jumper through the clamps.   (Figure is shown with panels removed.)

d.   Connect one end of jumper to ground stud located in generator skid base and connect the other end of jumper to the proper load terminal (Figure 4-1).

5-9.2.   Load Cables.

a.   Perform steps given in paragraph 5-9.1a thru c for the load cables.

b.   Connect one end of load cables to proper load terminals (Figure 4-1) and tighten clamps.   Connect other end of load cables to unit requiring power.

Section VI. OPERATOR MAINTENANCE PROCEDURES.

5-10   GENERAL.   This section contains information on the maintenance of the equipment that is the responsibility of the operators.

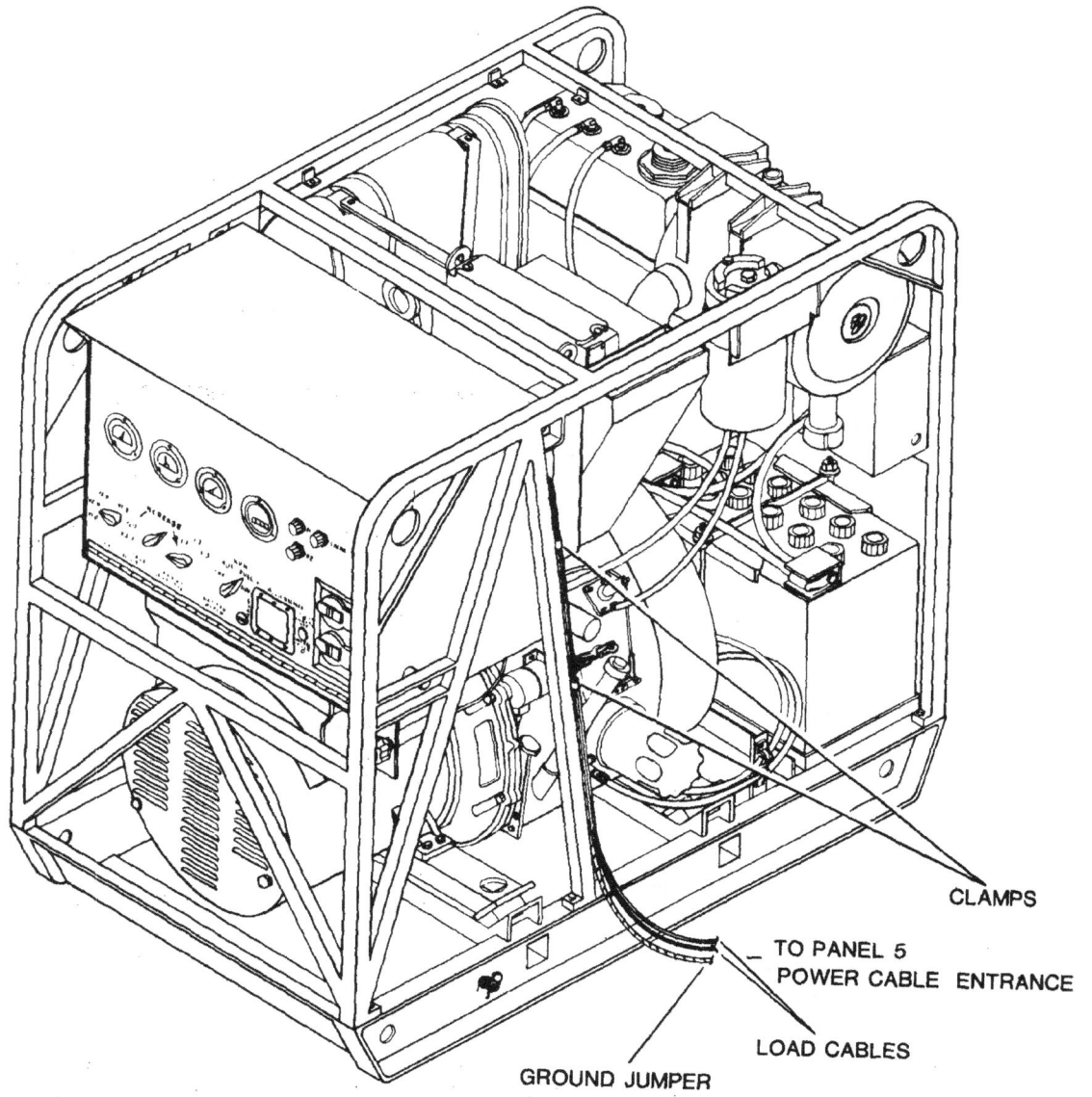

CLAMPS

TO PANEL 5
POWER CABLE ENTRANCE

LOAD CABLES

GROUND JUMPER

5-11.   **PANELS.**   (See Figure 5-8.)

a. Inspect surface of panels for damage such as cracks, dents, and missing hardware (items 1, 2, 3, 7, 12, 14, 15, 16, 17, and 22 thru 30). Notify higher echelon of maintenance for repair or replacement of panels or hardware.

NOTE

Operator is only allowed to remove Panel 4 and to open the service/access doors of the panels.

b. Inspect panel items such as fans (6, 18), various latches (7, 12, and 17), and sound insulation (32) for damage.   Notify higher echelon of main-tenance if above items need replacement or repair.

Section VII.   ORGANIZATIONAL MAINTENANCE PROCEDURES.

5-12. GENERAL. This section contains information on the repair and removal/ installation procedures of the ASK panels 1 thru 6, ASK items contained in the modified generator, gaskets, and information on the common and special tools required. Table 5-6 lists the items called out in the maintenance sections of chapter 3 and chapter 4 and the ASK panels which are to be removed or the access door to be opened in order to gain entry to the item.

CAUTION

Do not overtighten mounting screws when installing ASK panels. Damage to the rivet fas-teners located in generator frame can occur.

5-13. COMMON TOOLS AND EQUIPMENT. For common tools and equip Modified Tables of Organization and Equipment (MTOE) applicable ment refer to the e to your unit.

5-14. SPECIAL TOOLS. Special tools consist of a drill stop P/N V1577 which is listed and illustrated in the Repair Parts List TM 5-6115-615-24P. 1/16" - 1/4") and Special Tools

5-15. PANEL REPAIR. Repair of the ASK panels is limited to the replacement of a defective latch assembly or hinge.   Removal of panels, if necessary, to repair latch or hinge is given

CAUTION

Do not allow drill bit to extend beyond surface of material when drilling out pop rivets. Dam-age to the noise absorber insulation can occur.

a. Removal-Large Latch Assembly.

   (1) Loosen drill bit stop until opening becomes large enough to insert a No. 10 drill bit.

   (2) Insert bit into drill stop and tighten drill stop. Tapered part of drill stop should be towards bit tip; part of drill bit extending from tapered end of drill stop should be such that bit will not damage the sound insulation of panel when drilling. See Figure 5-8.

NOTE

As a quick check, place drill bit along edge of area to be drilled.   Bit tip should be flush with or less than the surface of panel.

   (3) Place drill bit stop into drill chuck and tighten.

   (4) Carefully drill out latch pop   rivets (2, Figure 5-9) from section of latch assembly (1).

b.   Installation

   (1) Place section of latch assembly   (1, Figure 5-9) into position on

   (2) Install 3/16" pop rivet (2) using pop   rivet gun.

CAUTION

Do not allow drill bit to extend beyond   surface of material when drilling out pop rivets. Damage to the noise absorber can occur.

c. Removal-Small Latch Assembly and Hinges. Perform same Removal pro-cedures used for the large latch assembly in paragraph 5-15a(1) thru (4) except use a No. 30 bit.

d.   Installation.   Perform same Installation procedures used for the large latch assembly in para-graph 5-15b(1) and (2), except use a 1/8" pop rivet.

DRILL BIT

DRILL STOP

INCORRECT   CORRECT

PANEL/ACCESS DOOR
EDGE

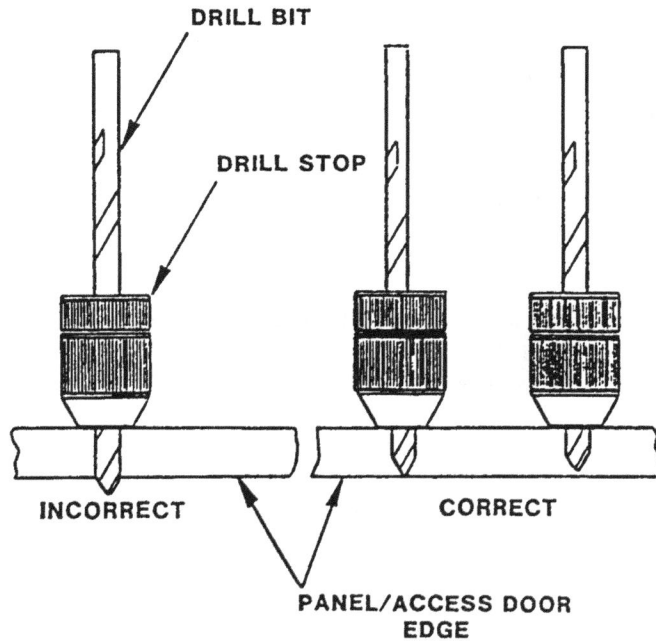

Figure 5-8. Panel Repair Setup.

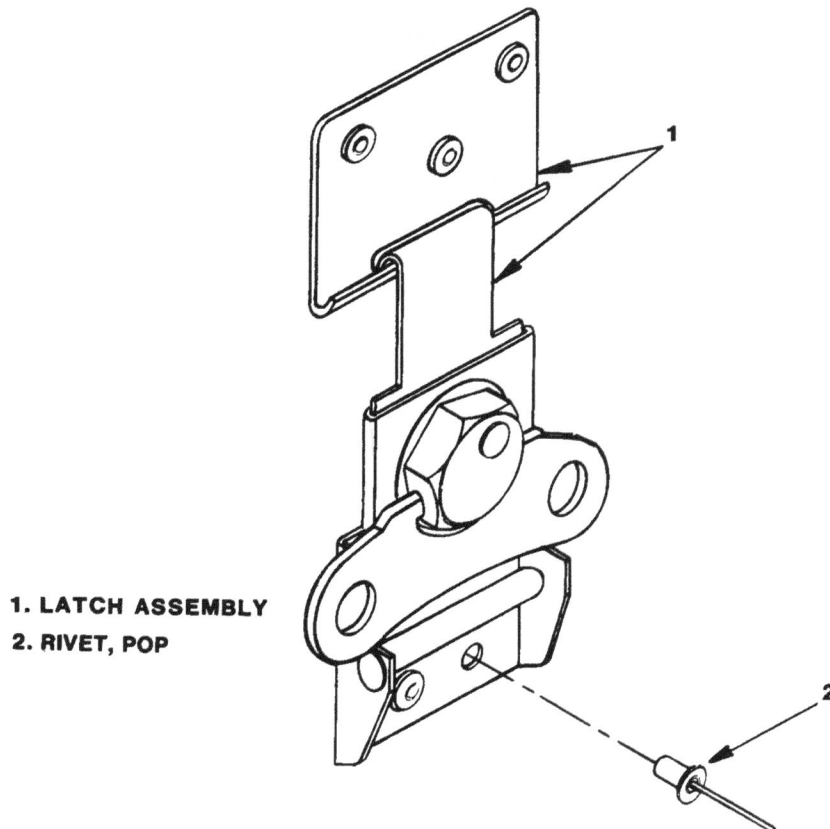

1. LATCH ASSEMBLY
2. RIVET, POP

5-16.   **REMOVAL/INSTALLATION.**   The following paragraphs 5-17 thru 5-33, contain information on the removal/ installation of the acoustic suppression kit (ASK) panels, along with the removal/installation of the remaining ASK items including any testing and repair.

5-17.   **PANEL 1.**   (See Figure 5-10).

   a. Removal.

      (1) Remove two screws (9), lockwashers (10), and washers (11) holding
          Panel 6 (5) to Panel 1 (31).
   (2) Remove screw (1) lockwasher (2), and washer (3) holding Panel 2

          to Panel 1.   (4)

      (3) Remove screw (25), lockwasher (26), and washer (27) holding Pane
          (20) to Panel 1.   5

      (4) Remove six screws (28), lockwashers (29), and washers (30) from
          Panel 1 and pull panel straight out.

   b.   Installation.
              (1) Place Panel 1 (31) into position of generator frame and install six

          washers (30), lockwashers (29), and screws (28).

      (2) Install washer (27), lockwasher (26) and screw (25) holding Panel 5
          (20) to Panel 1.

      (3) Install washer (3), Iockwasher (2), and screw (1) holding Panel 2
          (4) to Panel 1.

5-18. **PANEL 2.**   (See Figure 5-10.)

   a.   Removal.

      (1) Remove three screws (33), lockwashers (34) and washers (35) from
          Panel 2 (4).

      (2) Remove 11 screws (1), lockwashers (2), and washers (3) and remove panel.

   b.   Installation.

      (1) Place Panel 2 in position on generator frame and install 11 washers
              (3), lockwashers (2), and screws (1).

      (2) Install three washers (35), lockwashers (34), and screws (33).

Figure 5-10.Acoustic Suppression Kit, Exploded View (Sheet 1 of 2).

1. SCREW
2. LOCKWASH-ER
3. WASHER
4. PANEL 2
5. PANEL 6
6. FAN 1
7. LATCH

9. SCREW
10. LOCKWASHER

8. PANEL 4
1. WASHER
2. LATCH
3. PANEL 3
4. SCREW
5. LOCKWASH-ER
6. WASHER

19. SERVICE DOOR
20. PANEL 5

7. LATCH
8. FAN 2
2. HOSE
3. SCREW
4. LOCKWASHER
5. WASHER
6. SCREW
7. LOCKWASHER

29. LOCKWASHER
30. WASHER

8. WASHER
9. SCREW
10.   PANEL 1
11.   INSULA-TION
12.   SCREW
13.   LOCK-WASHER
14.   WASH-

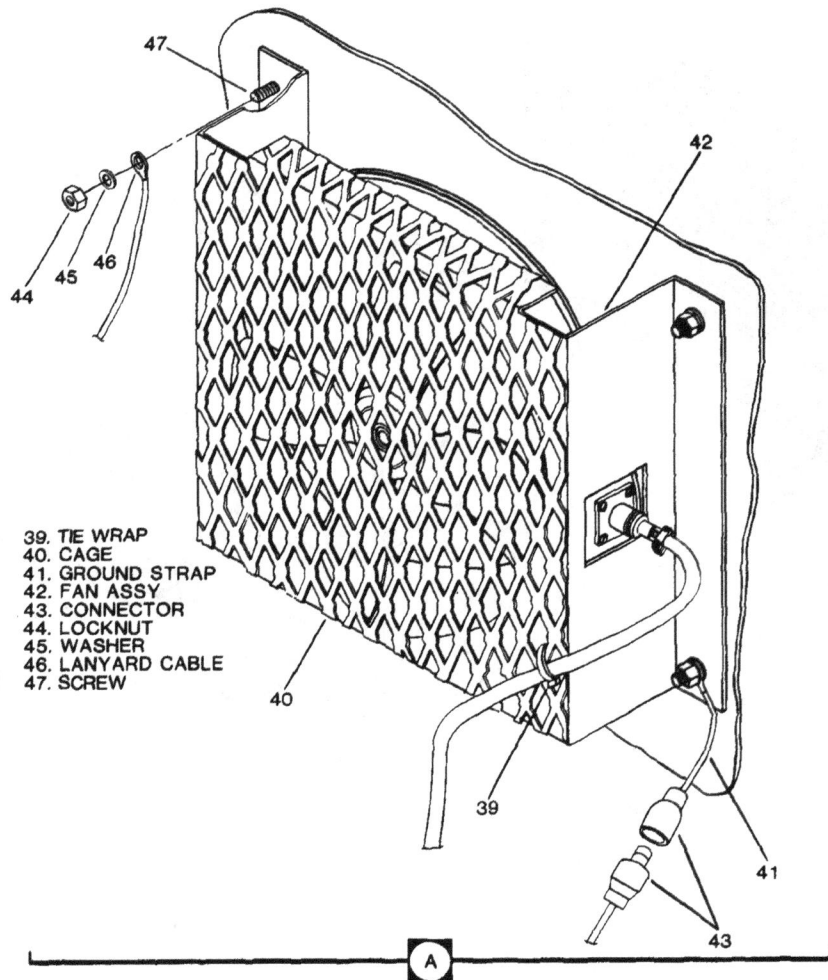

39. TIE WRAP
40. CAGE
41. GROUND STRAP
42. FAN ASSY
43. CONNECTOR
44. LOCKNUT
45. WASHER
46. LANYARD CABLE
47. SCREW

Figure 5-10. Acoustic Suppression Kit, Exploded View (Sheet 2 of 2).

MARINE CORPS TM 059266/065096-12
ARMY   TM 5-6115-615-12
NAVY   NAVFAC P-8-646-12
AIR FORCE   TO 35C2-3-386-31

5-19.   **PANEL 3.**   (See Figure 5-10.)

    a. Removal.

       (1) Loosen two large quarter-turn latches (12) by turning in a counter-clockwise direction.

       (2) Remove four screws (14), lockwashers (15), and flat washers (16) from Panel 3 (13) and remove panel.

    b. Installation.

       (1) Place Panel 3 in position on generator frame and install four flat washers (16), lockwashers (15), and screws (14).

       (2) Tighten two large quarter-turn latches (12) by turning in a clock-wise direction.

5-20.   **PANEL 4.**   (See Figure 5-10.)

    a, Removal. Loosen two large quarter-turn latches (12) from Panel 3 (13)

    and two large quarter-turn latches (7) from Panel 6 (5) by turning latches in a counterclockwise direction. Remove Panel 4 (8) from generator frame.

    b.   Installation.   Put Panel 4 into position on generator frame and tighten four large quarter-turn latches.

5-21.   **PANEL 5.**   (See Figure 5-10.)

    a. Removal.

       (1) Loosen two small   quarter-turn latches (17). counterclockwise open the service direction, and   door access (19) of Panel   5 (20).

       (2) Disconnect fan

       (3) Remove tie-wrap (39) securing harness to protective cage   (40).

       (4) Disconnect the ground strap (41) from fan mounting assembly   (42) by pulling apart male-female connector (43).

       (5) Remove the wrap(s) as necessary.

       (6) Remove lanyard cable assembly (46) by removing locknut (44) and washer (45).   Reinstall locknut and washer onto screw (47) and close service access door.

       (7) Remove four screws (22), lockwasher (23), and flat washers (24) which mount the engine-pipe assembly (21) to Panel 5.

(8) Remove three screws (36), lockwashers (37), and washers (38) from

(9) Remove 11 screws (25), lockwasher (26) and washers (27) and remove
panel .

b. Installation.

(1) Place Panel 5 (20) into position on generator frame and install 11
washers (27), lockwashers (26), and screws (25).

(2) Install three washers (38), lockwashers (37) and screws (36).

(3) Install lanyard cable assembly (46) by removing locknut (44) and washer (45) previously reinstalled on
fan mounting screw (47) in Removal procedure, step 6.

(4) Place lanyard onto screw and install washer and locknut.

(5) Reach in through open service access door (19) and position exhaust
pipe assembly (21) against Panel 5.

(6) Holding pipe assembly in place install four washers (24), lockwashers
(23), and screws (22).

(7) Reconnect fan wiring harness plug to Fan 2 (18).

(8) Replace tie-wrap (39) and secure wiring harness to protective cage
(40).

(9) Replace tie-wrap(s) removed in step (5) of Removal Procedure

(10) Reconnect ground strap (41).

(11) Close service door access by turning latches (17) in clockwise
direction.

5-22. PANEL 6.   (See Figure 5-10.)

a. Removal.
(1) Loosen two large quarter-turn latches (7) which secure Panel
to Panel 6 (5).

(2) Remove 10 screws (9), lockwashers (10), and washers (11) from   4 (8)
(3) Carefully lift panel to the point where the fan cable (41) can be
disconnected from Fan 1 (6).

MARINE CORPS TM 05926B/06509B-12
ARMY   TM 5-6115-615-12
NAVY   NAVFAC P-8-646-12
AIR FORCE TO 35C2-3-386-31

(4) Disconnect fan cable   assembly from fan and remove Panel 6 from
     generator frame.

b. Installation.

(1) Place Panel 6 (5) onto generator frame and position panel in order
    to reconnect fan cable assembly (41) to Fan 1 (6).

(2) Install 10 washers (11), lockwashers (10), and screws (9).

(3) Tighten two large quarter-turn latches (7) and secure Panel 4 to
    Panel 6.

5-23. BLIND RIVET FASTENER (See Figure 5-11.)

a. Removal.

(1) Drill through the rivet fastener with same size drill original hole; l/4-inch bit for small   that drilled (1) or "Q" bit
    rivet fastener for large rivet fastener (2).

(2) Punch out shank of fastener.

b.   Installation.

(1) Screw rivet fastener (1 or 2) onto pull-up stud (4) of blind hand riveter (3) as illustrated. Use a 10/32" hand
    riveter for small fastener (1) or a 1/4" hand riveter for large fastener (2). Beveled edge (5) of pull-up
    stud should extend just beyond the end of the fastener when fastener head rests against anvil (6) of hand
    riveter. If not...

(2) Adjust pull-up stud head (7) to shorten or lengthen pull-up stud
    and perform step (1) above.

(3) Place fastener into drilled-out hole on generator.
    When installing fastener, do not overtighten. Damage to fastener threading can occur.

CAUTION

(4) Using a wrench, hold hand riveter nut (8) in place while turning
    pull-up stud head (7) with Allen wrench handle (9).

1. RIVET FASTENER (SMALL)
2. RIVET FASTENER (LARGE)
3. HAND RIVETER
4. PULL-UP STUD
5. BEVELED EDGE
6. ANVIL
7. STUD HEAD
8. RIVETER NUT
9. WRENCH HANDLE

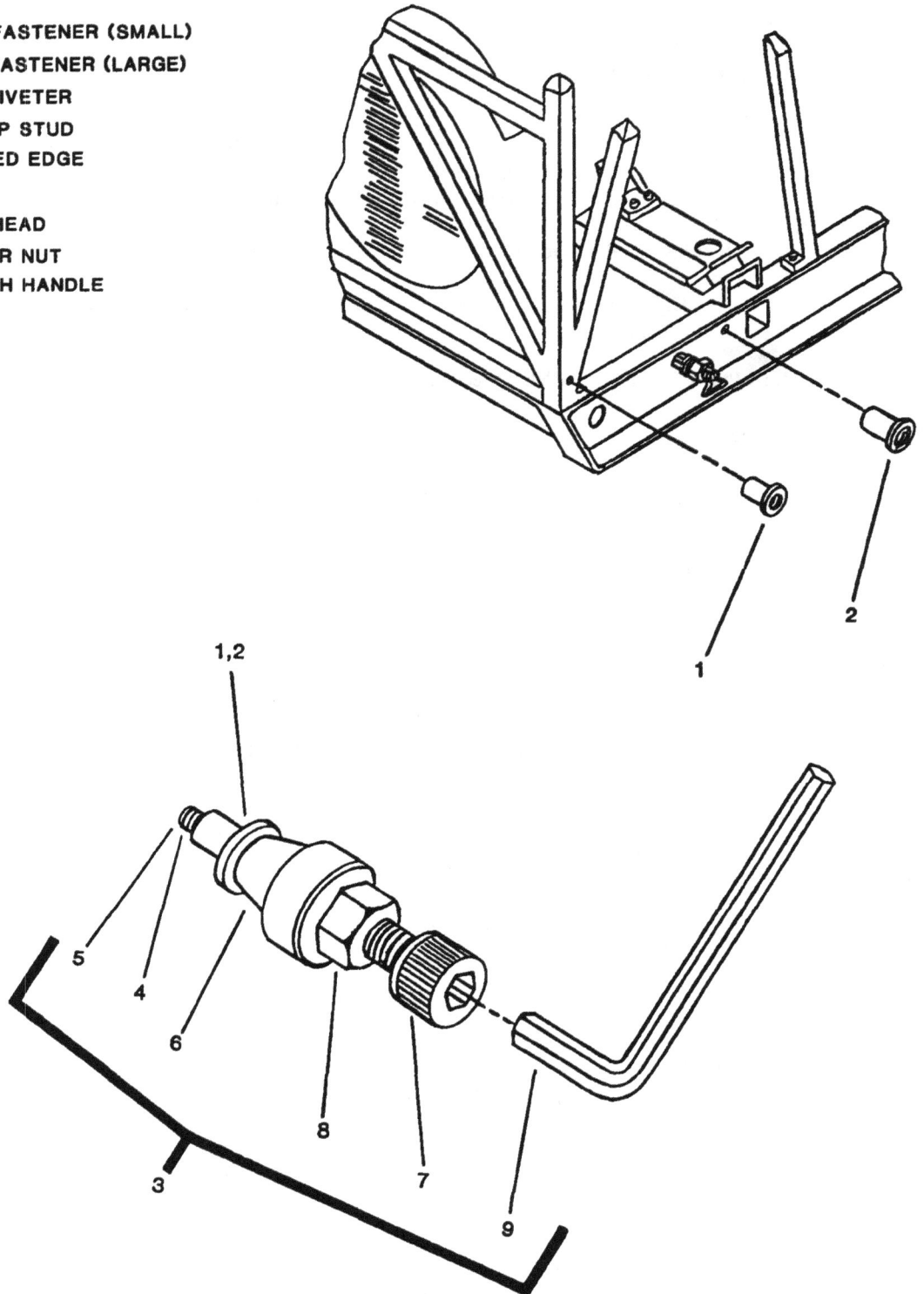

Figure 5-11.   Blind Rivet Fastener and Installation Tool.

5-24.   FAN 1.   (See Figure 5-12).

a.   Test. Using a multimeter, check continuity between 5-12) of Fan 1.   Pins D to pins (12, Figure between
C and D to B should read ohms; pins C to B, approximately 500 ohms (Pins are 200-300 labeled conse-
tively in a counterclockwise direction from the top cu-keyway).

b.   Repair.   Repair consists of replacing blown fuse or replacing or repair-
ing the wiring harness as described in paragraph 5-27.

c.   Removal.

(1) Perform Removal steps for Panel 6 given in paragraph 5-22a(1) thru
a(4).

(2) Remove four locknuts (1, Figure 5-12), eight washers (2), and four screws (3), which secure the fan assembly (4) to Panel 6, and remove fan mount.

(3) Loosen four locknuts (9) just enough to slide the four fan mount clips (10) away from fan rim.   Note
fan position with regard to fan plug.

(4) Remove fan (11) from mounting assembly.

d.   Installation

NOTE

Before installing fan into fan mounting assembly, ensure that blade rotation is such that air will blow
out of generator set.

(1) Place fan (11) into housing in same position as it was removed and
as shown in Figure 5-12.

(2) Slide four fan mount clips (10) forward and tighten locknuts (9)
securing fan in place.

(3) Place fan assembly into designated panel section; fan plug pointing
towards lifting point access (Figure 5-1), and align holes.

(4) Install four screws (3), eight washers (2), and four locknuts (1).

(5) Perform Installation steps for Panel 6 given in paragraph 5-22b(1)
thru (3).

1. LOCKNUT
2. WASHER
3. SCREW
4. FAN ASSY
5. LOCKWASHER
6. GROUND STRAP
7. CAGE
8. TIE WRAP

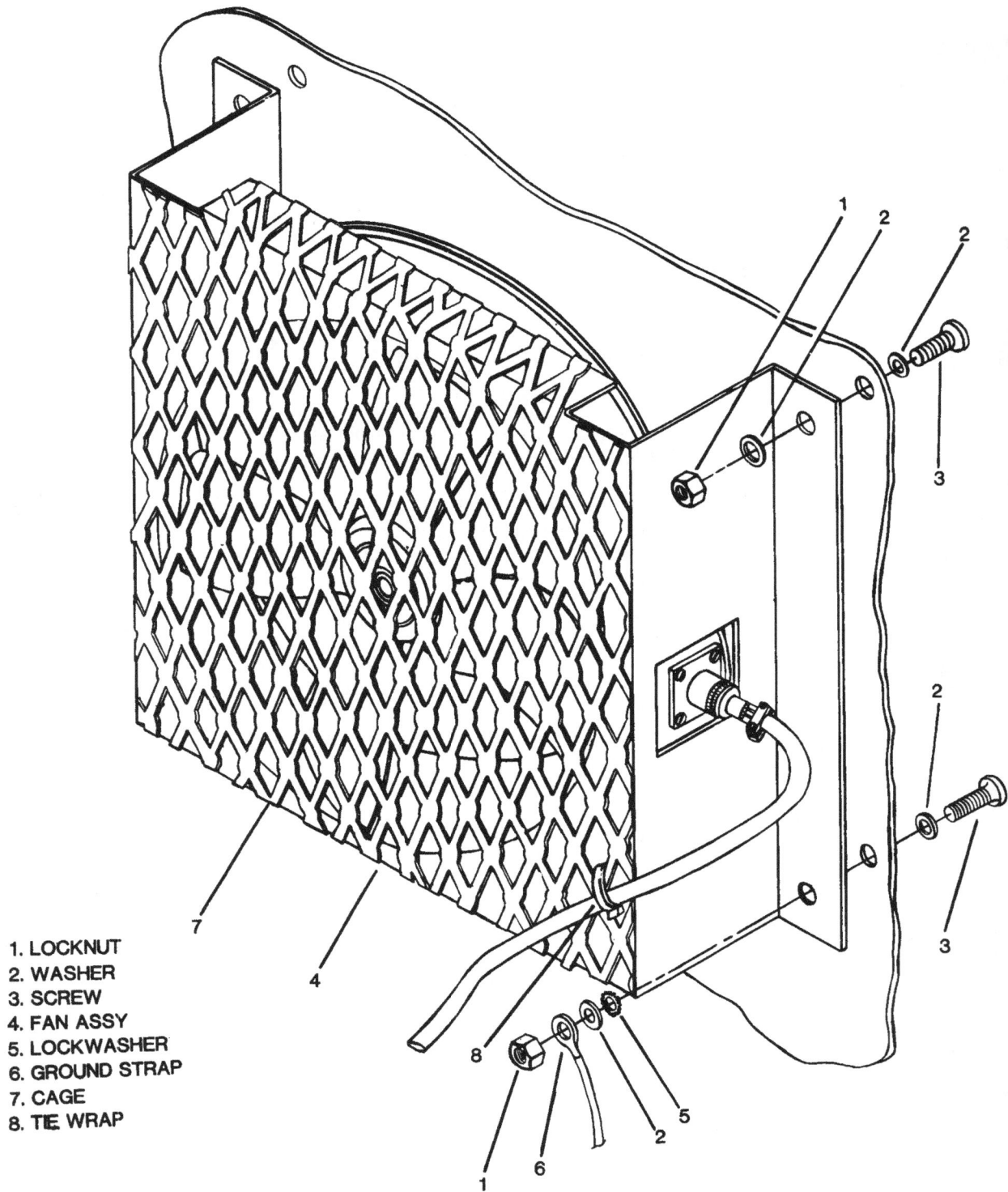

NOTE: ITEMS 5, 6, 7, & 8 NOT APPLICABLE TO FAN 1

Figure 5-12.   Fan Assembly (Sheet 1 of 2).

MARINE CORPS TM 05926B/06509B-12
ARMY   TM 5-6116-615-12
NAVY   NAVFAC P-8-646-12

9. LOCKNUT
10. CLIP
11. FAN
12. PIN A

TOP KEYWAY

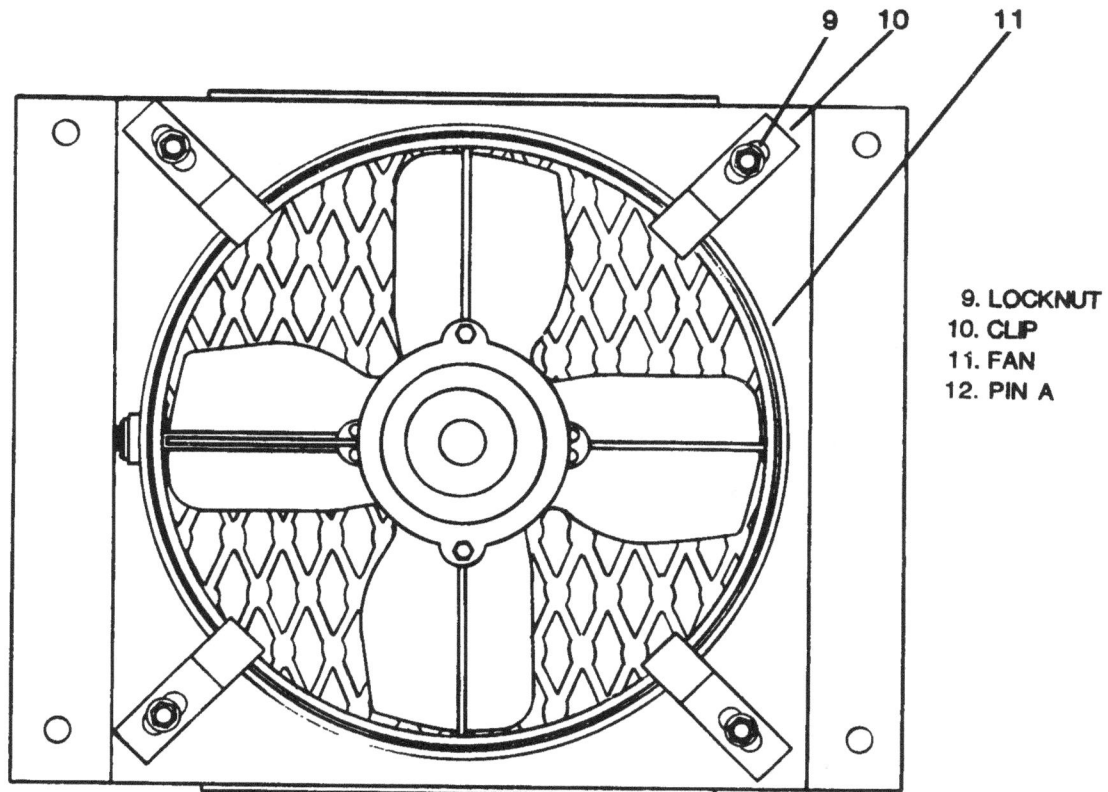

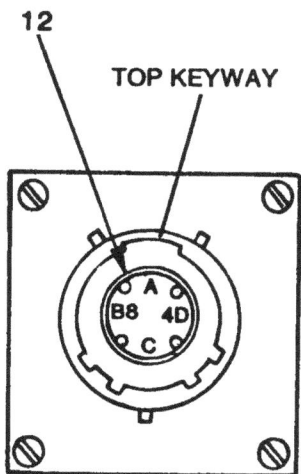

Figure 5-12.   **Fan Assembly (Sheet 2 of 2).**

Change 1

5-25.   FAN 2.   (See Figure 5-12.)

a.   Test._____Perform same test as given in paragraph 5-24a for Fan 1.

b._____Repair. Perform same repair instructions as given in paragraph 5-24b
     for Fan 1.

c. Removal.

(1) Perform Removal steps for Panel 5 given in paragraph 5-21a(1) thru
     (3).

(2) Remove ground strap (6) by removing locknut (1), lockwasher (5),
     washer (2), and screw (3).

(3) Remove the remaining three nuts (1), six washers (2), and three screws (3) which secure the fan assem-
     bly (4) to the service door access of Panel 5 and remove the fan assembly.

(4) Loosen four locknuts clips (10) (9) just enough to slide the four fan mount the fan rim.   Note fan posi-
     away from fan plug.   tion with regard to

(5) Remove fan (11) from

                                        mounting assembly and protective cage (7).

d.   Installation

NOTE

Before installing fan into fan mount, ensure that blade rotation is such that air will blow out
of generator set.

(1) Perform Installation steps given for Fan 1, paragraph 5-19d(1) thru
     (3) with Fan 2 plug pointing towards instrument panel.

(2) Connect ground strap (6) by installing screw (3) washer (2), lock-
     washer (5), and locknut (l).

(3) Install remaining three screws (3), six washers (2), and three lock-
     nuts (1).

(4) Replace tie-wrap

                                        (8) and secure wiring harness to protective cage (7). harness plug to fan.
(5) Reconnect wiring

5-26. FUSE BLOCK ASSEMBLY, MODIFIED GENERATOR. (See Figure 5-13.)

a. Test.

    (1) Open generator control panel (Figure 1-1) and check fuses.

    (2) perform steps b(3) and b(4) of Removal procedure below.

    (3) Using a multimeter, check for continuity between point of origin to point of termination.

b. Removal.

    (1) Open generator control panel (Figure 1-1).

    (2) Remove Panel 6 according to paragraph 5-22a(1) thru (4).

    (3) Remove black wire (6), A1-VR1-27, from terminal 27 on voltage regulator assembly.

    (4) Tag wires and disconnect the two insulated male tabs (7) of the fuse assembly from the insulated female receptacles (8) of the fan wiring harness.

    (5) Remove tie-wraps, as necessary.

    (6) Remove bracket (4) from voltage regulator assembly by removing four nuts (5) and pulling the voltage regular forward just enough to remove bracket.

c. Repair.   Repair consists of wrapping deteriorated insulation with electrical tape, replacing damaged wires or tabs, or replacing the fuse fuse holders as follows:

    (1) Perform Removal steps given in paragraph 5-26a above (1) thru (6).

(2) Remove fuse from fuseholder (1) and remove fuseholder by resoldering, tagging wires, and removing nut (3) and lockwasher (2).

    (3) Remove tie-wraps, as necessary.

    (4) Install fuseholder (1) into bracket and tighten with lockwasher (2) and nut (3).

    (5) Solder wires to fuseholder and reinstall fuse.

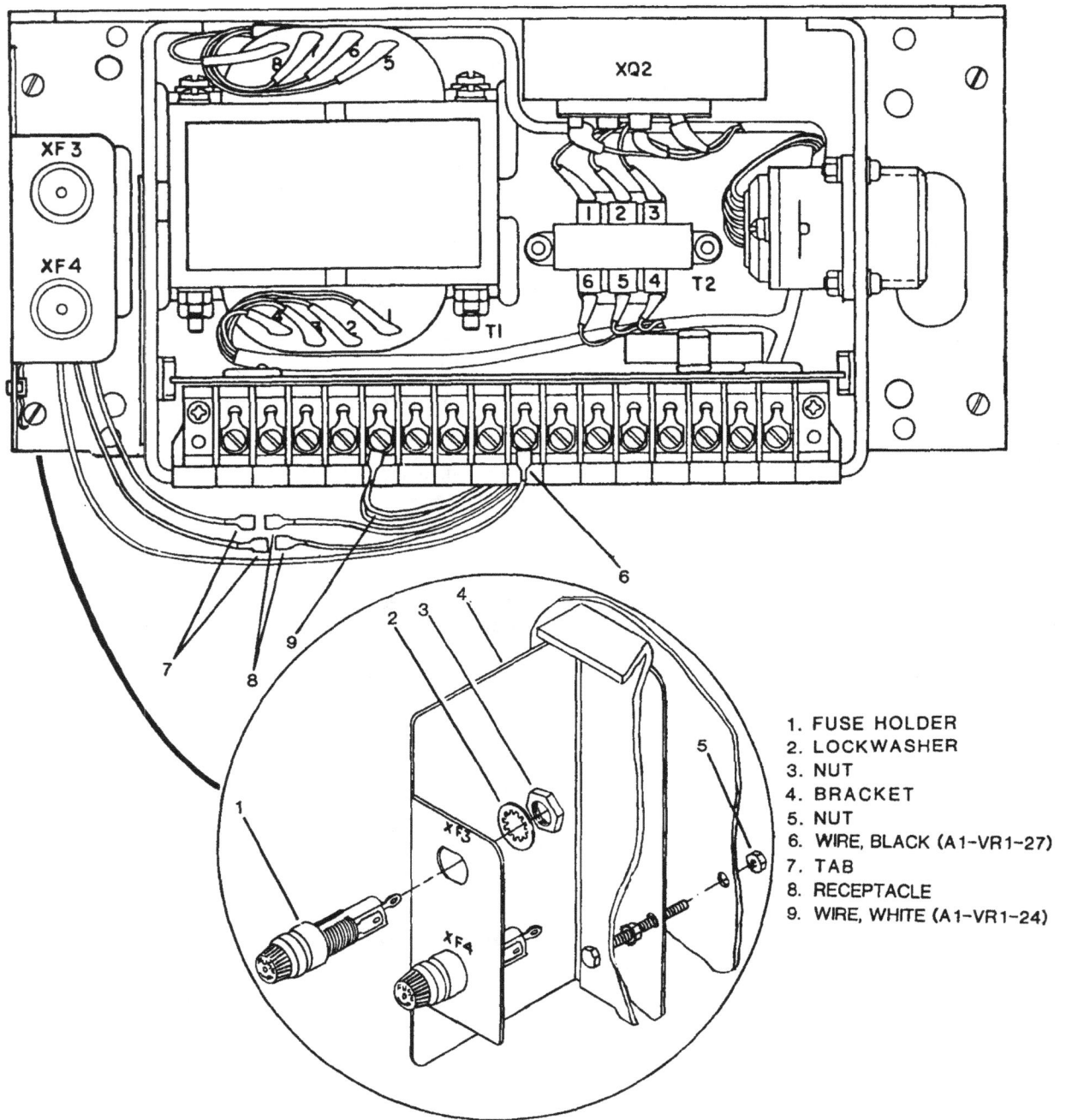

1. FUSE HOLDER
2. LOCKWASHER
3. NUT
4. BRACKET
5. NUT
6. WIRE, BLACK (A1-VR1-27)
7. TAB
8. RECEPTACLE
9. WIRE, WHITE (A1-VR1-24)

Figure 5-13.   Fuse Block Assembly.

d. Installation.

   (1) Place bracket (4) onto voltage regulator assembly and secure regulator assembly with four nuts (5).

   (2) Reconnect wiring removed in paragraph 5-26a(3) and (4) above.

   (3) Close generator control panel. **5-27.**   **FAN WIRING**

HARNESS, MODIFIED GENERATOR.

WARNING

To prevent electrical shock,discharge capacitors C3 and C4,
Figure 5-14, before attempting to test wiring harness.

a.   Inspection.

   (1) Inspect that all connections are tight.

   (2) Inspect harness for damaged wires, loose or missing ties, missing
insulation,or other visible damage.

b.   Test. Test wiring harness by measuring continuity of wires from point of origin to point of termination.
Continuity (low ohms) indicates a good wire.   No continuity (high ohms) indicates a broken
wire that must be replaced by replacing entire wire harness.
Disconnect battery cables before servicing fan wiring harness.   The high current output of the

WARNING

DC electrical system can cause arcing and/or burns if a short circuit occurs near the area
adjacent to the wiring harness.

c.   Repair.   Repair consists of wrapping deteriorated insulation wi trical tape or replacing
damages terminals, or open or shorted
tors as necessary.   Be sure to use the same size, color, and length of wire when making   th elec-
repairs.   Solder connections where applicable. Tag each wire and corresponding   capaci-

terminals at time of removal to ensure correct reassembly.   Repair the wiring harness by removing
capacitors C3 or C4 as follows:

Change 1

d. Removal.

(1) Perform Removal steps for Panel 2, 3, 5, and 6 paragraphs 5-18, 5-19,
5-21 and 5-22, respectively.

(2) Disconnect the negative (-) battery cable.

(3) Open generator control panel.

(4) Tag and disconnect white wire, A1-VR1-24, from terminal 24 of the regulator assembly A1-VR1 and
disconnect the two male tabs
(7, Figure 5-13) of the fuse block assembly from the two female receptacle(s) of the fan wiring harness.

(5) Remove ties and clamps securing wiring harness.

(6) Carefully remove harness from generator control panel bottom (7,
Figure 5-14).

WARNING

To prevent electrical shock, discharqe capacitor C3 before attempting to remove leads from
capacitor terminals.

(1) Capacitor C3. (See Figure 5-14.)

    (a) Removal

      2.   Slide terminal boot (1) away from capacitor C3 (2) wires from terminals.   and desolder

      3. Remove C3 from holding bracket (3)

(b) Installation
1.   Solder Leads to capacitor C3 (2) terminals: two black wires

on one terminal;red wire on remaining terminal.

    2.   Place C3 into holding bracket (3).

    3.   Slide terminal boot (1) back into place over terminals.

    4.   Replace Panel 2 according to paragraph 5-18b.

1. TERMINAL BOOT
2. C3
3. BRACKET
4. TERMINAL BOOT
5. C4
6. BRACKET
7. PANEL BOTTOM

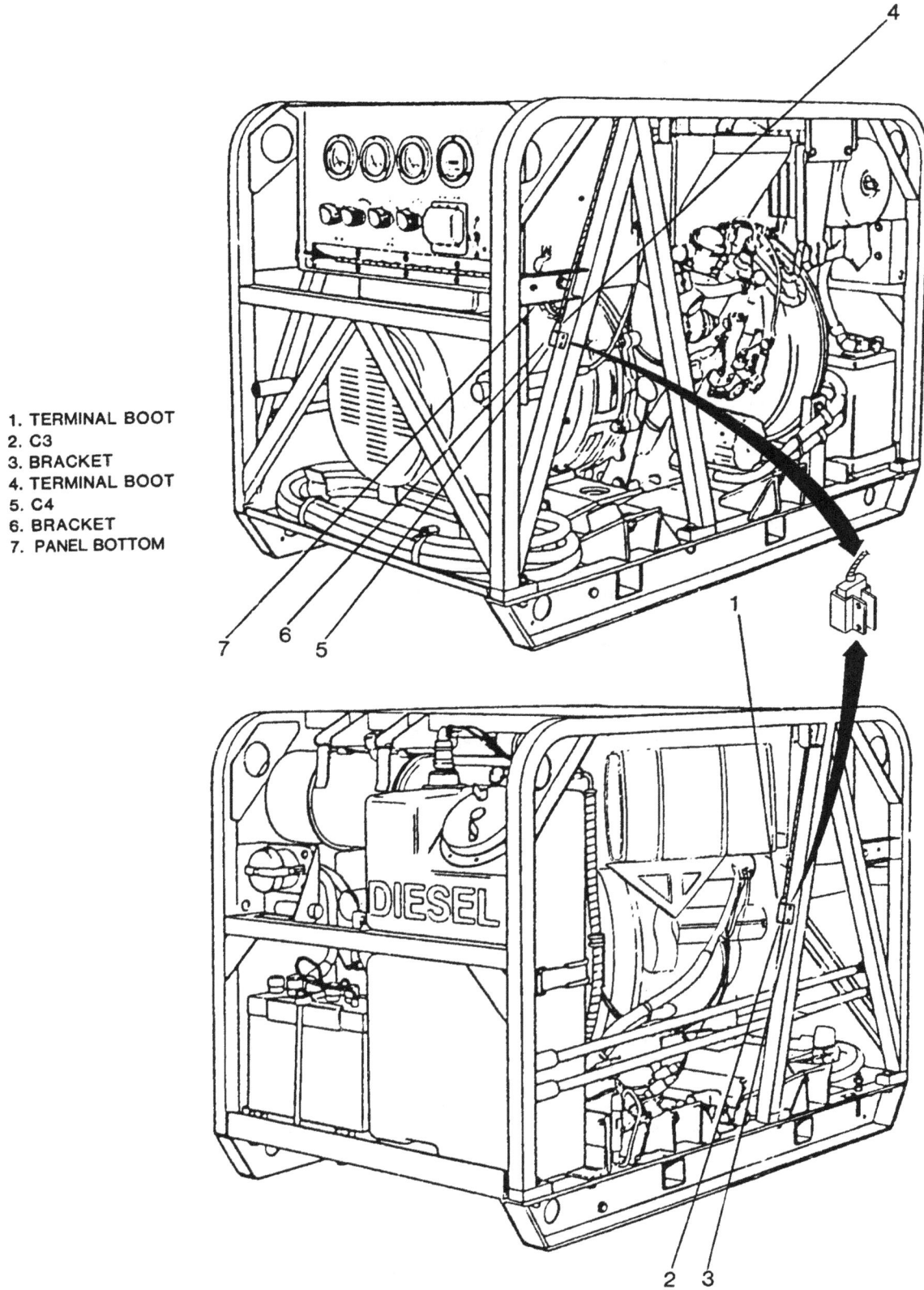

Figure 5-14.   Capacitors C3 and C4.

(2) Capacitor C4. (See Figure 5-14.)

WARNING

To prevent electrical shock, discharge before attempting to   capacitor C4 leads capacitor
remove leads from

—————   —————

2.   Slide terminal boot (4) away from capacitor C4 and desolder wires from terminals.

3.   Remove C4 from holding bracket (6).

(b) Installation.

1.   Solder leads to capacitor C4 (5) terminals: two black wires on
one terminal;red wire on remaining terminal.

2.   Place C3 into holding bracket (6).

3.   Slide terminal boot (4) back into place over terminals,

4.   Replace Panel 5 according to paragraph 5-21b.

e. Installation.

(1) Using tags of identification, attach leads to components.

(2) Secure harness with ties and clamps.

5-28. CAPACITOR BRACKET, MODIFIED GENERATOR. (See Figure 5-15.)

a. Removal.

(1) Perform Removal steps for Panel 5, given in paragraph 5-21a, for capacitor C4 bracket on Panel 2, paragraph 5-14b, for capacitor C3 bracket.

(2) Remove four screws (1, figure 5-15) from bracket (2) and remove
bracket from generator frame (3).

1. SCREW
2. BRACKET
3. FRAME
4. SCREW
5. LOCKWASHER
6. WASHER
7. CLAMP
8. SCREW
9. BONDING WIRE
10. WASHER
11. CLAMP
12. HOSE
13. CLAMP
14. EXHAUST PIPE
15. MODIFIED PIPE
16. SCREW
17. LOCKWASHER
18. LANYARD
19. NUT
20. SCREW
21. CLAMP
22. NUT
23. LOCKWASHER
24. EXTENDER

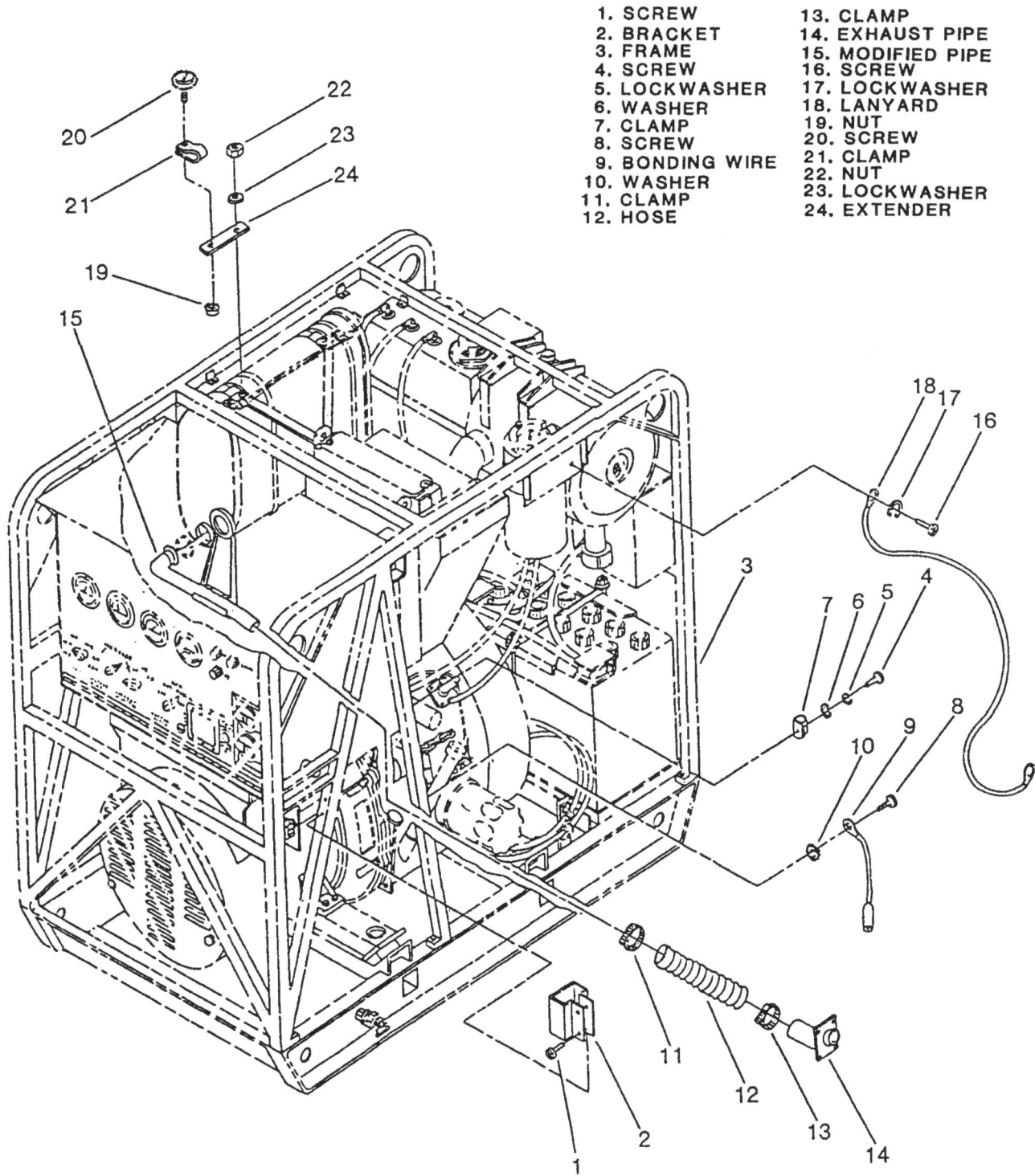

Figure 5-15.   Generator ASK Items, Right Front, Three-Quarter View.

MARINE CORPS TM 05926B/06509B-12
ARMY   **TM 5-6115-615-12**
NAVY   **NAVFAC P-8-646-12**
AIR FORCE   **TO 35C2-3-386-31**

b. Installation.

    (1) Place bracket (2) onto generator frame (3) and secure with four screws (1).

    (2) perform Installation steps for Panel 5, paragraph 5-21b, or Panel 2, paragraph 5-18b.

5-29. LOAD CABLE CLAMP, MODIFIED GENERATOR. (See Figure 5-15.)

a. Removal.

    (1) Open the service door access (19, Figure 5-10) for Panel 5.

    (2) Remove screw (4, figure 5-15), lockwasher (5), and washer (6) from frame (3) and remove clamp (7).

b. Installation.

    (1) Place clamp (7) onto frame (3) and secure with washer (6), lock-washer (5) and screw (4).

    (2) Close service door access for panel 5.

5-30. BONDING WIRE ASSEMBLY, MODIFIED GENERATOR. (See figure 5-15).

a. Removal.

    (1) Open the service access door (19, Figure 5-10) for Panel 5.

    (2) Remove screw (8, figure 5-15), wire (9), and washer (10) from frame (2).

b.   Installation.

    (1) Place washer (10) and wire (9) onto frame (2) and secure with screw (8).

    (2) Close service access door for Panel 5.

5-31. MODIFIED EXHAUST PIPE, FLEXIBLE EXHAUST TUBE, AND EXHAUST PIPE ASSEMBLY, MODIFIED GENERATOR.   (See figure 5-15).

a. Modified Exhaust Pipe.

    (1) Removal.

        (a) Perform Removal steps for Panel 6 given in paragraph 5-22a.

        (b) Perform Removal steps for Panel 2 (if necessary) given in paragraph 5-18a.

(c) Loosen clamp (11) and remove flexible exhaust tube (12).

(d) Perform Removal steps for EXHAUST PIPE and CLAMP given in paragraph 4-55a(1) and (2).

(e) Remove modified exhaust pipe (15).

(2) Installation.

(a) Perform Installation steps given in paragraph 4-55b.

(b) Place flexible hose (12) onto pipe (15) and tighten clamp (11).

(c) Perform Installation steps for Panel 2 given in paragraph 5-18b.

(d) Perform Installation steps for Panel 6 given in paragraph 5-22b.

b.   Flexible Exhaust Tube.

(1) Removal.

(a) Perform Removal steps for Panel 5 given in paragraph 5-21a.

(b) Perform Removal steps for Panel 6 given in paragraph 5-22a.

(c) Loosen clamp (11) and remove tube (12) from modified pipe (15).

(d) Loosen clamp (13) and remove exhaust pipe assembly (17) from tube (12).

(2) Installation.

(a) Place tube (12) onto modified pipe (15) and tighten clamp (11).

(b) Place other end of tube (12) onto exhaust pipe assembly (14) and tighten clamp (13).

(c) Perform Installation steps for Panel   5 given in paragraph 5-21b.

(g) Perform Installation steps for Panel   6 given in paragraph 5-22b.

c.   Exhaust Pipe Assembly.

(1) Removal.

(a) Perform Removal steps for Panel 5 given in paragraph 5-21a.

(b) Loosen clamp (13) and remove exhaust pipe assembly (14) from flexible exhaust tube (12).

(2) Installation.

    (a) place exhaust pipe assembly (14) into flexible exhaust tube (12) and tighten clamp (13).

    (b) Perform Installation steps for Panel 5 given in paragraph 5-21b.
_____

5-32. LANYARD CABLE ASSEMBLY, MODIFIED GENERATOR. (See Figure 5-15.)

a. Removal.

(1) Open the service door access (19, Figure 5-10) for Panel 5.

(2) Remove locknut (38) and washer (39) securing lanyard to Fan 2 Mounting and remove one end of lanyard. Reinstall locknut and washer back onto Fan 2 screw.

(3) Completely remove lanyard cable (18, Figure 5-15) by removing screw (16) and lockwasher (17).

b. Installation.

(1) Install one end of lanyard cable (18, Figure 5-15) by removing lock-nut (38, Figure 5-10) and washer (39) previously reinstalled on fan mounting screw in Removal procedure, step (b) above.

(2) Place lanyard cable onto screw and install washer (39) and locknut

(3) Holding service door access in place, secure other end of lanyard cable by installing lanyard (18, Figure 5-15), lockwasher (17), and screw (16).

5-33. GASKETS.

a.   Removal.   Remove gasket material from panel area by scrapping or peeling.

b. Installation.

(1) Mark edge of panel holes with white chalk on side of panels where new gasket is to be installed.

(2) Cut gasket to correct length and peel off protective adhesive strip.

(3) Lay gasket on panel area to be mounted and then remove gasket. Hole markings should appear on gasket.

                                  ng a 5/16-inch punch.

(4) Punch out hole markings usi

(5) Reinstall gasket onto pane

5-34. CABLE CLAMP EXTENDER, MODIFIED GENERATOR. (See Figure 5-15.)

    a.   Removal.

      (1) Remove cable clamp (21) from cable clamp extender (24) by removing nut (19) and screw (20). Remove wiring harness from clamp.

      (2) Remove nut (22) and lockwasher (23) and remove clamp clamp extender (24) from upper muffler bracket.

    b.   Installation.

      (1) Place cable clamp extender (24) on to upper muffler bracket and secure extender with lockwasher (23) and nut (22).

      (2) Place wiring harness into cable clamp (21) and install clamp onto extender using screw (20) and nut (19).

5-35.   OIL DRAIN ADAPTER, HOSE, AND CLAMP, MODIFIED GENERATOR. (See Figure 5-16.)

    a. Removal.

      (1) Loosen clamp (1) and remove hose (2) from oil drain adapter (3).

      (2) Remove oil-drain adapter from drain valve (4)

    b.   Installation.

      (1) Place oil drain adapter (3) onto drain valve (4) and tighten adapter.

      (2) Install hose (2) onto adapter (3) and tighten clamp (1).

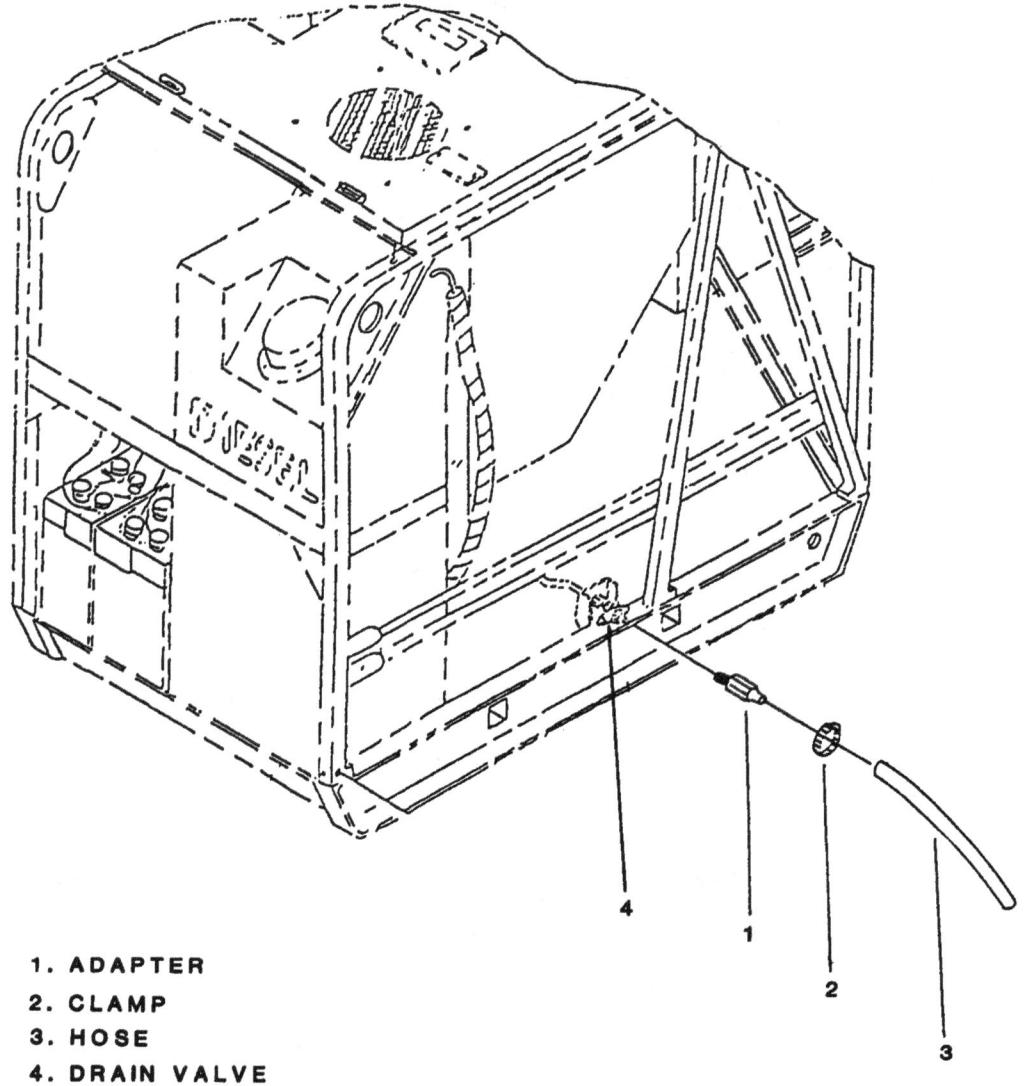

1. ADAPTER
2. CLAMP
3. HOSE
4. DRAIN VALVE

Figure 5-16.   Oil Drain Adapter, Hose, and Clamp.

MARINE CORPS TM 05926B/06509B-12
ARMY   TM 5-6115-61.5-12
NAVY   NAVFAC P-8-646-12
AIR FORCE TO 35C2-3-386-31

Table 5-6.Maintenance Item/Panel Removal List.

NOTE

Operator/crew personnel authorized to only remove Panel 4 or open the various service/access doors of the panels.

Item	Chapter	Page	Paragraph	Access/ Removal	Para- graph
Air Cleaner	4	4-47	4-42	Panels 4, 5 (Service Door), and 6	5-20, 5-21, and 5-22
Air Filter	3	3-31	3-29	Panel 5 Service Door	5-21
Auxiliary Fuel Pump	3/4	3-27/4-37	3-24/4-35	Panel 2	5-18
Battery	3/4	3-18/4-19	3-16/4-22	Panels 3, 4, and Panel 5 (Service Door)	5-19, 5-20, 5-21
Battery Cables-	3/4	3-20/4-20	3-17/4-23	Panel 4 and Panel 5 (Service Door)	5-20 and 5-21
Slave Recept.					
Battery Frame-	3/4	3-18/4-16	3-15/4-19	Panel 4, 5, and 6	5-20, 5-21, and 5-22
Tray & Holdown					
Control Box Assy	4	4-67	4-58	Panels 2, 3, 5, and 6	5-18, 5-19, 5-21, and 5-22
Control Box Assy					
Switches and Meters	4	4-66; 4-69 thru 4-85	4-57; 4-59 thru 4-70	Panel 3	5-19

Table 5-6.   Maintenance Item/Panel Removal List, Continued.

Item	Chapter	Page	Paragraph	Access/ Removal	Para- graph
Engine Assy	3	3-31	3-28	All Panels	5-17 thru 5-22
Engine Exhaust	3	3-35	3-33	Panel 2, 5, and 6	5-18, 5-21, and 5-22
Engine Mounting Brackets	4	4-14	4-17	Panels 2 and 5	5-18 and 5-21
Exhaust Pipe and Clamp	4	4-64	4-55	Panels 2, 5, and 6	5-18, 5-21, and 5-22
Frame	3	3-16	3-12	All Panels	5-17 thru 5-22
Fuel Filter Assy	3/4	3-29/4-40	4-25/4-36	Panels 5 and 6	
Fuel Injection Pump	4	4-46	4-40	Panel 5 Service Door	5-21
Fuel Injector	4	4-46	4-41	Panel 6	5-22
Fuel Lines-Valves	3/4	3-31/4-44	3-27/4-39	Panels 2 and 4; Panel 5 Service Door	5-18, 5-20, and 5-21
Fuel Tank	3	3-29	3-26	Panels 2, 3, and 4	5-18, 5-19, and 5-20
Fuel Transfer Pump	4	4-34	4-34	Panel 5 Service Door	5-21
Generator Assy	4	4-28	4-28	Panels 1, 2, and 5	5-17, 5-18, and 5-21
Glow Pump	4	4-59	4-50	Panel 6	5-22

Table 5-6.   Maintenance Item/Panel Removal List, Continued.

Item	Chapter	Page	Paragraph	Access/ Removal	Para-graph
Governor Linkage	3/4	3-33/4-49	3-31/4-44	Panel 5	5-21
Grounding Assy	3/4	3-24/4-18	3-21/4-20	Panel 2	5-18
Intake Manifold	4	4-58	4-49	Panels 4, 5 (Service Door), and 6	5-20, 5-21, and 5-22
Lifting Eye	3/4	3-16/4-15	3-13/4-18	Panel 6	5-22
Load Terminal Board	4	3-25/4-29	3-22/4-30	Panel 5	5-21
Muffler	4	4-62	4-54	Panels 2 and 6	5-18, 5-22
Oil Cooler	4	4-61	4-52	Panels 3 and 4	5-19, 5-20
Oil Filter	3/4	3-35/4-54	3-32/4-46	Panel 5	5-21
Oil Pan-Drains	3/4	3-33/4-60	3-30/4-51	Panel 2	5-18
Rocker Arms	4	4-48	4-43	Panel 6	5-22
Skid Base-Ground Stud	3	3-16	3-14	Panels 1 thru 5	5-17 thru 5-21
Solenoid, Fuel cutoff	4	4-41	4-37	Panel 5 Service Door	5-21
Starter	4	4-55	4-48	Panels 2 and 4	5-18 and 5-20
Throttle and Bracket	4	4-65	4-56	Panels 1 and 5	5-17 and 5-21

MARINE CORPS TM 05926B/06509B-12
ARMY   TM 5-6115-615-12
NAVY   NAVFAC P-8-646-12

APPENDIX A

1.   PAINTING:

     T.O.35-1-3   Painting and marking of USAF Aerospace Ground
                                     Equipment.

2.   RADIO SUPPRESSION:

     MIL-STD-461   Radio Interference Suppression.

3.   MAINTENANCE :

     T.O.1-1-1   Cleaning of Aerospace Equipment.
     T.O.1-1-2   Corrosion Control and Treatment for Aerospace
                                       Equipment.
     T.O.35-1-11   Organization, Intermediate and Depot Level
                                       Maintenance for FSC 6115 Equipment.
     T.O.35-1-12   Components and Procedures for Clening Aerospace Ground Equipment.
                                       Repair/Replacement Criteria for FSC 6115 Aerospace Ground Equipment.
     T.O.35-1-26   USAF Equipment Registration Number System Applicable to FSC 6115 Equip-
                                       ment.
     T.O.35-1-524

                                       Electric Motor and Generator Repair.
                                       Electric Power Generation in the Field.
     TM 5-764   Organizational, Intermediate (field) Direct Support and General Support and
     TM 5-766   Depot Maintenance Repair Parts Lists.
     TM 5-6115-615-24P (A)
     SL-4-05926B/
           06509B-24P/2 (MC)   Processing and Inspection of Aerospace Ground Equipment for Storage
                                         and Shipment.

4.   SHIPMENT AND STORAGE: Processing and Inspection of Non-Mounted, Non-Aircraft Gasoline and Diesel
     T.O.35-1-4   Engines for Storage and Shipment.

     T.O.38-1-5

                                         procedures for Destruction of Equipment to Prevent Enemy Use.

5.   DESTRUCTION OF MATERIEL:

     TM 750-244-3

6.   MAINTENANCE   AND RECORDS:
     FORMS
                                       The Army Maintenance Management System.
     DA PAM 738-750   Consolidated Index of Army Forms and Publications. Marine Corps Index of
     DA PAM 310-1   Authorized Publications for
     SL-1-2   Euipment support.
                                       Air Force Maintenance Forms and Records.
     AFM 66-1   Marine Corps Forms and Records Procedures.
     TM 4700.15/1   Recommendations for Changes and Improvements for Technical publications.
     NAVMC Form 10772   Recommendations for Changes and Improvements for Technical Publications.
                                       Recommendations for Changes and Improvements for Technical Publications.
     AFTO Form 22

APPENDIX B

AUTHORIZED ITEMS LIST

This page intentionally left blank.

MARINE CORPS TM 05926B/06509B-12
ARMY   TM 5-6115 -615-12
NAVY   NAVFAC P-8-646-12
AIR FORCE

APPENDIX C

MAINTENANCE ALLOCATION CHART

Section 1.   INTRODUCTION

1.   **GENERAL.**

    a.   This section provides a general explanation of all maintenance and repair functions authorized at various maintenance levels.

    b.   Section II designates overall responsibility for the performance of maintenance functions on the identified end item or component.The implementation of the maintenance functions upon the end item or component will be consistent with the assigned maintenance functions.

    c.   Section III lists the tools and test equipment required for each maintenance function as referenced from Section II.

    d.   Section IV contains supplemental instructions, explanatory notes and/or illustrations required for particular maintenance functions.

    e. **EXPLANATION OF COLUMNS IN SECTION II.**

    a.   Group Number.   Column 1.   The assembly group is a numerical group assigned to each assembly in a top down breakdown sequence. The applicable assembly groups are listed on the MAC in disassembly sequence beginning with the first assembly removed in a top down disassembly sequence.

    b.   Assembly.   Group.   Column 2. This column contains a brief description of the components of each assembly group.

    c.   Maintenance Functions.   Column 3.   This column lists the various maintenance functions (A through K) and indicates the lowest maintenance category authorized to perform these functions. The symbol designations for the various maintenance categories are as follows:

C-   Operator or crew
O-   Organizational maintenance
F-   Direct support maintenance
H-   General support maintenance
D-   Depot maintenance

The maintenance functions are defined as follows:

    A-Inspect.   To determine serviceability of an item by comparing its physical, mechanical, and electrical characteristics with established standards.

    B-- Test. To verify serviceability and to detect electrical or mechanical failure by use of test equipment.

    C-- Service. To clean, to preserve, to charge, and to add fuel, lubricants, cooling agents, and air.   If it is desired that elements, such as painting and lubricating, be defined separately, they may be so listed.

D– Adjust. To rectify to the extent necessary to bring into proper operating range.

E — Align. To adjust specified variable elements of an item to bring to optimum performance.

F-- Calibrate. To determine the corrections to be made in the readings instruments or test equipment of used in precise measurement. Consists of the comparison of two instruments, one of which is a certified standard of know

accuracy, to detect and adjust any discrepancy in the accuracy of the instrument being compared with the certified standard.

G-- Install. To set up for use in an operational environment such as an emplacement, site, or vehicle.
H-- Replace. To replace unserviceable items with serviceable like items.

I-- Repair Those maintenance operations necessary to restore an item to serviceable condition through correction of material damage *or* a specific failure. Repair may be accomplished at each category of maintenance.

J -- Overhaul. Normally, the highest degree of maintenance performed by the Army in order to minimize time work in process is consistent with quality and economy of operation. It consists of that maintenance necessary to restore an item to completely serviceable condition as prescribed by maintenance standard in technical publications for each item of equipment. Overhaul normally does not return an item to like new, zero mileage, or zero hour condition.

K-- Rebuild. The highest degree of material maintenance. It consists of restoring equipment as nearly as possible to new conditions in accordance with original manufacturing standards. Rebuild is performed only when required by operational considerations or other paramount factors and then only at the depot maintenance category. Rebuild reduce to zero the hours or miles the equipment, or component thereof, has been in use.

d. Symbols. The uppercase letter placed in the appropriate column indicates the lowest level at which that particular maintenance function is to be performed.

e. Tools and Equipment. Column 4. This column is provided for referencing by code, the special tools and test equipment, (Section III) required to perform the maintenance functions (Section II).

f. Remarks. Column 5. This column is provided for referencing by code, the remarks (Second IV) pertinent to the maintenance functions.

3. EXPLANATION OF COLUMNS IN SECTION III.

a. Reference Code. This column consists of a number and a letter separated by a dash. The number references the T and TE requirements column on the MAC. The letter represents the specific maintenance function the item is to be used with. The letter is representative of columns A through K on the MAC.

b. Maintenance Category. This column shows the lowest level of maintenance authorized to use the special tool or test equipment.

C-2

c. Nomenclature. test   of the tool or

d. Tool Number. This column lists the manufacturer's code and National Stock Number of tools and test equipment.

4.   **EXPLANATION OF COLUMNS IN SECTION IV.**

a. Reference Code. This column consists of two letters separated by dash, both of which are references to Section II. The first letter references column 5 and the second letter references a maintenance function, column 3, A through K.

b. Remarks. This column lists information pertinent to the maintenance function being performed, as indicated on the MAC, Section II.

SECTION II — MAINTENANCE ALLOCATION CHART

(1) GROUP NO.	(2) ASSEMBLY GROUP	(3) MAINTENANCE FUNCTIONS											(4) TOOLS AND EQUIPMENT	(5) REMARKS
		A INSPECT	B TEST	C SERVICE	D ADJUST	E ALIGN	F CALIBRATE	G INSTALL	H REPLACE	I REPAIR	J OVERHAUL	K REBUILD		
01	**DC ELECTRICAL AND CONTROL SYSTEM**													
	Batteries	C	O	C,O					O	F			1-B	
	Battery Cables	C		O					O	O				
	Battery Frame and Tray	C							O					
	Battery Hold Down	C	O						C					
	Slave Receptacle	C	O						O	O			4-B, 5-B	
	Voltage Regulator Battery Chg.	C	O						O	O			5-B	
	Engine Wiring Harness	O							O					
02	**FRAME**													
	Frame	C								F				
	Grounding Rod	C		C										
	Grounding Strap	C		C										
03	**ELECTRICAL POWER GENERATION SYSTEM**													
	Generator Assy	O	F						F	F			5-B	
	Bearings	F	F						C,O					
	Rectifier Assy Rotating Diode	F							C					
	Generator Fan	F							F					
	Rotor Assy	F	F						F				5-B	
	Housing Assy	F	F						F					
	Exciter = Parts Rotor	F							F					
04	**FUEL SYSTEM AND OIL DRAIN**													
	Fuel Pump, Transfer	C	O	O					O	O				
	Fuel Pump, Auxiliary	C	O	O					O	O				
	Filters	C		O					O					

SECTION II — MAINTENANCE ALLOCATION CHART

(1) GROUP NO.	(2) ASSEMBLY GROUP	A INSPECT	B TEST	C SERVICE	D ADJUST	E ALIGN	F CALIBRATE	G INSTALL	H REPLACE	I REPAIR	J OVERHAUL	K REBUILD	(4) TOOLS AND EQUIPMENT	(5) REMARKS
04	FUEL SYSTEM AND OIL DRAIN (Cont)													
	Fuel Tank	C	O,F	C					F				5-B	
	Float Switches	O	O						O				7-B & D	No Repair in -34
	Fuel Injection Pump	O	F	F	F				F	F				
	Fuel Injector, Nozzle	F	F		F				F					
	Fuel Line, Flexible	C							O					
	Oil Drain Valve	C							O					
05.	ENGINE													
	Engine Assembly	C		C					F					
	Air Cleaner Assy	C		C					O					
	Valves	F							F					
	Rocker Arms	O							F					
	Cylinder Head Assy	F	F						F	H			13-I	
	Governor, Mechanical (D)	O			F				O	F			12-D	
	Governor, Linkage	C		O	O				O	O				
	Valve Seats	F							H				15-I	
	Valve Guides	F							F					
	Valve Lifters	F							F					
	Valve Springs	F							F					
	Push Rods	F							F					
	Camshaft	F							F					
	Piston	F							F				10-A	
	Piston Pin	F							F				10-A	
	Piston Rings	F							F				10-A, 12-D	
	Connecting Rod	F							F				10-A	
	Rod Bearings	F							F				10-A	
	Cylinder Barrel	F							F				10-A	
	Main Bearings	F							F				10-A	
	Oil Pump	F	F						F	F			12-D	

SECTION II — MAINTENANCE ALLOCATION CHART

(1) GROUP NO.	(2) ASSEMBLY GROUP	(3) MAINTENANCE FUNCTIONS											(4) TOOLS AND EQUIPMENT	(5) REMARKS
		A INSPECT	B TEST	C SERVICE	D ADJUST	E ALIGN	F CALIBRATE	G INSTALL	H REPLACE	I REPAIR	J OVERHAUL	K REBUILD		
05	ENGINE (Cont)													
	Oil Filter	C							O					
	Camshaft Bearings	H							H				10-A	
	Crankcase	H							H				10-A	
	Intake Manifold	O							O					
	Flywheel Assembly	F							F					
	Timing Gears	H							H					
	Oil Pan	C							F					
	Crankshaft Assembly	H							H	H				
	Engine Mounts	O							O	F				
	Starter Assy	O							O					
	Starter Solenoid								F					
	Oil Filter Adapter	O		O					F					
	Glow Plug		O,F						O					
	Oil Cooler	O	O						F					
	Low Fuel Solenoid, Shutdown	O	O						O					
06	ENGINE EXHAUST													
	Muffler	C							O					
	Clamps	C							O					
	Exhaust Pipes & Ducts	C							O					
	Lifting Eye	C							O					
07	ENGINE CONTROLS													
	Governor Speed Control	C							O					
08	GENERATOR CONTROLS and INSTRUMENTS													
	Control Box Assembly	C	O						O	O			4-B	
	Voltmeters	C	O						O					
	Current Indicating Meter	C	C						O					

SECTION II — MAINTENANCE ALLOCATION CHART

(1) GROUP NO.	(2) ASSEMBLY GROUP	(3) MAINTENANCE FUNCTIONS											(4) TOOLS AND EQUIPMENT	(5) REMARKS
		A INSPECT	B TEST	C SERVICE	D ADJUST	E ALIGN	F CALIBRATE	G INSTALL	H REPLACE	I REPAIR	J OVERHAUL	K REBUILD		
08	GENERATOR CONTROLS and INSTRUMENTS (Cont)													
	Rheostat	C	O						O					
	Frequency Meter	C	F						O					
	Frequency Transducer	C	F						O					
	Tachometer	C	O						O					
	Rectifier Bridge Assy	C	O						O	O			5-B	
	AC Circuit Breaker	C	O						O				5-B	
	DC Control Circuit Breaker	C	O						O	O			5-B	
	Switches	C	O						O				5-B	
	Control Wiring Harness	C,O	O						O	F				
	Starter Relay	F	F		F				F					
	Voltage Regulator	O	F						F					
	Current Transformer	O							O					
	Load Terminal Board	O	O						O					
	Convenience Outlet	O							O					
	Convenience Outlet Fuses	C	O		F				F	F			5-B	
	Hourmeter		F						F					
	Engine Control PCB								F					
09	SKID BASE													
	Skid Base Assembly	C							F					
	Engine/Generator Mounts	F							F					
	Skid Base	C							F					
	Ground Stud	C							O					
10	ACOUSTIC SUPPRESSION KIT													
	Panels	C,O							C,O	O			17-I,19-I,20-I	
	Fan Assembly	C,O	O						O	O			5B	A-H, W-I

MARINE CORPS TM 05926B/06509B-12
ARMY   TM 5-6115-615-12
NAVY   NAVFAC P-8-646-12

SECTION II — MAINTENANCE ALLOCATION CHART

(1) GROUP NO.	(2) ASSEMBLY GROUP	(3) MAINTENANCE FUNCTIONS A INSPECT	B TEST	C SERVICE	D ADJUST	E ALIGN	F CALIBRATE	G INSTALL	H REPLACE	I REPAIR	J OVERHAUL	K REBUILD	(4) TOOLS AND EQUIPMENT	(5) REMARKS
10	ACOUSTIC SUPPRESSION KIT (cont'd)													
	Fan Wiring Harness	O	O						O	O			5B	X-I
	Fuse Block Assembly	O	O						O	O			5B	Y-I
	Bracket, Capacitor	C							O					
	Power Cable Clamp	C							O					
	Bonding Wire Assembly	O							O					
	Modified Exhaust Pipe	O							O					
	Flexible Exhaust Tube								O					
	Exhaust Tube Assembly								O				17-I	
	Oil Drain Adapter, Hose, and Clamp	C							O					
	Gaskets	O							O					
	Rivet Fasteners	O												

SECTION III. TOOLS, TEST AND SUPPORT EQUIPMENT REQUIREMENTS

REFERENCE CODE	MAINTENANCE CATEGORY	NOMENCLATURE	NSN
1-B	0	Tester, Battery, Electrolyte solution	6630-00-171-5126
4-B	F	Multimeter	6625-00-553-0142
5-B	0	Ohmmeter	6625-00-581-2466
7-B & D	F	Tool, Test Set, Diesel Injector	4910-00-317-8265
10-A	H	Micrometers, Inside & outside	
11-H	H	Puller Kit, universal	5180-00-701-8046
12-D	0	Gage, thickness	5210-00-221-1999
13-I	F	Grinding machine, valve face	4910-00-540-4679
15-I	F	Grinding Kit, Valve seat	4910-00-473-6437
17-I	0	Tool Kit, Blind Rivet	5180-01-201-4978
19-I	0	Drill Stop	
20-I	0	Drill, Electric	5130-00-889-8994
21-I	0	Soldering Gun	3439-00-618-6623

Section IV. REMARKS

REFERENCE CODE	REMARKS
A-A	Visual insepction
C-A	Visual inspection
D-C	Lubriate hinges and latches
E-I	Welding and staighten
G-B	Operational Test
H-B	Pressure Test after repair
I-B	Continuity Ted
J-K	Fabricate new harness
K-B	Continuity Test
M-B	Continuity Test
O-K	Fabricate new harness
P-I	Repair by replacement of components
Q-B	Operational Teat
R-D	Zero adjustment
S-B	Continuity Test
U-B	Continuity Test
V-B	Operational Test
A-H	Operator to replace Panel 4 only
W-I	Repair by replacement of latch assembly
X-I	Repair by replacement of capacitors or receptacles
Y-I	Repair by replacement of fuses or tabs

MARINE CORPS TM 05926B/06509B-12
ARMY   TM 5-6115-615-12
NAVY   NAVFAC P-8-646-12
AIR FORCE   TO 35C2-3-386-31

INDEX

Subject	Paragraph, Table Number

INDEX

INDEX

INDEX

MARINE CORPS TM 05926B/06509B-12
ARMY   TM 5-6115-615-12
NAVY   NAVFAC P-8-646-12
AIR FORCE   TO 35C2-3-386-31

INDEX

Subject

Paragraph,
Table

INDEX

ZONE	LTR	DESCRIPTION	DATE	APPROVED
	A	FIRST ARTICLE CONFIGURATION AUDIT	7/21/86	
	B	REVISED PER ONAN ER NO 28768	86/9/24	
A-3	C	ADD PART LABEL "AIKI"	86/10/8	
	D	REVISED PER ONAN ER NO 29370	87/1/15	
	E	REVISED PER ONAN ER NO 29475	87/1/29	
	F	REVISED PER ONAN ER NO 29519	87/2/9	
	G	SERVICE SIGN-OFF	3/12/87	

NOTES:

1. FOR 3KW, 60 Hz ASSEMBLY. SEE 84-016.
2. FOR 3KW, 400 Hz ASSEMBLY. SEE 84-021.
3. THE 400 Hz MODEL DOES NOT CONTAIN THE AC RECEPTACLE A2J1 OR THE FUSES A2F1 OR A2F2.
4. FRONT PANEL SWITCH A2S3 HAS JUMPERS INSTALLED:
   - BETWEEN A2S3-12 AND A2S3-15 AND A2S3-27
   - BETWEEN A2S3-16 AND A2S3-17 AND A2S3-14
   - BETWEEN A2S3-22 AND A2S3-23 AND A2S3-24
   - BETWEEN A2S3-25 AND A2S3-26 AND A2S3-13
5. FRONT PANEL SWITCH A2S1 HAS JUMPER INSTALLED BETWEEN A2S1-13 AND A2S1-14.
6. FOR 3KW, 60 Hz AND 400 Hz CONTROL WIRING DIAGRAM SEE 84-13194. FOR 3KW, 60 AND 400 Hz ENGINE WIRING DIAGRAM. SEE 84-13307.
7. FOR INTERPRETATION OF DRAWING, SEE 84-13110. NOTE 1.

8. CONTROL BOX SWITCH A1S1 HAS JUMPERS INSTALLED
   - BETWEEN A1S1-2 AND A1S1-4
   - BETWEEN A1S1-3 AND A1S1-7
   - BETWEEN A1S1-6 AND A1S1-8 AND A1S1-10 AND A1S1-14
   - BETWEEN A1S1-9 AND A1S1-11
   - BETWEEN A1S1-18 AND A1S1-20 AND A1S1-22 AND A1S1-24

A2S3 VOLTAGE SELECTOR SWITCH

POSITION	CONTACTS CLOSED
V1-0	11 AND 12 AND 15, 21 AND 22
V2-0	11 AND 13, 21 AND 23
V3-0	11 AND 14, 21 AND 24
V1-2	11 AND 12 AND 15, 21 AND 25
V2-3	11 AND 15, 21 AND 26
V3-1	11 AND 17, 21 AND 27

A2S2 CURRENT SELECTOR SWITCH

POSITION	CONTACTS CLOSED
I-1	11 AND 12
I-2	11 AND 13
I-3	11 AND 14

A1S1 VOLTAGE RECONNECTION SWITCH

POSITION	CONTACTS CLOSED
120/208 3Ø	1 AND 2, 5 AND 6, 7 AND 8, 9 AND 10, 17 AND 18
120	1 AND 2, 3 AND 4, 11 AND 12, 13 AND 14, 19 AND 20
240 3Ø	3 AND 4, 9 AND 10, 21 AND 22
120 1Ø	3 AND 4, 11 AND 12, 13 AND 14, 15 AND 16, 23 AND 24

A2S1 MASTER SWITCH

POSITION	CONTACTS CLOSED
PRE-HEAT	11 AND 15
OFF	NONE
RUN/AUX.	11 AND 13 AND 14 AND 17
FUEL	11 AND 13 AND 14
RUN	11 AND 13 AND 14 AND 15,
START	25 AND 27 AND 28

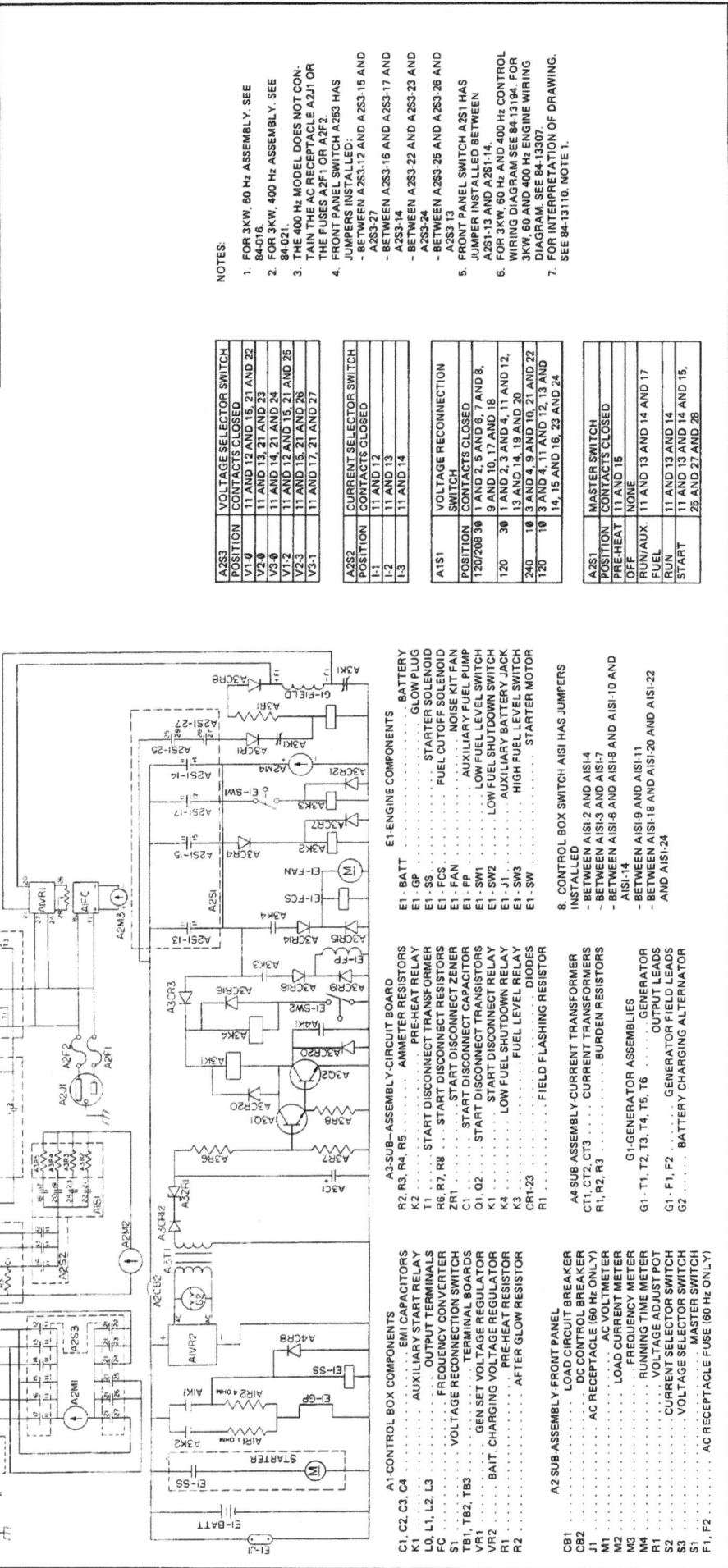

A1-CONTROL BOX COMPONENTS
- C1, C2, C3, C4 ... EMI CAPACITORS
- K1 ... AUXILIARY START RELAY
- L0, L1, L2, L3 ... OUTPUT TERMINALS
- FC ... FREQUENCY CONVERTER
- S1 ... VOLTAGE RECONNECTION SWITCH
- TB1, TB2, TB3 ... TERMINAL BOARDS
- VR1 ... GEN SET VOLTAGE REGULATOR
- VR2 ... BATT. CHARGING VOLTAGE REGULATOR
- R1 ... PRE-HEAT RESISTOR
- R2 ... AFTER GLOW RESISTOR

A2-SUB-ASSEMBLY-FRONT PANEL
- CB1 ... LOAD CIRCUIT BREAKER
- CB2 ... DC CONTROL BREAKER
- J1 ... AC RECEPTACLE (60 Hz ONLY)
- M1 ... AC VOLTMETER
- M2 ... LOAD CURRENT METER
- M3 ... FREQUENCY METER
- M4 ... RUNNING TIME METER
- R1 ... VOLTAGE ADJUST POT
- S2 ... CURRENT SELECTOR SWITCH
- S3 ... VOLTAGE SELECTOR SWITCH
- S1 ... MASTER SWITCH
- F1, F2 ... AC RECEPTACLE FUSE (60 Hz ONLY)

A3-SUB-ASSEMBLY-CIRCUIT BOARD
- R2, R3, R4, R5 ... AMMETER RESISTORS
- K2 ... PRE-HEAT RELAY
- T1 ... START DISCONNECT TRANSFORMER
- R6, R7, R8 ... START DISCONNECT RESISTORS
- ZR1 ... START DISCONNECT ZENER
- C1 ... START DISCONNECT CAPACITOR
- Q1, Q2 ... START DISCONNECT TRANSISTORS
- K1 ... LOW FUEL SHUTDOWN RELAY
- K4 ... FUEL LEVEL RELAY
- K3 ... FUEL LEVEL RELAY
- CR1-23 ... DIODES
- R1 ... FIELD FLASHING RESISTOR

A4-SUB-ASSEMBLY-CURRENT TRANSFORMER
- CT1, CT2, CT3 ... CURRENT TRANSFORMERS
- R1, R2, R3 ... BURDEN RESISTORS

G1-GENERATOR ASSEMBLIES
- G1 ... T1, T2, T3, T4, T5, T6 ... GENERATOR OUTPUT LEADS
- G1 ... F1, F2 ... GENERATOR FIELD LEADS
- G2 ... BATTERY CHARGING ALTERNATOR

E1-ENGINE COMPONENTS
- E1 ... BATT ... BATTERY
- E1 ... GP ... GLOW PLUG
- E1 ... SS ... STARTER SOLENOID
- E1 ... FCS ... FUEL CUTOFF SOLENOID
- E1 ... FAN ... NOISE KIT FAN
- E1 ... FP ... AUXILIARY FUEL PUMP
- E1 ... SW1 ... LOW FUEL LEVEL SWITCH
- E1 ... SW2 ... LOW FUEL SHUTDOWN SWITCH
- E1 ... J1 ... AUXILIARY BATTERY JACK
- E1 ... SW3 ... HIGH FUEL LEVEL SWITCH
- E1 ... SW ... STARTER MOTOR

Figure FO-1. Electrical Schematic, Models MEP-016B and MEP-021B.

FO-1/(FO-2 blank)

ZONE	LTR	DESCRIPTION	DATE	APPROVED
	A	FIRST ARTICLE CONFIGURATION AUDIT	9/12/86	
	B	REVISED PER ONAN ER NO 28768	86/9/24	
	C	REVISED PER ONAN ER NO 29,475	87/1/29	
	D	REVISED PER ONAN ER NO 29,519	87/2/9	
	E	SERVICE SIGN-OFF	3/12/87	

A2S1	MASTER SWITCH
POSITION	CONTACTS CLOSED
PRE-HEAT	11 AND 15
OFF	NONE
RUN/AUX FUEL	11 AND 13 AND 14 AND 15 AND 17
RUN	11 AND 13 AND 14
START	11 AND 13 AND 14 AND 15, 25 AND 27 AND 28

NOTES:

1. FRONT PANEL SWITCH A2S1 HAS A JUMPER INSTALLED BETWEEN A2S1-13 AND A2S1-14.

2. FOR 3KW, 28 VDC ASSEMBLY, SEE DRAWING 84-026.

3. FOR 3KW, 28 VDC CONTROL WIRING DIAGRAM, SEE DRAWING 94-13193.

4. FOR 3KW, 28 VDC ENGINE/GENERATOR WIRING DIAGRAM SEE DRAWING 84-13305.

5. FOR INTERPRETATION OF DRAWING SEE 84-13110, NOTE 1.

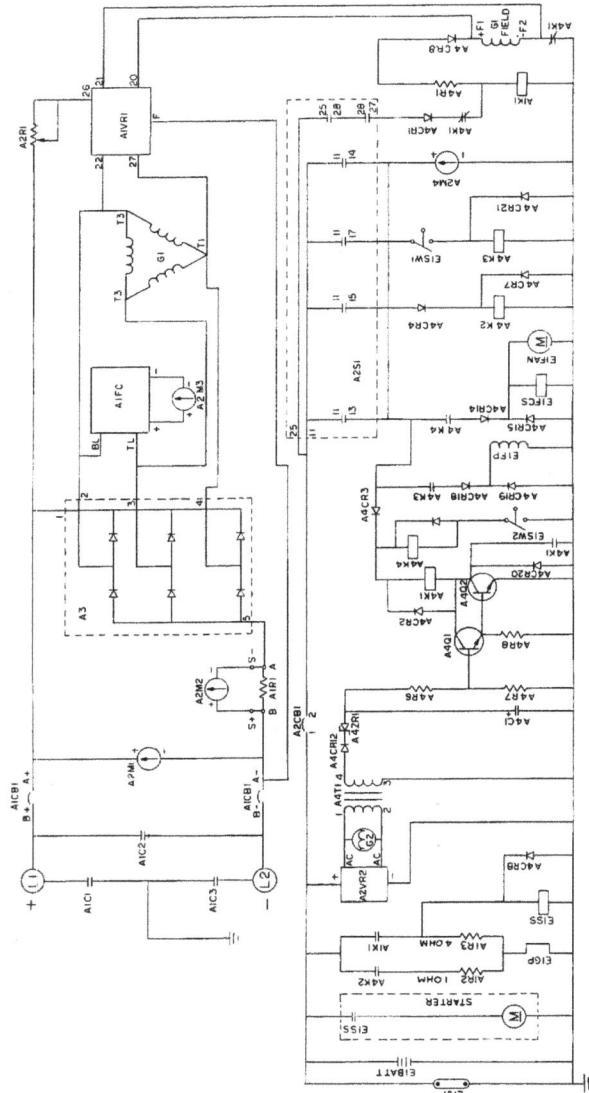

A1 .... CONTROL BOX COMPONENTS
A1, C1, C2, C3 .... EMI CAPACITOR
K1 .... AUXILIARY START RELAY
L1, L2 .... OUTPUT TERMINALS
FC .... FREQUENCY CONVERTER
VR1 .... GENERATOR SET VOLTAGE REGULATOR
CB1 .... LOAD CIRCUIT BREAKER
R1 .... LOAD CURRENT SHUNT
R2 .... PRE-HEAT RESISTOR
R3 .... AFTER GLOW RESISTOR

A2-SUB-ASSEMBLY-FRONT PANEL
CB1 .... DC CIRCUIT BREAKER
M1 .... DC VOLTMETER
M2 .... DC LOAD CURRENT METER
M3 .... RPM METER
R1 .... VOLTAGE ADJUST POT
S1 .... MASTER SWITCH
VR2 .... BATTERY CHARGING VOLTAGE REG.
TB3 .... TERMINAL BOARD
M4 .... RUNNING TIME METER

A3-SUB-ASSEMBLY-RECTIFIER STACK

E1 .... ENGINE COMPONENTS
BATT - BATTERY
GP - GLOW PLUG
SS - STARTER SOLENOID
FCS - FUEL CUTOFF SOLENOID
FAN - NOISE KIT FAN
FP - AUXILIARY FUEL PUMP
SW1 - LOW FUEL LEVEL SWITCH
SW2 - LOW FUEL SHUTDOWN SWITCH
J1 - AUXILIARY BATTERY JACK
SW3 - HIGH FUEL LEVEL SWITCH
GENERATOR ASSEMBLIES
G1-T1, T2, T3 .... GENERATOR OUTPUT LEADS
G1-F1, F2 .... GENERATOR FIELD LEADS
G2-BATTERY CHARGING ALTERNATOR

A4-SUB-ASSEMBLY-CIRCUIT BOARD

K2 .... PRE-HEAT RELAY
T1 .... START DISCONNECT TRANSFORMER
R6, R7, R8 .... START DISCONNECT RESISTORS
ZR1 .... START DISCONNECT ZENER
C1 .... START DISCONNECT CAPACITOR
Q1, Q2 .... START DISCONNECT TRANSISTORS
K1 .... START DISCONNECT RELAY
K4 .... LOW FUEL SHUTDOWN RELAY
K3 .... FUEL LEVEL RELAY
CR1-23 .... DIODES
R1 .... FIELD FLASHING RESISTOR

Figure FO-2. Electrical Schematic, Model MEP-026B.

FO-3/(FO-4 blank)

DEPARTMENT OF THE NAVY
HEADQUARTERS UNITED STATES MARINE CORPS
WASHINGTON, D.C. 20380-0001

NORMAL

MI-96-13200/00
1 APRIL 1999

U.S. MARINE CORPS MODIFICATION INSTRUCTION

INSTALLATION INSTRUCTIONS FOR

MODIFICATION OF GENERATOR SET,

DIESEL ENGINE DRIVEN, TACTICAL, SKID MOUNTED

TUBULAR FRAME, 3KW, 3 PHASE,

60 HZ, 120/208 VOLTS AC, MODEL MEP-O16B

NSN 6115-01-150-4140

1. <u>PURPOSE OF MODIFICATION INSTRUCTIONS.</u> The purpose of this MODIFICATION INSTRUCTION (MI) is to repower the diesel engine driven, skid mounted, tubular frame generator sets to a more reliable diesel engine driven, skid mounted, tubular frame generator sets for improved generator efficiency and logistic support.

2. <u>EQUIPMENT IN USE (INCLUDING EQUIPMENT IN SUPPLY OR MAINTENANCE ACTIVITIES AND ADMINISTRATIVE STORAGE).</u> All equipment in use or storage will be modified. This MI is to be applied to all existing Model MEP- 016B generator sets in the field.

3. <u>Technical Manuals Affected.</u> TM 05926B/06509B/3,SL-4-05926B/06509B-24P/2,TM 5-6115-615-24P, TM 5-6115-615-34,NAVFAC P-8-646-24P,NAVFAC P-8-646-34,TO 35C2-3-386-34,TO 35C2-3-386-32.

4. <u>Major Items Affected</u>.

Description	NSN	TAM No.
Generator Set, Diesel Engine Driven Tactical, Skid Mounted, Tubular Frame 3kw, 3 Phase, 60hz, 120/208 Volts AC Model Mep-016B	6115-01-150-4140	B0730

<u>New Model</u>

Mep-016E	6115-01-456-5046	B0730

5. <u>Material Affected</u>.

   a) <u>Material Required</u>

Description	NSN	Part Number	Quantity
Kit, Retrofit, Powerpack	2920-01-418-0970	GSA-247	1 Each

   b) <u>Material Discarded</u>

Description	NSN	Part Number	Quantity
Engine, Diesel, 3KW	2815-10-274-6803	Q106DL10399	1 Each

6. <u>MI KIT/PARTS AND THEIR DISPOSITION</u>.

   a) Kit/Parts Required To Accomplish This MI. The new parts required to accomplish this MI are contained in a modification kit composed of the items listed in Table 1. One kit is required for each Model MEP-016B generator set. The National Stock Number (NSN) for the kit is 2920-01-418-0970.

   b) Distribution and Issue Instructions. Modification kit/parts will be requisitioned from Commander General, Bldg. 3700, Code 837-1, MCLB, Albany, Ga. 31704. The requisition will specify the nomenclature and Serial Number of each Model Mep 016B generator set and the number of this MI.

   c) Bulk and Consumable Materials. Fuel-diesel, Oil-10W30, Loctite and Teflon tape.

   d) Parts Disposition. Parts will be disposed of in accordance with current Marine Corps directives

TABLE 1. MODIFICATION

ITE	QTY	DESCRIPTION	PART NUM	NATIONAL STOCK NUM
1	1	Engine,Diesel,3kw Generator	GS-2515	N/A
2	1	Oil Drain Adapter	CW-0071	4730-01462-8782
3	1	Stop Lever Bracket	GS-85	2815-01-463-1855
4	1	Throttle Solenoid Return Spri	GS-291	5360-01-462-9713
5	1	Engine Mount Bracket	GS-2470	2815-01-463-1826
6	1	Key, Machine	GS-2476	5315-01-462-9949
7	1	Drive Hub	GS-2477	2990-01-463-1829
8	1	Throttle Solenoid	GS-2479	2990-01-463-1860
9	1	Intake Air Hose, Diesel Engin	GS-2481	4720-01-463-0865
10	1	Intake Air Hose, Air Filter Si	GS-2483	4720-01463-0864
11	1	Adapter, Intake Hose	GS-2503	4720-01-463-0861
12	1	Coil Timer Assembly	GS-2491	5950-01-462-9335
12a	1	Coil Timer(w/o terminal ends	Cage 7838! P/N SA-46	5950-01-462-9335
13	1	Bracket Assembly	GS-2493	2815-01-463-1864
14	1	Fuel Hose, Pump To Filter	GS-2495	4720-01-463-0858
15	1	Fuel Hose, Filter To Injection	GS-2497	4720-01-463-0855

TABLE 1. MODIFICATION KIT (continued)

ITEM	QTY.	DESCRIPTION	PART NUMBER	NATIONAL STOCK NUMBER
16	1	Oil Drain Hose	GS-2499	4720-01-463-0856
17	1	Exhaust Pipe	GS-2501	2815-01-463-1888
18	1	Gasket, Exhaust Pipe	183671-13211	2990-01-464-6872
19	1	Wiring Harness	GS-2505	2815-01-463-1885
20	1	Jumper Wire Assembly	GS-2507	2815-01-463-1882
21	1	Fuel Filter Drain Hose	GS-2509	4720-01-463-0857
22	1	Speed Lever Modification	GS-2517	2815-01-463-1875
23	1	Throttle Solenoid Pusher	GS-3607	2990-01-463-1857
24	1	Flywheel Housing	GS-3609	2815-01-463-0819
25	1	Fuel Pump	MS51321-2-24N1	2910-00-930-9367
		Hardware required for Throttle Solenoid		
26	1	Screw, Cap, Hexagon Head	B1821BH025C100N	5305-00-225-3843
27	1	Washer, Flat	NAS1149F0463P	5310-00-141-1795
28	1	Washer, Lock	MS35338-44	5310-00-582-5965
29	1	Hex Spacer	H10375	5340-01-464-1079

7. SPECIAL TOOL; TOOL KITS; JIGS; TEST, MEASUREMENT, AND DIAGNOSTIC EQUIPMENT (TMDE); AND FIXTURES REQUIRED. A General Mechanic's Tool Kit and a Common Number One Tool Kit is required.

8. MODIFICATION PROCEDURES.

9. First part of the Modification Instructions covers the Generator Set Disassembly (pages 1-17).The second part of the Modification instructions covers the Retrofit Kit Installation Instructions (pages 1-23).

WARNING

. Remove watches, rings, and all other jewelry while working on or near this equipment. Wearing these items could result in injury or death to personnel or damage to equipment.

. Operation of this equipment presents a noise hazard to personnel in the area. The noise level exceeds the allowable limits for unprotected personnel. Wear earmuffs or earplugs that were fitted by a trained professional.

10. Instruction Plate Removal.

    a) Removal

    Remove instruction data plate by pushing out or prying out four rivets.

    Note 1: Maintain old data plate until old serial number is stamped on new data plate.

    b) Installation

    Install new data plate and secure with four rivets.

11. Instruction to Maintenance Management

    a) The Maintenance Management shall inscribe the old serial number as the new serial Number on the new data plate.

12. Generator Set Function Check (use the Test Data Sheet)

    a) Load bank installation:

        i)   Connect load bank to generator set.
        ii) Set all load switches to the off position.

13. Start generator set in accordance with generator starting procedure.

14. **After generator set is sufficiently warmed up , perform the following functional checks:**

    a) Voltage
    b) Frequency
    c) Percentage load-using load bank, load generator set as follows:
        -25%
        -50%
        -75%
        -100%
    d) Load duration shall be no longer than five minutes.
    e) Record test results on the Test Data Sheet

TEST DATA SHEET

FUNCTIONAL AND OPERATIONAL
TEST

Item _____   Date_____

_____

_____   Operator_____

Serial No._____

1. Electrical Control Box Functional Test:	YES	NO	COMMENTS
. voltage selector switch	_____	_____	_____
. current selector switch	_____	_____	_____
. frequency adjust	_____	_____	_____
. percent load meter	_____	_____	_____
. voltage meter	_____	_____	_____
. frequency meter	_____	_____	_____
. voltage adjust	_____	_____	_____
. circuit breaker	_____	_____	_____

2. Operational Test (Load Bank Hookup) :	YES	NO	COMMENTS
. voltage at 28/120/240	_____	_____	_____
. frequency 60/400 Hz	_____	_____	_____
. load duration			
25%	_____	_____	_____
50%	_____	_____	_____
75%	_____	_____	_____
100%	_____	_____	_____

3. Comments:_____

_____

_____

_____

I.   GENERATOR SET DISASSEMBLY

1. Remove ground rods (1), if applicable, from cage by loosening 1/2 inch
   bolt on clamp (2) with a wrench or socket. Removal of clamp is not
   required.

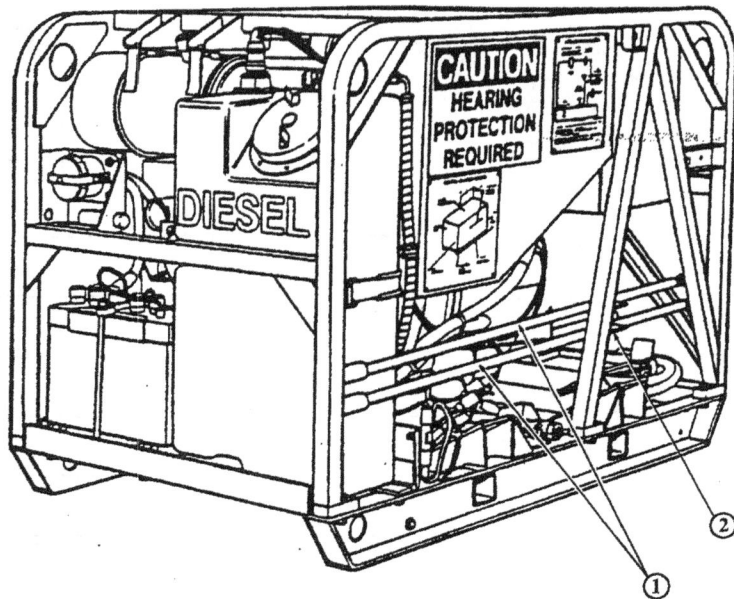

2. Prepare engine for disassembly by draining fluid:
   A. oil - Turn valve (1) 90°. After draining, close valve.

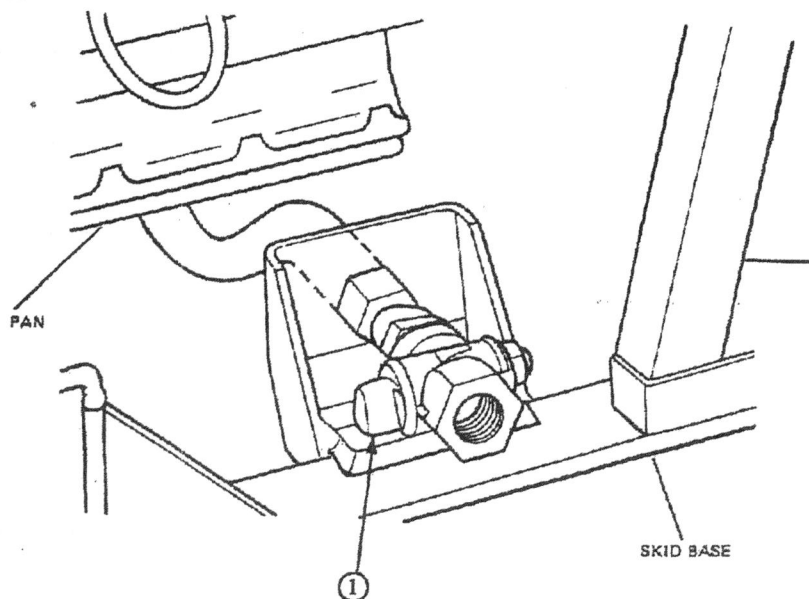

fuel - Remove fuel drain cap from petcock from base of fuel tank using a 5/8" open end wrench. Open petcock and drain fuel.

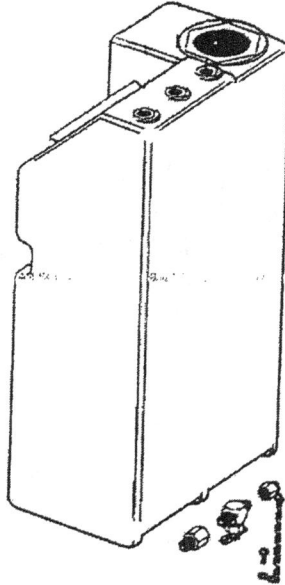

After draining fuel tank close petcock.

Note: Use proper containers and clean-up procedures.

3. Remove engine access cover (1) by removing two screws (2) with flat tip screwdriver.

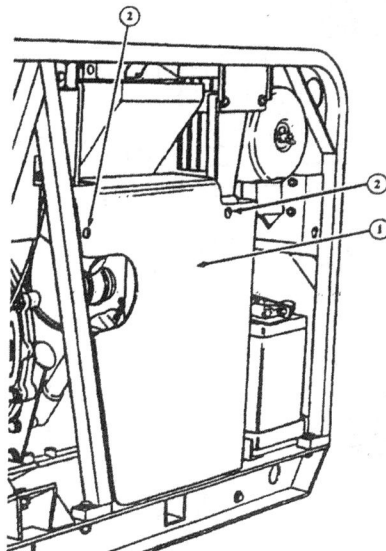

4. Disconnect battery cable (1) using 9/16 inch open end wrench. Remove battery bracket using 1/2 inch wrench. Remove the bracket (2) and battery (3) from generator set.

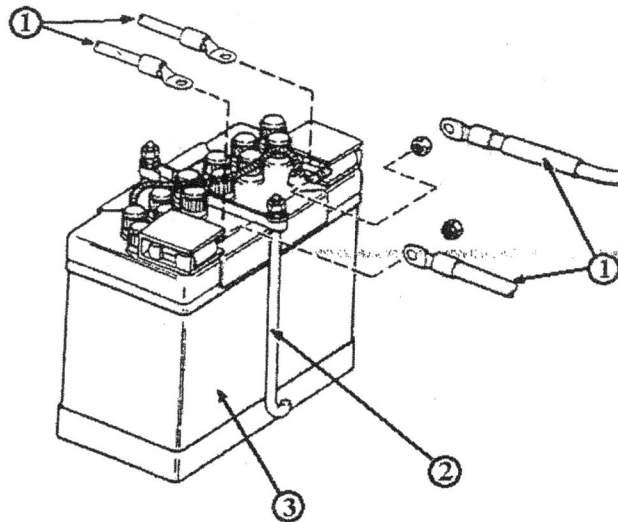

5. Open control panel (1) by using a 7/16 inch wrench to remove two bolts (2) (one bolt on each side of control panel), then use a flat tip screwdriver to unlock three screws (3).

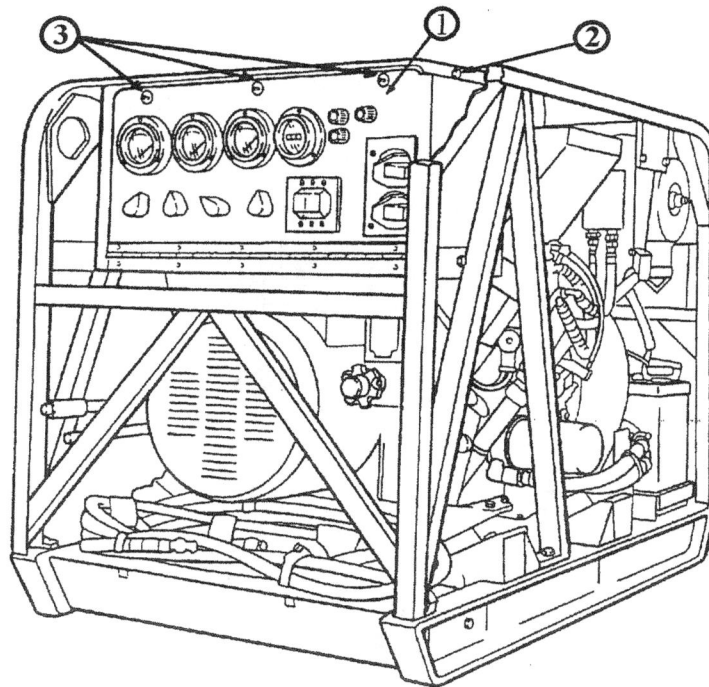

6. Locate cap screw holding hand throttle bracket. Remove hand throttle and bracket using flat tip screwdriver and 7/16 inch wrench. Replace nut and bolt on bracket for future use.

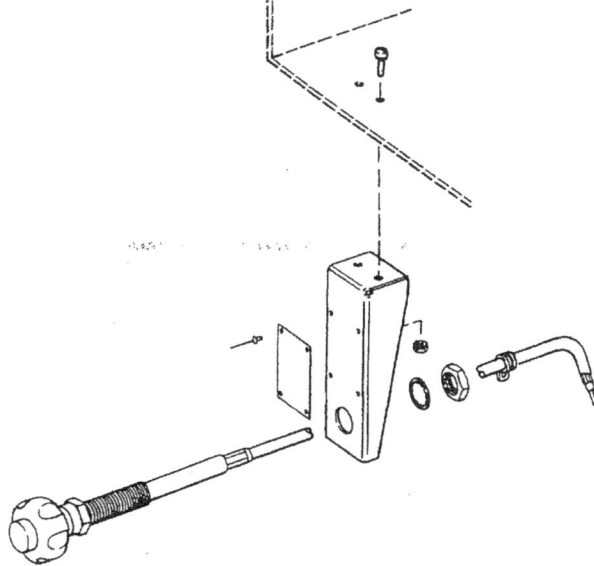

7. Drain fuel filter by pushing -up on drain fuel spout (1) and unscrew bleeder valve (2). Drain into proper container.

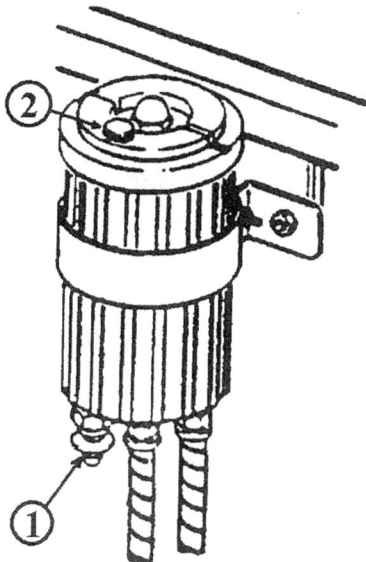

8. Remove two fuel lines (1) from fuel transfer pump (2) using a 5/8 inch and a 9/16 inch wrench.

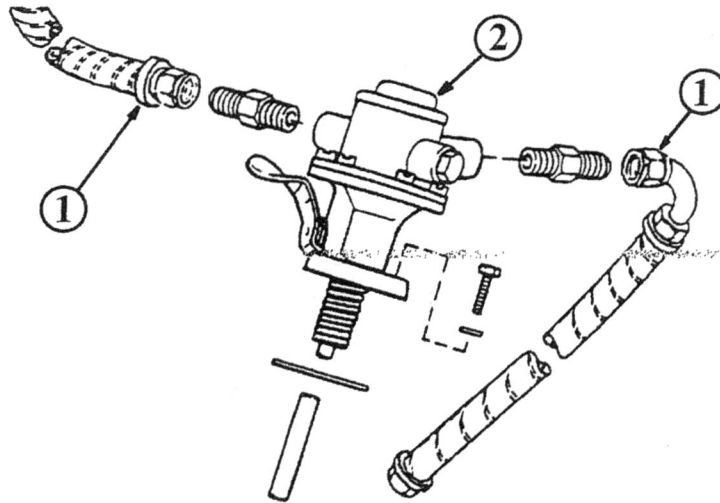

9. Remove fuel line (1) from fuel injection pump (2) using a 9/16 inch open end wrench. Use proper container to capture excess fuel and clean up spill.

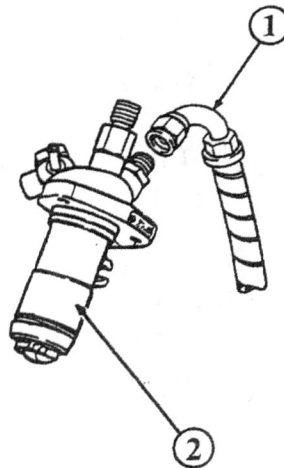

10. Loosen hose clamp (1). Remove fuel return line (2) from fuel injection pump (3) using a flat tip screwdriver. Retain fuel line and clamp for reassemble.

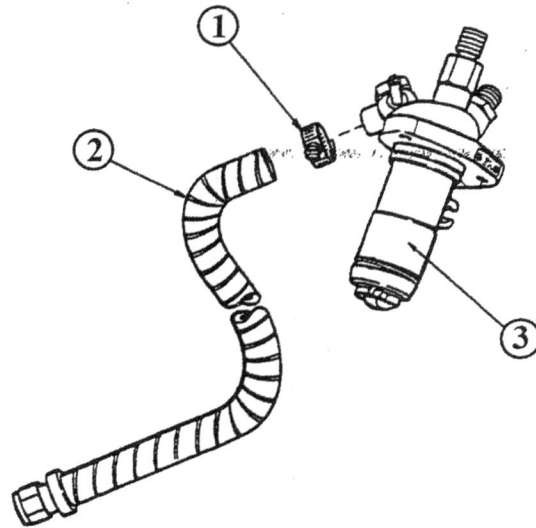

11. Remove two fuel lines (1) from fuel filter (2) using a 9/16 inch open end
wrench. Discard after removal.

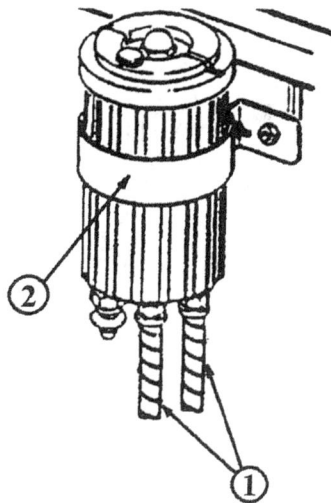

12. Disconnect ground wires (1) and (3) from grounding terminal (5) using two 7/16 inch wrench. Replace nut and washer on grounding terminal.

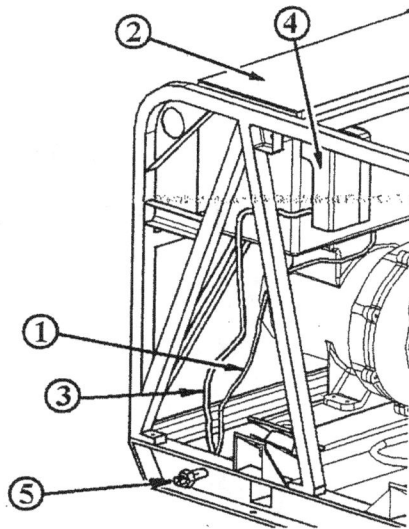

13. Disconnect hose clamp (1) at air cleaner (2) using flat tip screwdriver.

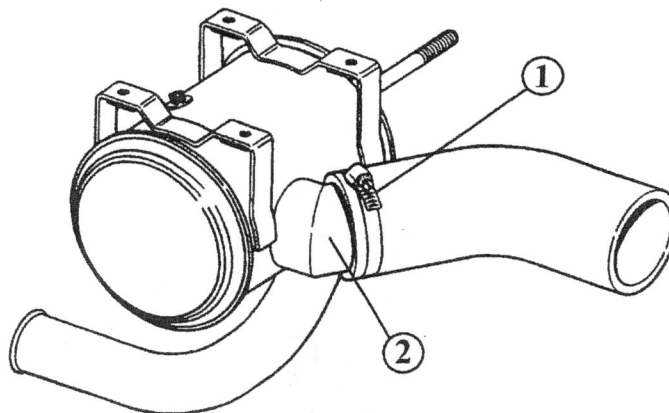

14. Disconnect glow plug wire male (1) from female plug (2) and identify and mark glow plug connector (1) for reassemble.

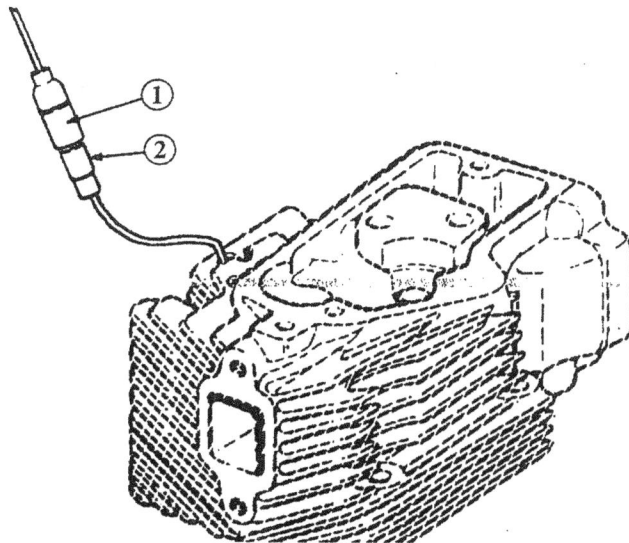

15. Disconnect T1 through T6 wires from terminal board TB1 using a small flat tip screwdriver. Disconnect F1 and F2 wire from terminal board TB2 using a small flat tip screwdriver. Individually mark and tag wires, if necessary.

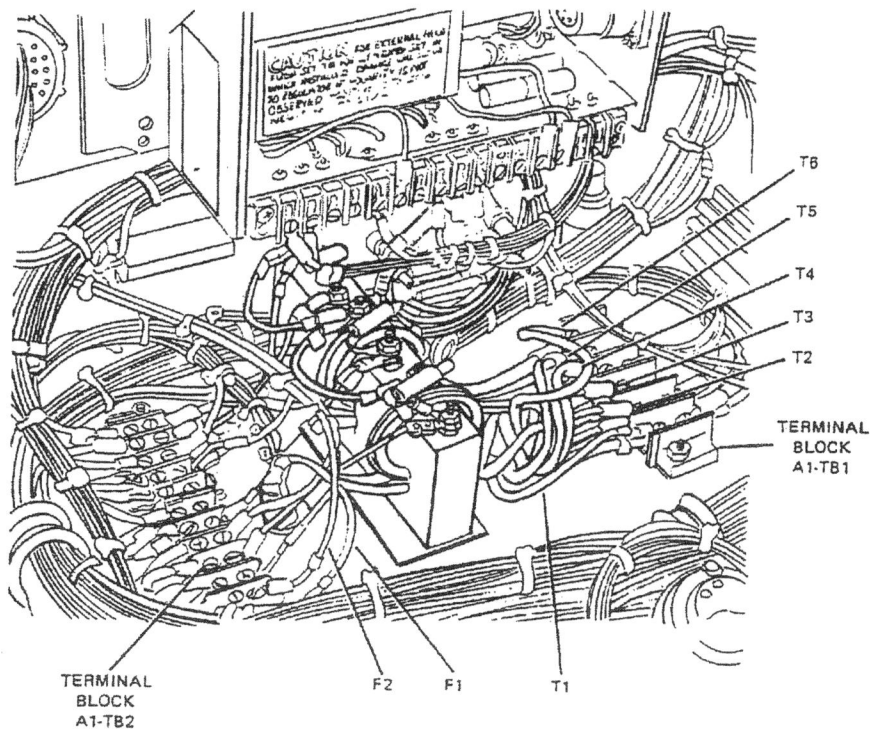

16. Disconnect positive battery wire (1) and small wire (2) from starter solenoid (3) using a 8mm and 12mm wrench. Identify and tag small positive
wire for reassemble. Disconnect ignition wire (4) from starter solenoid using an phillips screwdriver or a 5/16 inch wrench. Clip terminal lug (5) from ignition wire (4). Identify and tag for reassembly.

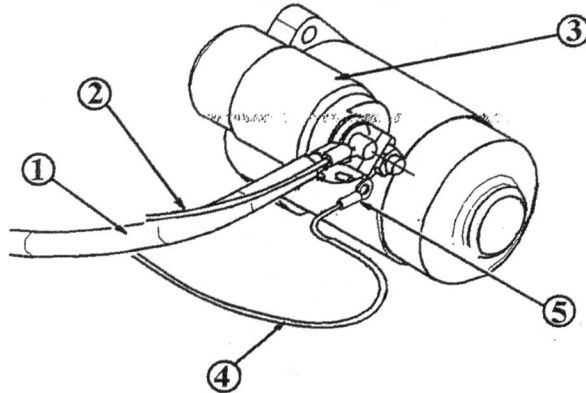

Note: There is no need to remove starter solenoid and starter solenoid negative wire.

17. Disconnect battery ground wire (1) and small wire (2) on starter (3) using a 15mm wrench. Remove and retain both battery cable and small wire. Replace nut and washer on the starter.

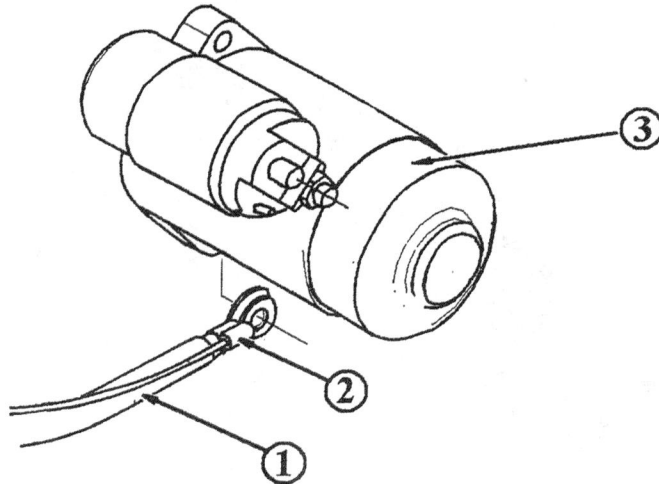

18. Unplug auxiliary fuel pump wire male plug (1) from female plug (2).

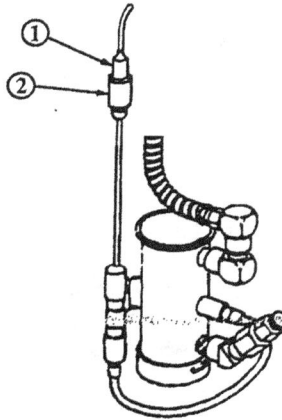

19. Disconnect the two male wire connectors from engine stator underneath starter from the two female connectors in existing wiring harness located near auxiliary fuel pump.

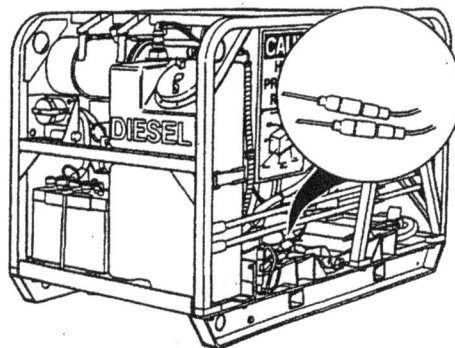

20. Disconnect oil drain line (1) from drain valve (2) using a 11/16 inch open end wrench.

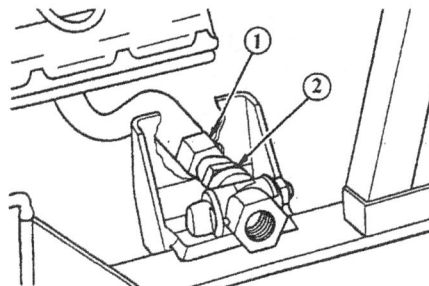

21. Disconnect fuel line (1) on top of auxiliary fuel pump (2) using a 9/16 open end wrench.

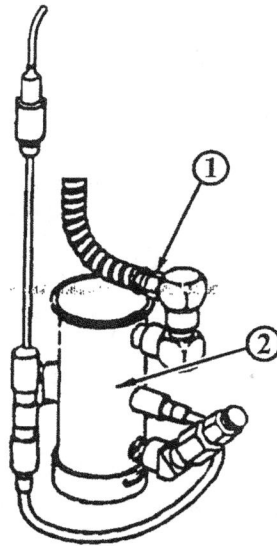

22. Disconnect fuel solenoid wires by using a small flat tip screwdriver to depress retaining tab (1) to unplug wire plug (2) from solenoid. (Be sure to pull wire free of engine so the cage can be removed.) Remove wire clamp (4) using a 10mm wrench or socket. Replace wire clamp after disassembly.

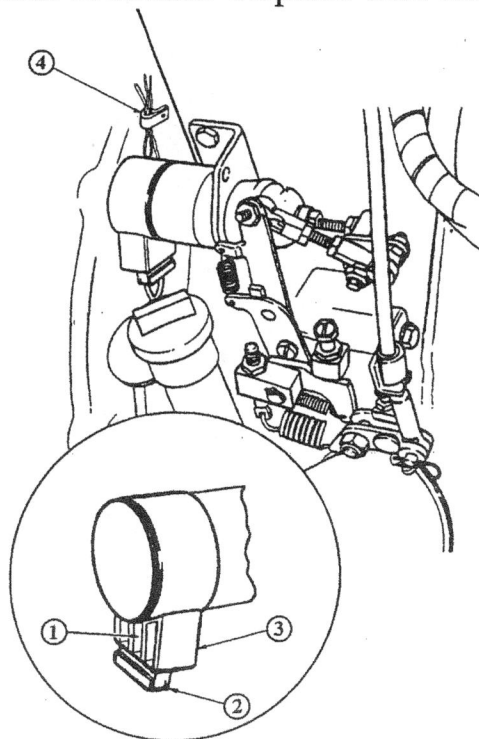

23. Remove screw (1) from fuel tank bracket (2) beside fuel tank using a flat tip screwdriver.

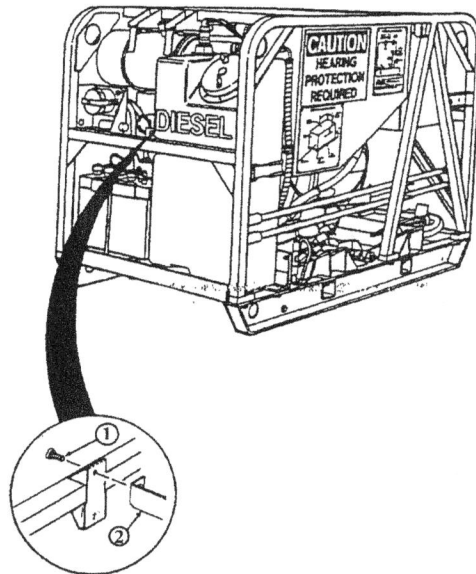

24. Remove six bolts (1), washers (2) and nuts (3) from cage (4) using two 1/2 inch box ends wrench. Retain bolts, washers and nuts for reassemble.

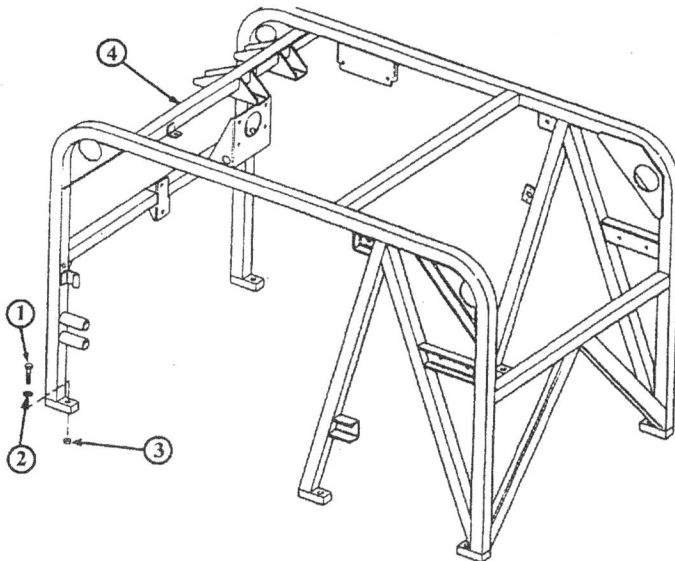

25. Lift cage off frame.

Caution should be taken to ensure wires that was disconnected in control box are not broken.

26. Remove two screws (1) from air baffle plate (2) using a flat tip screwdriver. Replace two screws on baffle plate.

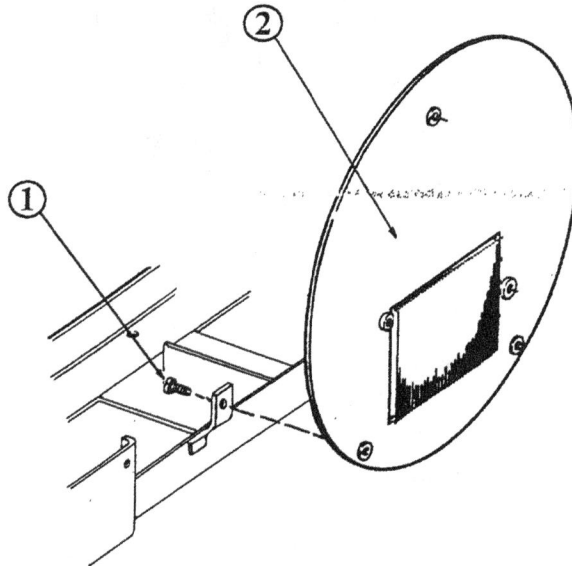

27. Remove two screws (1) from oil-cooler bracket (2) using a 7/16 inch wrench. Replace two screws on bracket.

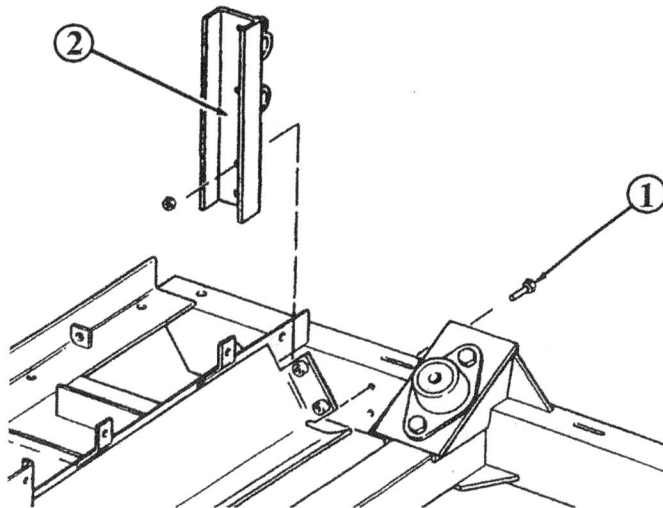

28. Remove two motor mount bolts (1) using a 9/16 inch box end wrench and socket. Remove four bolts (2) from generator mounting bracket (3) using an 11/16 inch wrench and 3/4 inch wrench. Retain bolts, washers, and nut for reassembly.

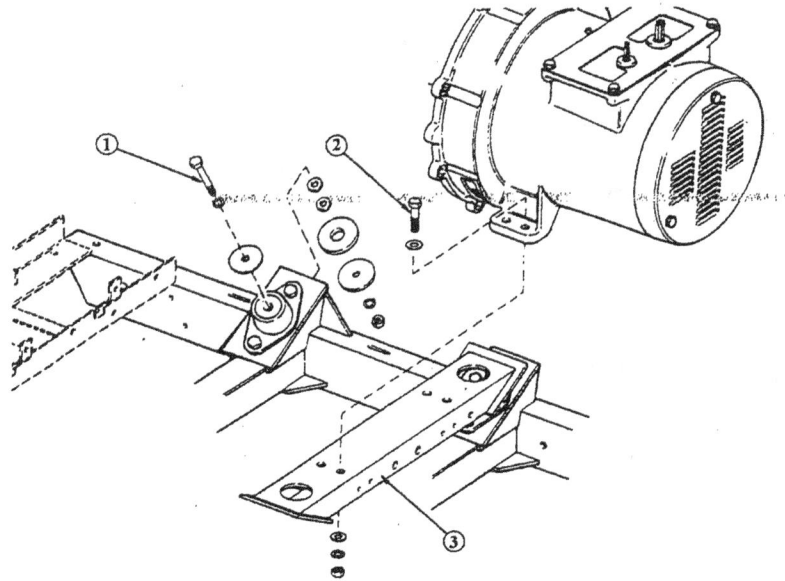

29. Remove engine/generator (1) from base (2).

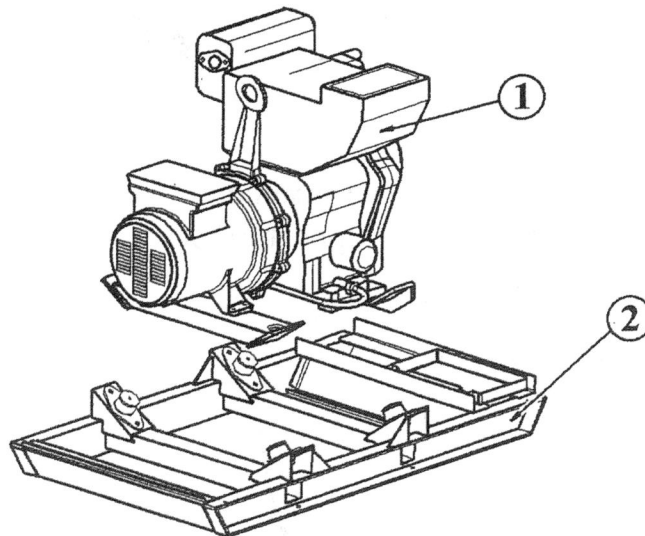

30. Remove three capscrews (1) on housing end cover (2) of generator using a 7/16 inch wrench.

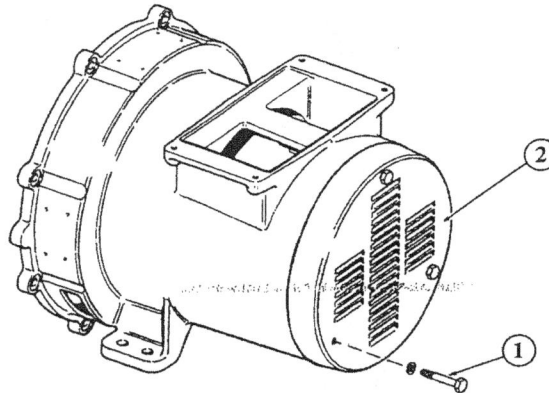

31. Remove eight bolts (1), lockwashers (2), and washers (3), from generator housing (4) assembly using a 13mm wrench or socket. Retain hardware for reassemble.

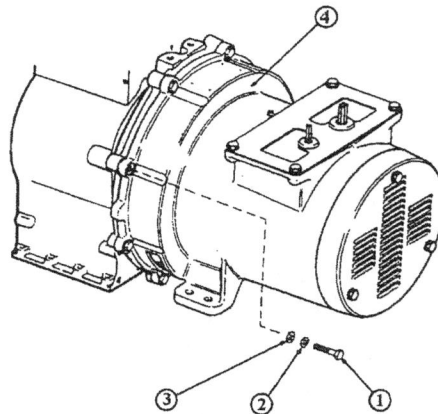

32. To support engine during generator removal, remove bolt (1) from engine support (2). Rotate engine support to the down position. Install bolt (1) to lock engine support in place.

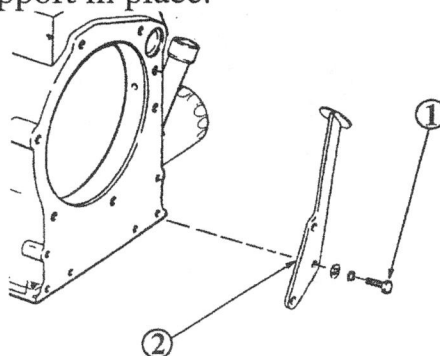

33. Utilizing a two jaw gear pulley at rear of generator housing assembly bearing, disconnect the housing from the generator rotor assembly.

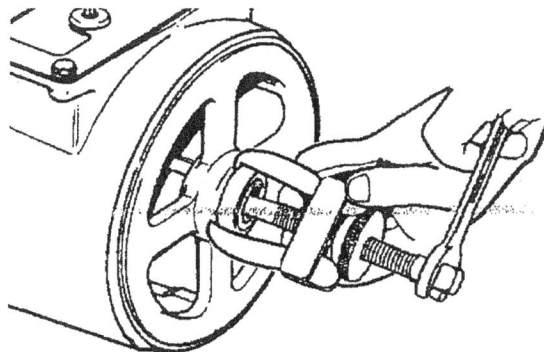

34. Remove eight capscrews (1) and washers (2) from armature coupling plate (3), using a 7/16 inch wrench. Retain hardware for reassemble.

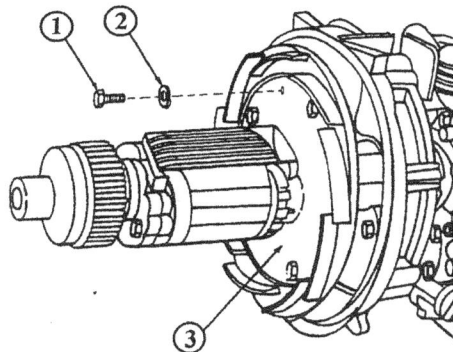

35. Remove rotor fan (4) by removing four bolts (1), lockwashers (2), and washers (3) using a 16mm wrench. Replace bolts, lockwashers, and washers on engine. Retain rotor fan for reassemble.

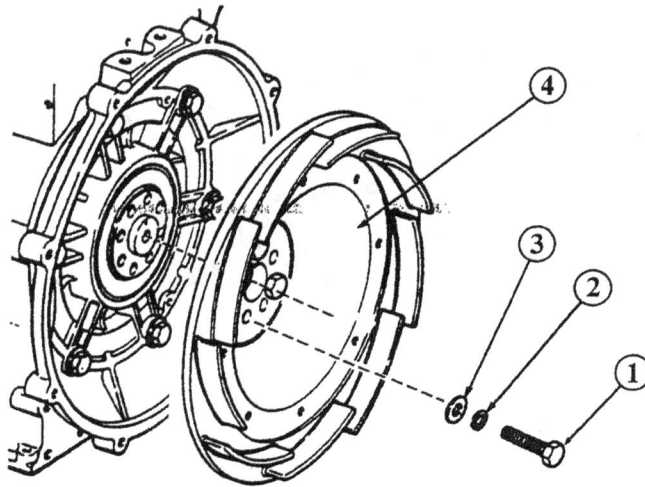

36. Transfer engine to secure area and remove engine lifting eye (1) by removing two capscrews (2), lockwasher (3) and spacer plate (4) using a 16mm socket. Retain hardware for reassemble on new engine.

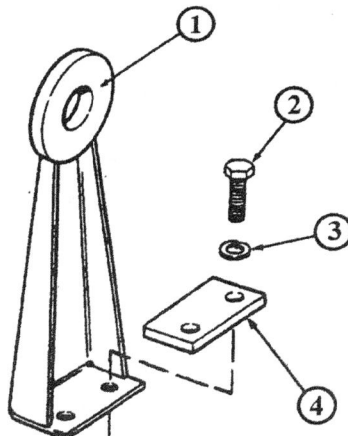

II. RETROFIT KIT INSTALLATION INSTRUCTIONS

1. Open the engine container and remove the box containing the retrofit hardware. Install lifting eye (1) with two capscrews (2), lockwasher (3) and spacer (4) using a 16-mm socket. Remove the engine from the container.

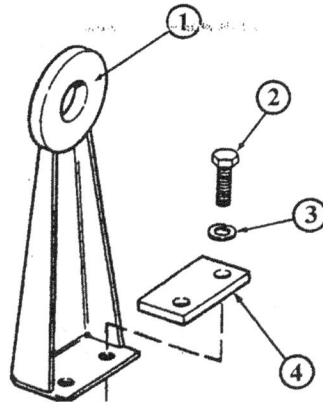

Figure 1

NOTE: Clean four (4) mating surfaces.

  a. Inside surface of rotor fan (1)
  b. Back surface of rotor fan (2)
  c. Outer edge of coupling plate (3)
  d. Outer edge of generator housing assembly (4)

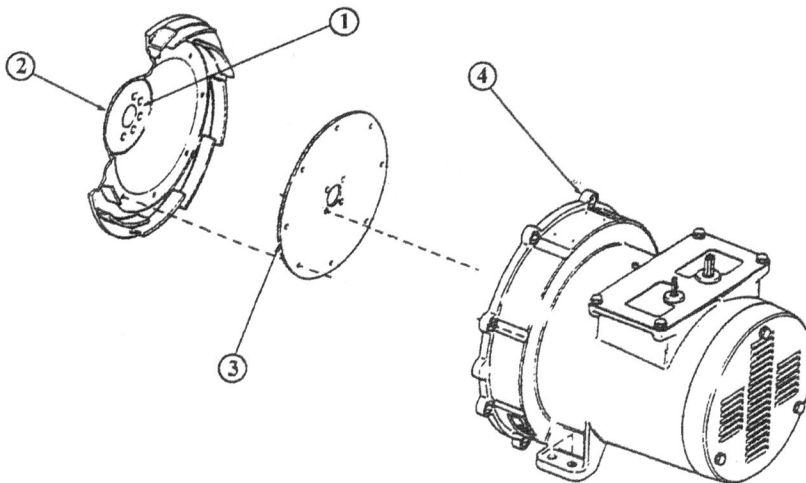

Figure 2

2. Install the fan (1) on crankshaft adapter (2). Use four new bolts 3/8-x 1/2 inch (3), lock washers (4), and flat washer (5) provided in kit. Use locktite on the bolt threads. Tighten using a 9/16-inch socket. The torque setting is 51 to 55 foot-pounds.

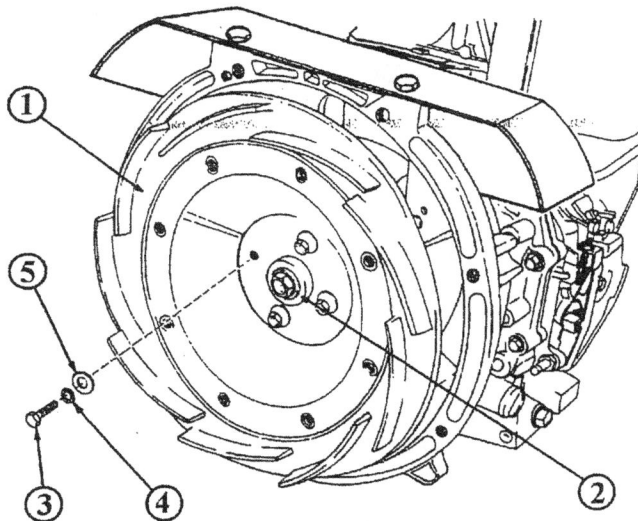

Figure 3

3. Install generator rotor assembly (1) with eight existing 1/4-x 3/4-inch bolts (2) using a 7/16-inch socket. Torque to 8 foot pounds.

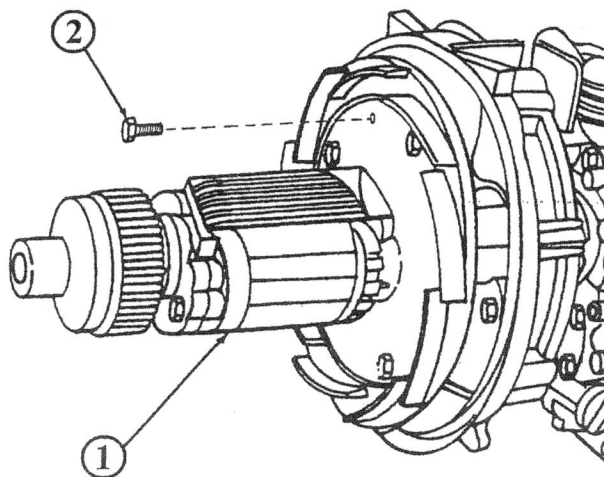

Figure 4

4. Install the generator housing assembly (4) over generator rotor assembly. Use eight existing bolts (1), lockwasher (2) and flat washer (3) and tighten with 13 mm wrench or socket. Use rubber mallet if necessary to seat housing. Note make sure the motor mount and the generator mount are aligned.

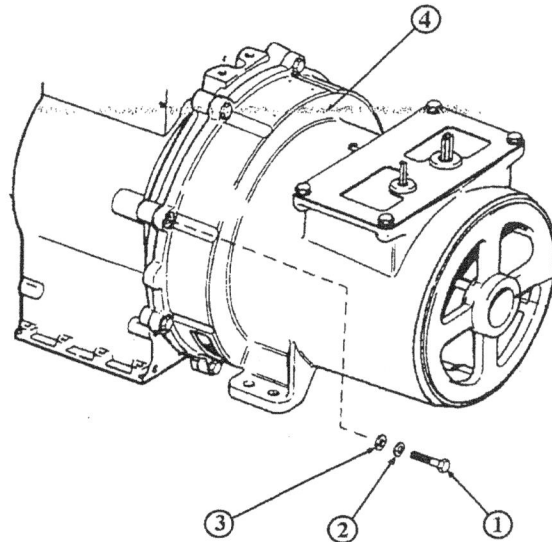

Figure 5

5. Install housing end cover (1) on generator housing (2) with three existing 7/16-inch bolts (3) and lockwasher (4). Make sure grill is facing down toward bottom of generator.

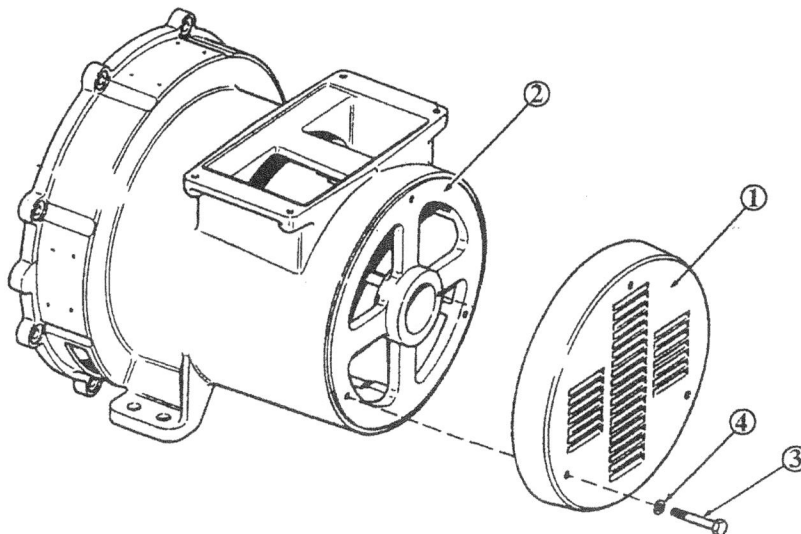

Figure 6

6. Install engine (1) and generator (2) onto base skid (3). Attach the engine (1) using two existing bolts (4), lockwasher (5), washer (6), rubber mount (7), washers 8, rubbers spacer (9) and nut (10) using a 9/16 inch box end wrench and socket. For the generator, use four existing bolts (11), washers (12), lockwashers (13), and nuts (14). Tighten using a 5/8 inch wrench or socket on the bolt and an 11/16 inch wrench or socket on the nut. *Get bolt pattern from S/L.*

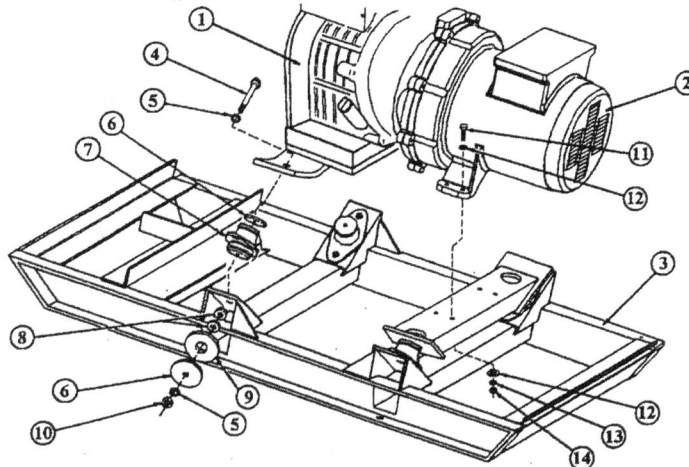

Figure 7

7. Install the new fuel pump assembly (1), P/N GS-2493 provided in kit, on the eng/gen skid assembly (2). Use the two holes that held the oil-cooler hoses bracket with two 1/4 x 1 1/4 inch bolts, two flat washers and two lock washers provided in kit, using a 7/16 inch wrench or socket.

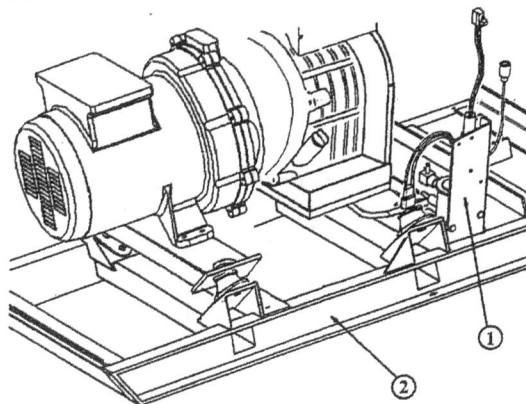

Figure 8

8. Remove the oil pan plug (1) on fuel pump side using a 17-mm wrench. **CAUTION: Some oil may drip**. Remove the gasket (2) from plug and it install on the hex fitting (3), P/N M16-3/8 NPT/ CW-0071 supplied in kit. Install hex plug in engine using 1-inch wrench.

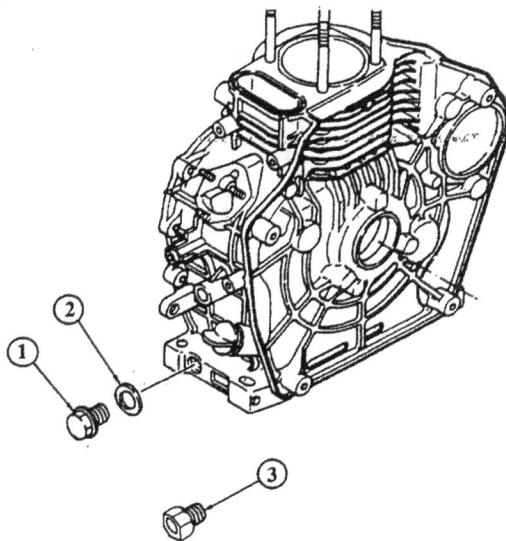

Figure 9

9. Install the oil drain hose (1) P/N GS-2499 supplied in the kit into hex fitting (2) on the engine, (use thread sealer or tape) and tighten with a 11/16 inch wrench.

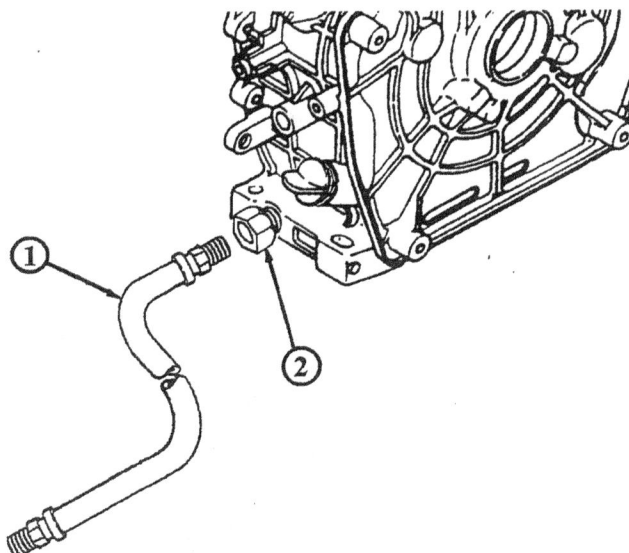

Figure 10

10. Connect other end of the oil drain hose (1) to the bulkhead of oil drain cock (2) on the other side of eng/gen skid assembly using an 11/16-inch wrench.

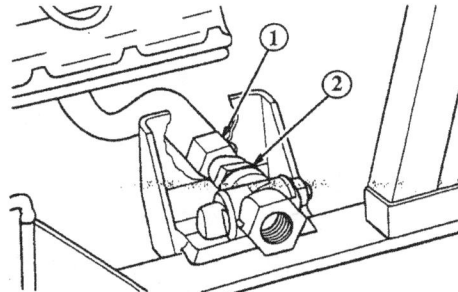

Figure 11

11. Remove the protective plug (1) from the exhaust muffler (2).

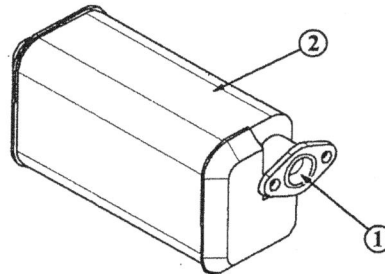

Figure 12

12. Using the new exhaust gasket (1) P/N 183671-13210 provided in the kit, then install the new exhaust pipe (2) (P/N GS-2501) provided in the kit, (down and out) with two (2) socket head bolts 8mx25mm and two (2) lock washers (8mm). Use a 6-mm allen wrench to tighten.

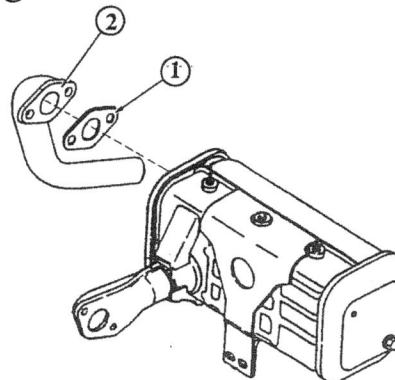

Figure 13

13. Install the new outlet fuel line (2) (P/N GS-2497) provided in the kit, onto the fuel filter (1) on cage assembly. **NOTE:** Outlet is the center fitting on fuel filter. Tighten using a 9/16-inch wrench.

14. Install the new inlet fuel line (3) (P/N GS-2495) provided in the kit, onto the fuel filter (1) on cage assembly. **NOTE:** Inlet is the remaining flare fitting on the fuel filter. Tighten using a 9/16-inch wrench.

15. Install the new fuel filter drain hose (4) (P/N GS-2509) provided in the kit, onto the fuel filter drain spout using a hose clamp and tighten with a flat head screwdriver. **NOTE:** this line is used to prevent fuel from spilling on fuel pump assembly during purging of system.

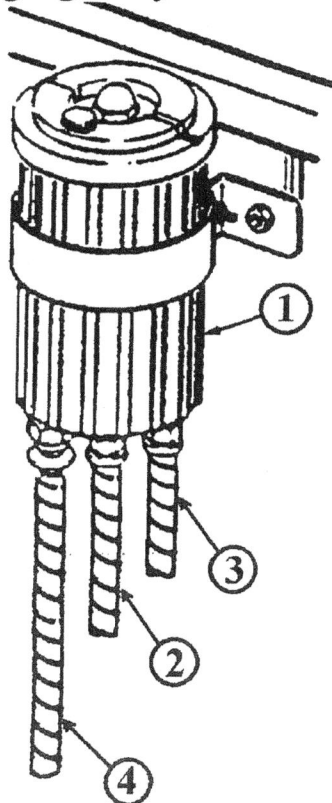

Figure 14

16. Install the main wiring harness (1), P/N GS-2505, provided in the kit into the inside surface of fuel pump assembly, P/N GS-2493, (previously installed). Use two hex head screws, nuts, lock washer, and washer, provided in the kit. Tighten with 3/8 and 5/16 inch wrenches.

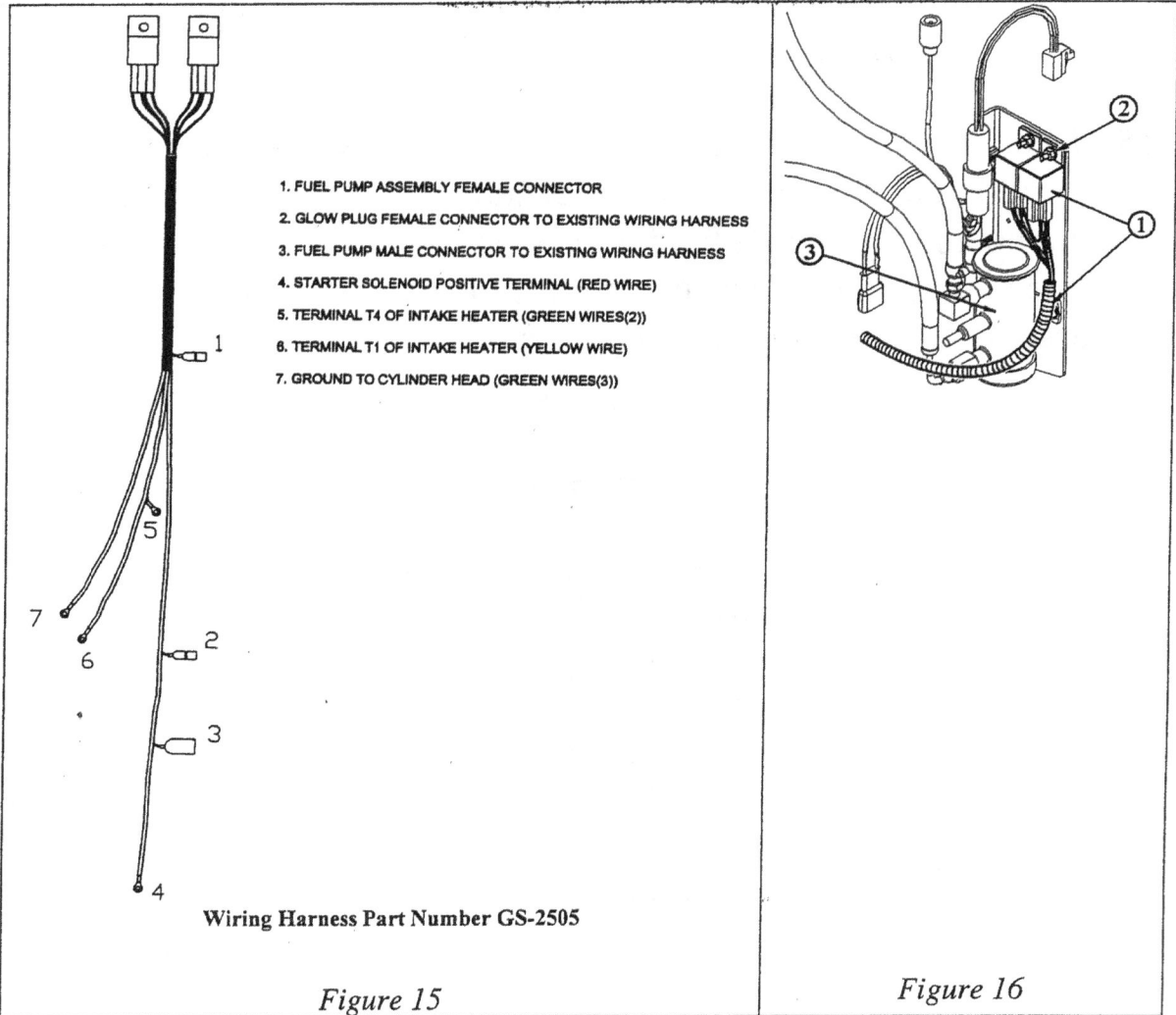

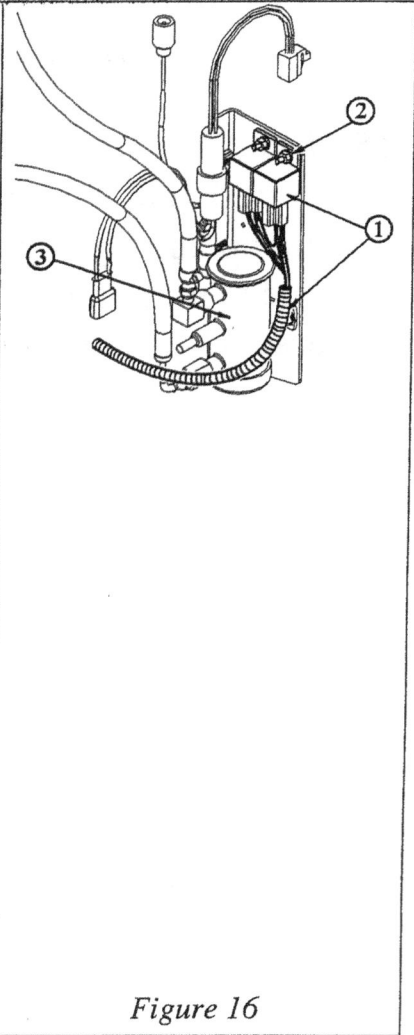

1. FUEL PUMP ASSEMBLY FEMALE CONNECTOR

2. GLOW PLUG FEMALE CONNECTOR TO EXISTING WIRING HARNESS

3. FUEL PUMP MALE CONNECTOR TO EXISTING WIRING HARNESS

4. STARTER SOLENOID POSITIVE TERMINAL (RED WIRE)

5. TERMINAL T4 OF INTAKE HEATER (GREEN WIRES(2))

6. TERMINAL T1 OF INTAKE HEATER (YELLOW WIRE)

7. GROUND TO CYLINDER HEAD (GREEN WIRES(3))

Wiring Harness Part Number GS-2505

Figure 15

Figure 16

17. Route the wiring harness between the air intake manifold and the pull start assembly to starter solenoid.

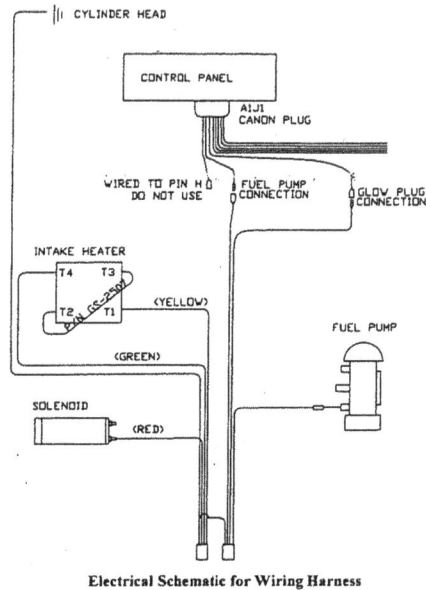

Electrical Schematic for Wiring Harness

Figure 17

18. Install **red** wire (1) (**HOT**) from the wiring harness to the positive post on the starter solenoid (2). Place washer and nut on terminal stud. Do not tighten.

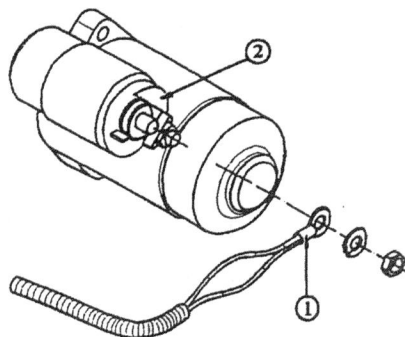

Figure 18

19. Looking from the fuel tank end, locate the intake heater head (1) with terminal marked T1 through T4. (Hardware installed).

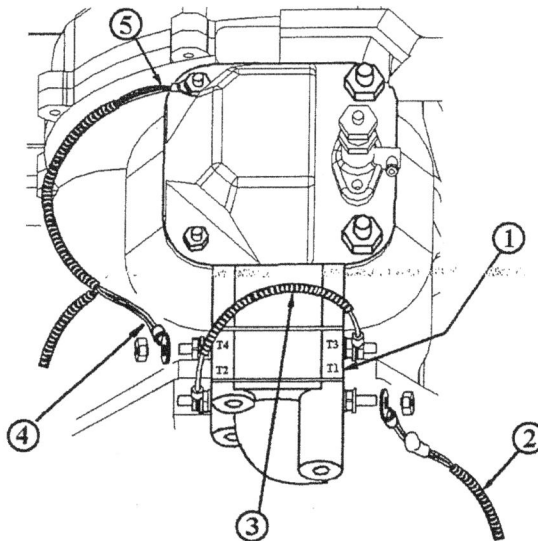

Figure 19

20. Install the **yellow** hot wire (2) (with boot) from the wiring harness to terminal T1 of the intake heater. Tighten using a 10mm wrench.

21. Install jumper cable (3) (wire), P/N GS-2507, provided in the kit over the **top** of the intake heater to terminals T2 and T3. Tighten using a 10mm wrench.

22. Install the **green** ground wire (4) from the wiring harness (with two wires) to terminal T4.

23. Install the **green** ground wire (5) from the wiring harness (with three wires) to the top of the cylinder head using one (1) bolt, 8mm x 20mm and one flat washer and lock washer. Use a 13mm wrench or socket to tighten.

NOTE: SHOCK HAZARD

24. Install the cage (1) on the eng/gen skid assembly (2) using six existing bolts (3), lockwasher (5), washer (4), and nuts (6).

Tighten with a 1/2-inch wrench. If the fuel tank was removed, make sure chain for the drain cap is installed. Also thread wires, T1 through T6 and F1 and F2, to the control box. Make sure all hoses and wires are free of obstruction and not bound in reassemble.

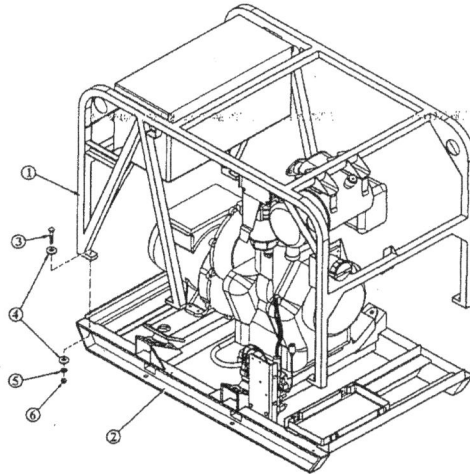

Figure 20

25. Connect F1 and F2 wires to TB2 terminal board inside control cabinet. Connect T1 through T6 wires to TB1 terminal board inside control cabinet.

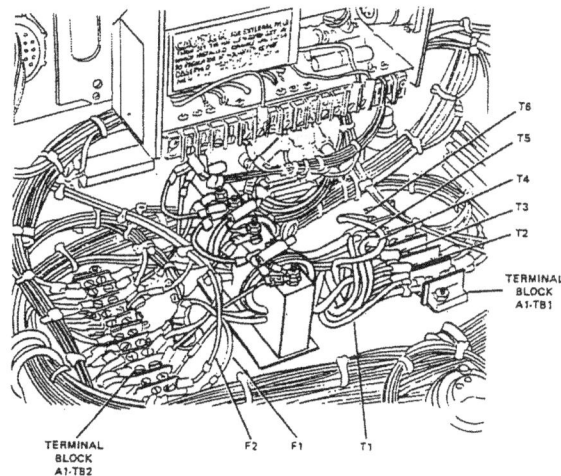

Figure 21

26. Connect the fuel line to the auxiliary fuel pump using 9/16-inch open-end wrench.

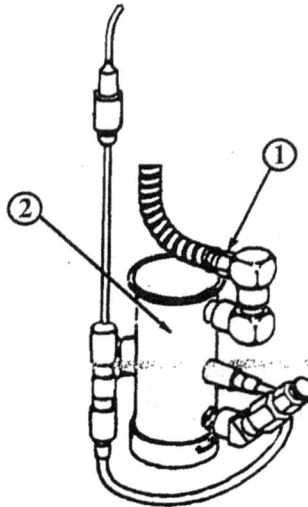

Figure 22

27. Connect two male wire connectors (1) from the engine stator assembly underneath the starter (2) to the two female wire connectors (3) from existing wiring harness located near the auxiliary fuel pump.

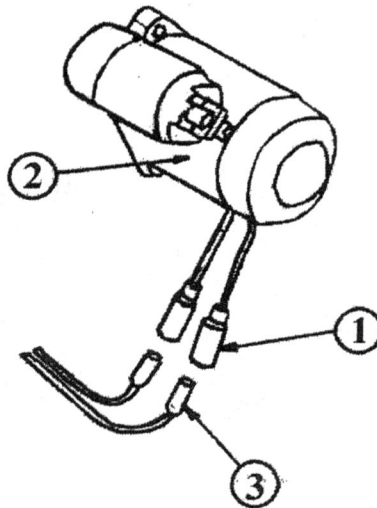

Figure 23

28. Connect the fuel suction line (1) from the fuel tank (2) to the bottom port (inlet) of new the electric fuel pump (2) using a 9/16-inch open-end wrench.

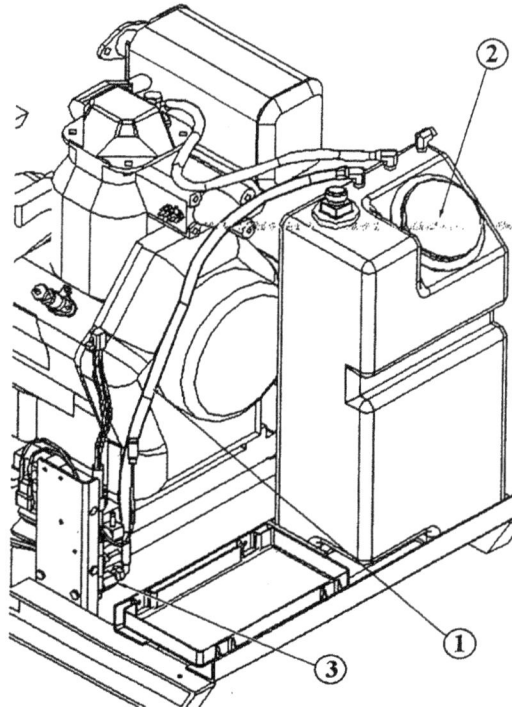

Figure 24

29. Connect the new outlet fuel line (1) from the fuel filter (2) to the injector (3) with a hose clamp using a flat tip screwdriver. (The outlet fuel line from fuel filter is the middle line).

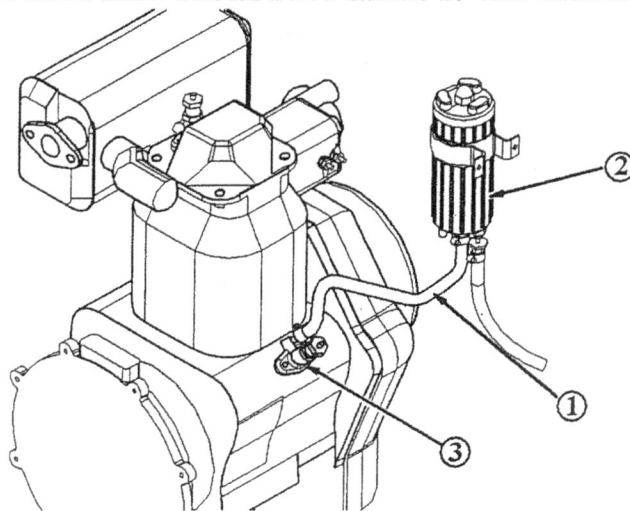

Figure 25

30. Connect the new inlet fuel line from fuel filter to top port (outlet) of electric fuel pump using a 9/16-inch open-end wrench. (The inlet fuel line from fuel filter is the outside line).

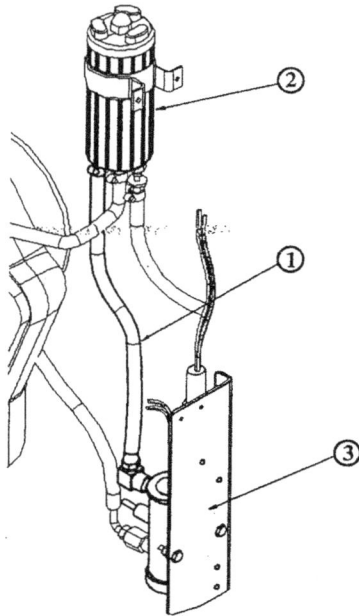

Figure 26

31. Connect the fuel return line (1) from the fuel tank (2) with a hose clamp onto the fuel injection (3) on cylinder head using a flat head screwdriver. Secure the hose in place by attaching the hose to the cylinder head using one bolt, 8mm x 20mm, flat washer, lock washer and 5/8 inch loom clamp provided in kit.
· Use a 13mm wrench or socket to tighten.

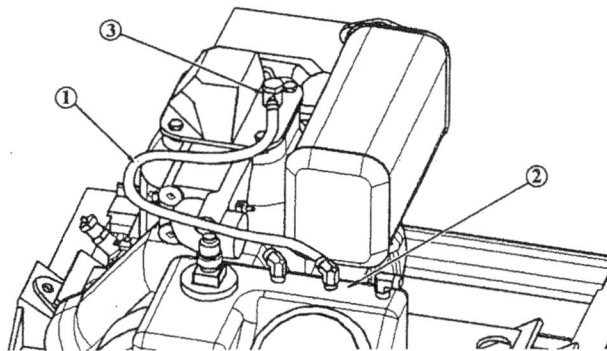

Figure 27

32. Connect the positive battery cable (1) and the small positive wire (2) to the large terminal lug on the starter solenoid terminal (3). Tighten using a 13-mm wrench or socket.

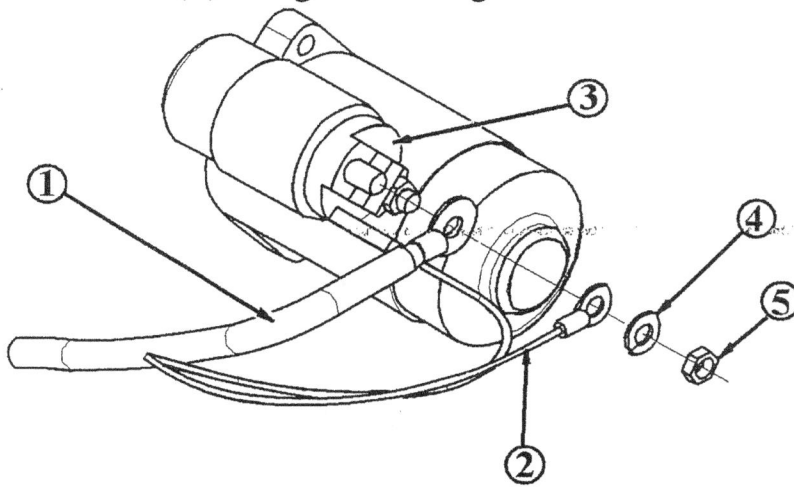

Figure 28

33. Cut the terminal lug (1) off of the ignition wire (2) (identified by number K14012 on wire) from starter solenoid (3) and install new female terminal lug and plastic cover (4), P/N DC925745 provided in the kit onto the ignition wire.

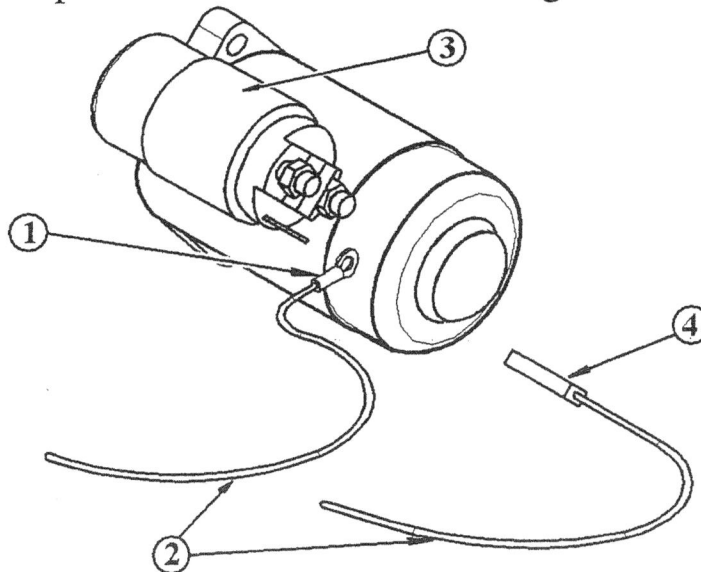

Figure 29

34. Connect the new female terminal lug installed in step 33 (1), P/N DC925745 to the tab on starter solenoid (2).

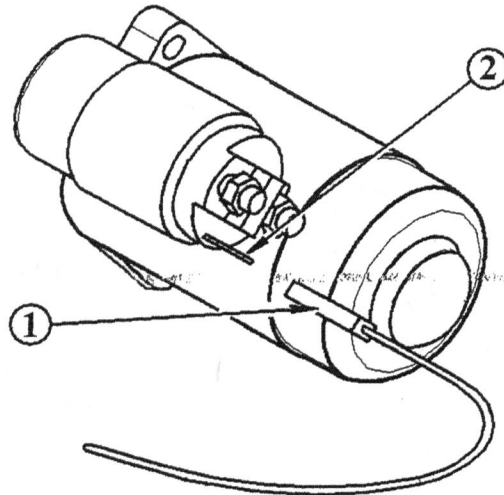

Figure 30

35. Install the battery ground cable (1) and the small ground wire (2) between two washers onto the left rear motor mount bolt (3). Using a 11/16 inch wrench to tighten. (Hardware on engine)

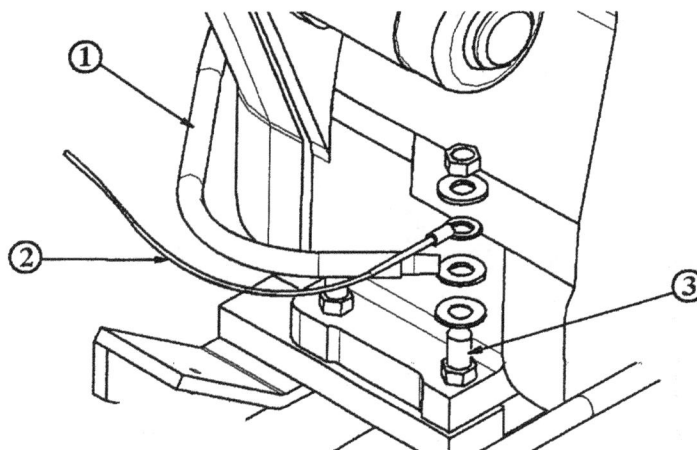

Figure 31

36. Install the male connector on auxiliary fuel pump to the female connector on the wiring harness.

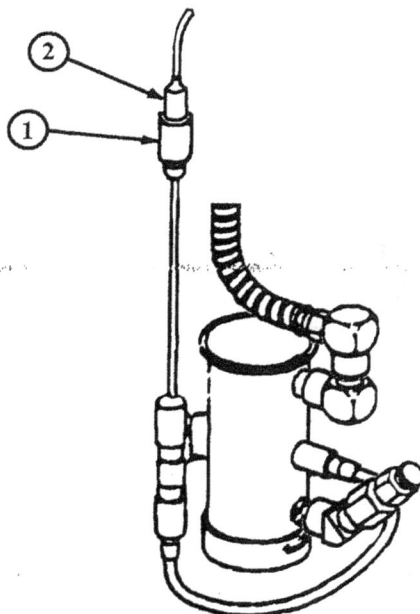

Figure 32

37. Connect the ground wire (1) from the control panel box (2) and the ground wire (3) from the load terminal box (4) to the grounding terminal (5) on the eng/gen skid base assembly using a 7/16 wrench.

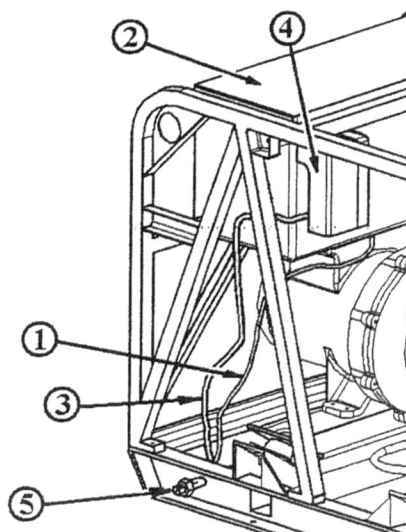

Figure 33

38. Connect the female connector on the wire marked "FUEL PUMP" (1) from the existing wiring harness to male connector on the fuel pump wire in new wiring harness.

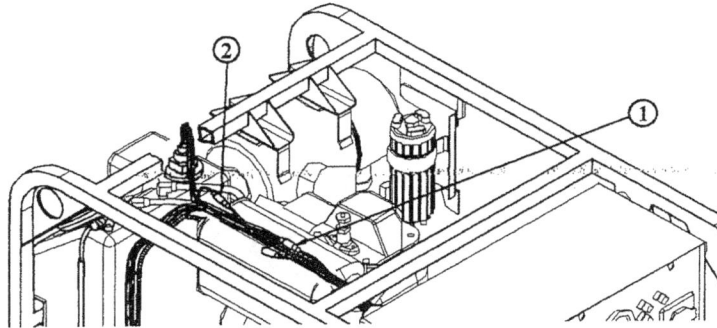

Figure 34

39. Connect the male connector on the wire marked "GLOW PLUG" (2) from existing wiring harness to female connector on the remaining white wire in the new wiring harness. (This wire runs to the relays mounted on the new fuel pump bracket.)

40. Install the female plug (1) from new wiring harness to the male plug (2) on the fuel pump assembly (3).

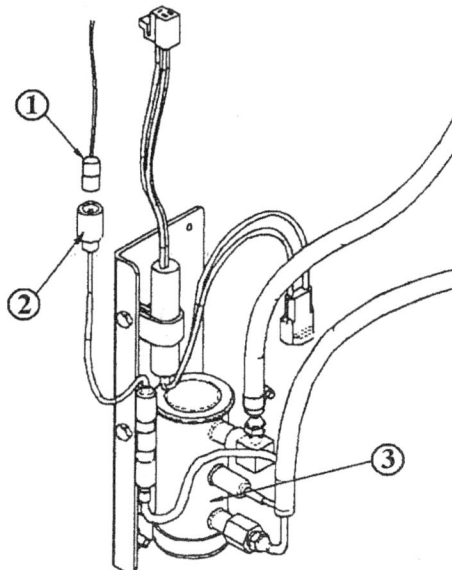

Figure 35

42. Install the WHITE female connector (1) from the command timer to the males plug (2) on the run-stop solenoid (3). Install the GRAY female connector (4) to the male connector on the existing wiring harness.

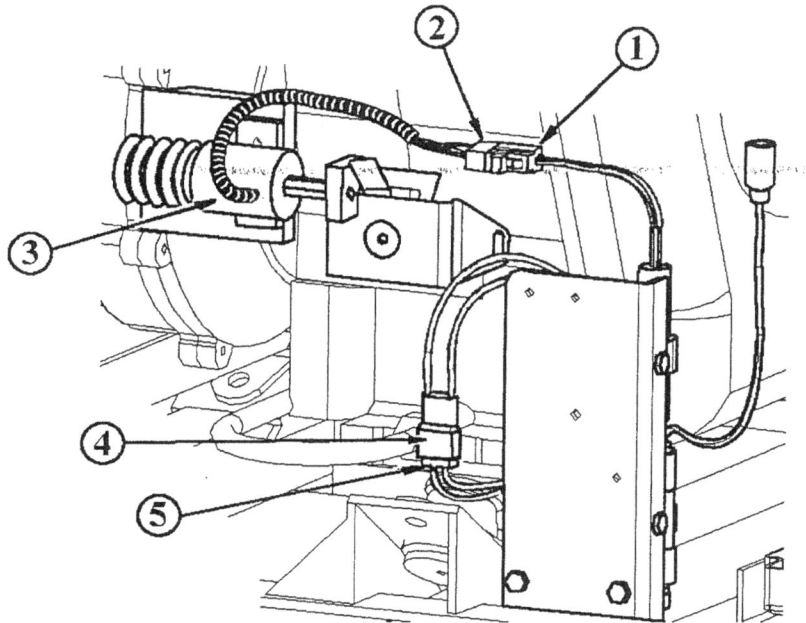

Figure 36

43. Install the cannon plug (1) from the existing wiring harness onto the top of fuel tank (2).

Figure 37

43. Utilize six (6) tee straps to tie wires as needed.

44. Remove the protective plastic plug (1) from the intake manifold (2). Install intake hose (3), P/N GS-2481 provided in the kit, to the engine intake manifold (2). Install intake hose (4), P/N GS-2483, with the large end to the air cleaner (5). Use three (3) hose clamps, number 32 (provide in the kit), on intake connection and aluminum connector (6). Use hose clamp number 40 (provided in kit), on air cleaner. Couple the hoses together with the aluminum connector, P/N GS-2803 provided in the kit. Tighten hose clamps with screwdriver or socket.

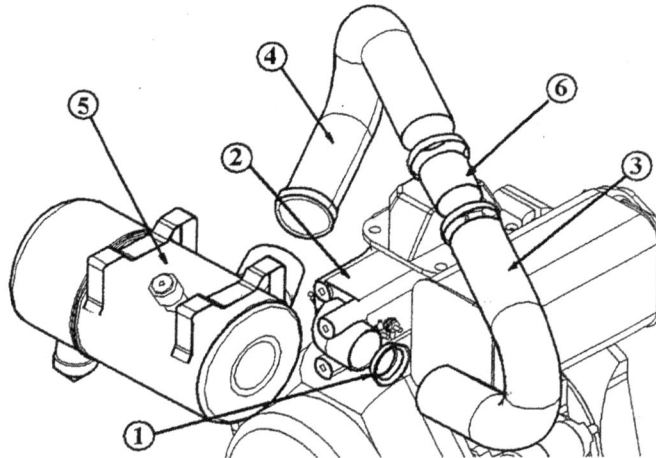

Figure 38

45. Add oil to engine.

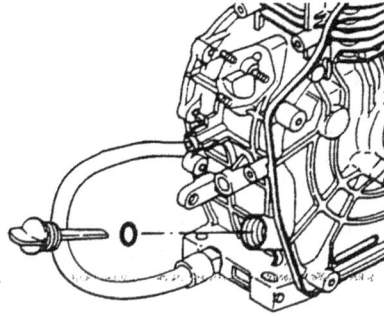

Figure 39

46. Replace battery and connect cables. Insure battery cover terminals are used.

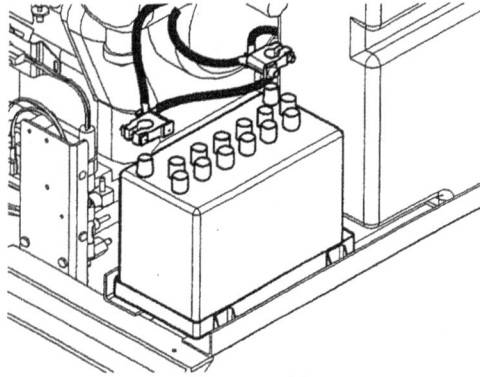

Figure 40

47.. Replace ground rods.

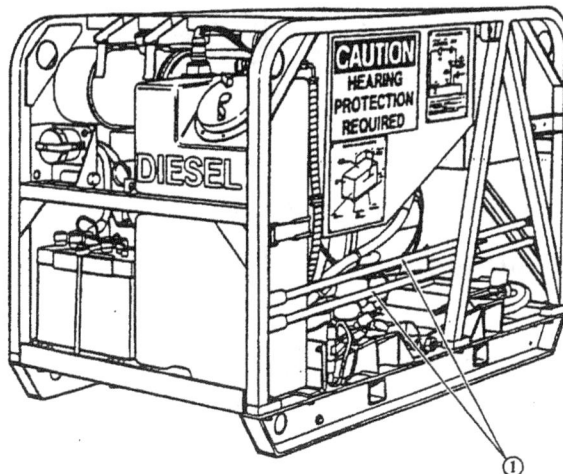

Figure 41

PURGING SYSTEM:

48. Turn the master switch on, open the bleeder vent on the fuel filter until fuel comes out. Close the bleeder and turn off the master switch. **CAUTION: FUEL SPILL.** Loosen the hose clamp on the fuel line at the injector pump, remove the fuel line, turn the master switch on and bleed the fuel line. **CAUTION: FUEL SPILL.** Connect the fuel line back and tighten the hose clamp.

    NOTE: If problems exist with the purging procedures described above, use the procedures in the Technical Manual.

49. Start the engine and run for approximately five (5) minutes. Shut the engine off and check the oil level. Add oil if required.

ADJUSTMENT OF _____ HERTZ (_ Hz)

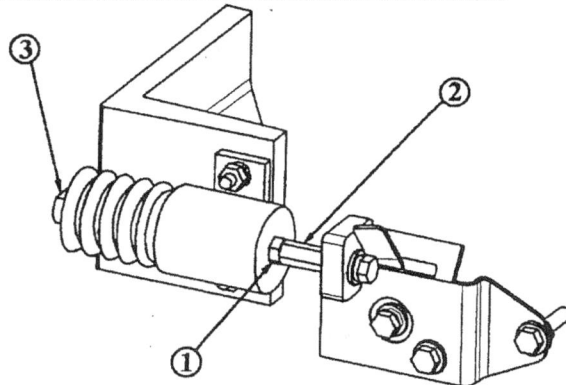

Figure 42

NOTE: This procedure is required because the hand throttle adjustment is no longer used.

TOOLS REQUIRED: One 3/8, 7/16 and 1/2 wrench.

50. Start the engine.

51. Loosen the jam nut (1) on the run-stop solenoid. By placing a 7/16-inch wrench on the jam nut (1) and 3/8-inch wrench on the long hex head stud (2).

52. Hold the long hex head stud with a 3/8-inch wrench and place a 1/2-inch wrench on the square end (3) of the run/stop solenoid shaft on rear of boot. Turn the 1/2 wrench to adjust .HERTZ (_ !Hz) to 61-62   .hertz's:
    a. Counterclockwise to increase _ Hz
    b. Clockwise to decrease _ .Hz.

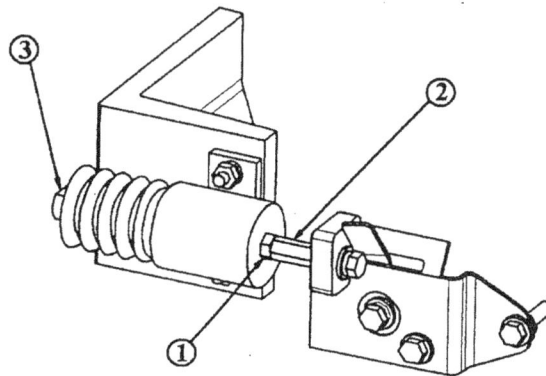

Figure 43

53. Place a 7/16-inch wrench on the jam nut and a 3/8-inch wrench on the long hex head stud and tighten jam nut.

54. Turn the engine off, then on quickly.

55. Let the engine run three to five minutes till warm.

LOAD BANK PROCEDURES

56. Perform load bank procedure I/A/W procedures in Technical Manual and record results as required.

57. Install the side cover plate. Attach the hose and wire clamps.

www.ingramcontent.com/pod-product-compliance
Lightning Source LLC
Chambersburg PA
CBHW080416030426
42335CB00020B/2467